David Woodring

Advancing Higher Education with Human-Centric AI Partnerships

A Sustainable Guide for Integrating Generative AI

palgrave
macmillan

David Woodring
Rogers, AR, USA

ISBN 978-3-032-05204-9 ISBN 978-3-032-05205-6 (eBook)
https://doi.org/10.1007/978-3-032-05205-6

This Palgrave Macmillan imprint is published by the registered company Springer Nature Switzerland AG.
The registered company address is: Gewerbestrasse 11, 6330 Cham, Switzerland

Competing Interests The author has no competing interests to declare that are relevant to the content of this manuscript.

About This Book

A distinctive feature of this guide is its focus on empowering users to move beyond theoretical understanding to practical application. Higher Education Institutions (HEIs) worldwide are entering an era of significant transformation driven by the integration of Generative Artificial Intelligence (GenAI). While these powerful tools offer immense potential for innovation across teaching, learning, assessment, research, and administration (TLARA), their effective and ethical adoption presents complex challenges. GenAI systems often reflect the cultural norms and biases embedded within their training data, primarily from Western contexts. Global deployment without careful adaptation risks biased outcomes and cultural irrelevance, demanding context-sensitive strategies. Furthermore, many HEIs lack the specific expertise needed to navigate the intricacies of AI design, implementation, and governance, making it difficult to mitigate risks related to academic integrity, data privacy, algorithmic bias, and the need for widespread AI literacy.

This textbook addresses these critical challenges by providing a comprehensive, adaptable, and sustainable guide for HEIs globally. It offers a structured, customizable four-phase rollout plan—encompassing Implementation, Piloting, Evaluation, and Scaling—designed to guide stakeholders through the complexities of integrating GenAI responsibly and effectively. The book caters to diverse user journeys within the institution, including those of undergraduate students, graduate students, full-time and part-time faculty, student- and institution-supporting staff, administrators, institutional leaders, and dedicated "Smart Teams" tasked with guiding AI initiatives.

Structure and Approach

The book is organized to build foundational understanding before moving into practical implementation:

- **Part 1: Foundations:** Establishes the context of GenAI within the evolving landscape of higher education, explores the science of human-AI collaboration, and introduces core theoretical frameworks essential for ethical and effective integration. This includes foundational concepts and core human-centric frameworks—such as HCAS (Human-Centric AI Synergy) and RSA (Regulation Synergy & Augmentation)—while emphasizing ethical considerations from the outset.
- **Part 2: The Four-Phase Rollout Plan:** Details the adaptable four-phase strategy (Implement, Pilot, Evaluate, Scale), providing step-by-step guidance, practical considerations, and customizable activities for each phase. This part transitions into a detailed, phased implementation strategy, guiding institutions through research, stakeholder engagement, strategic planning, piloting, evaluation, scaling, and long-term governance. Crucially, this book introduces the TLARA (Teaching, Learning, Assessment, Research, Administration) structure as an organizational scaffold. Starting in Chap. 5 and continuing throughout Part 2, readers are guided to use this structure to develop their own custom, context-specific ethical frameworks, policies, tools, resources, and solutions. This process emphasizes a holistic, multi-level systems approach, prompting consideration of AI's impact across different levels (individual, departmental, institutional), within various spaces (classrooms, labs, administrative offices), and as it evolves over time within unique sociocultural contexts. To facilitate this practical development, each chapter in Part 2 is complemented by a User Experience (UX) Workbook and a Personalized Adaptive Learning Session (PALS) Workbook.

Supplemental Resources

Each chapter is accompanied by downloadable resources designed for practical application:

- **The Human-GenAI Collaboration & Synergy (HAICS) Guide:** Provides in-depth explanations of the book's core conceptual frameworks—including Human-Centric AI Synergy (HCAS), Regulation, Synergy, & Augmentation (RSA), Enhancing Human-Centric Skills (EHCS), Enhancing Academic Outcomes (EAO), and Authentic High-Impact Practices (AHIPs)—along with the operationalizing five-component Human-Centric AI Synergy & Regulation (HCASR) process. This guide serves as a vital reference throughout the book.
- **User Experience (UX) Workbooks:** Offer structured activities and templates aligned with each chapter, guiding users through practical planning, design, assessment, and implementation tasks using human-centered design principles.
- **Personalized Adaptive Learning Session (PALS) Workbooks:** Provide guided learning experiences, leveraging GenAI itself as a partner to help users research concepts, customize frameworks, and apply strategies tailored to their specific roles and contexts. The PALS Workbooks feature a recurring section titled "Long-Term Scaling, Sustainability & AI Governance: Iteratively Developing Your Custom TLARA-Informed AI Guidelines." This standardized activity, which includes a customizable prompt, positions the user's GenAI model as an expert Higher Education AI Transformation Specialist and Strategic Advisor. It guides users to systematically apply the chapter's core topics to refine their "Living Document" of TLARA-informed guidelines, ensuring that their strategies for AI integration are not only ethically sound and effective but also deeply aligned with their institutional mission, stakeholder needs, and prepared for long-term sustainability and responsible scaling.

Methodology: Integrating Frameworks and Human-AI Partnership

This guide employs a unique methodology integrating principles from Design Thinking (Brown, 2019), Change Management (Kotter et al., 2021), and Human-Centric AI (Shneiderman, 2022), operationalized through the book's core frameworks (HCAS, RSA, EHCS, EAO, AHIPs) and the practical HCASR process detailed in the HAICS Guide. A central tenet is the concept of leveraging GenAI as a strategic partner.

How AI Was Used in Creating This Book

In the spirit of the human-AI partnerships advocated within these pages, GenAI tools were utilized as strategic partners throughout the research, drafting, and refinement process of this textbook. This engagement was explicitly guided by the Human-Centric AI Synergy & Regulation (HCASR) framework and its underlying principles (HCAS, RSA), leveraging AI for tasks like initial literature exploration (HCASR Component 3) and drafting assistance, while human authors focused on critical analysis, ethical judgment, framework development, contextualization, and ensuring the final content align with the book's human-centric values. The process encountered many of the "design-induced challenges" discussed herein, particularly with early model limitations and the complexities of applying AI to nuanced knowledge work in higher education. These practical experiences directly informed the guidance offered, aiming to model the responsible and synergistic collaboration this book seeks to foster.

A portion of the functional diagrams and figures in this book were created through an iterative nonlinear process of human-AI synergy, as presented in the core human-centric frameworks for this text. As the author, I developed the core concepts and intellectual structure for each figure. Generative AI tools, such as ChatGPT 4o, o3, and o4, and Google Gemini 2.5 Pro were then used as a tool to render initial visual drafts, which were subsequently subjected to significant manual refinement, editing, and compositing over several months to create the final original works presented here.

How to Use This Book

This textbook is designed for flexibility. Readers can engage with it sequentially to follow the entire four-phase process or focus on specific chapters or frameworks relevant to their immediate needs or role. The downloadable HAICS Guide and the chapter-specific UX and PALS workbooks are essential companions, providing the tools to translate concepts into action. Individuals can use the resources for personal development or skill-building, while teams and institutional leaders can use the frameworks and workbooks to guide strategic planning, pilot design, evaluation efforts, and large-scale implementation initiatives. The ultimate goal is to empower users to ethically and effectively partner with GenAI, creating customized tools, solutions, resources, and outcomes aligned with their unique needs, goals, objectives, and sociocultural contexts.

References

Brown, T. (2019). Change by design: How design thinking transforms organizations and inspires innovation (Rev. and updated ed.). Harper Business.
Kotter, J. P., Akhtar, V., & Gupta, G. (2021). Change: How organizations achieve hard-to-imagine results in uncertain and volatile times. Wiley.
Shneiderman, B. (2022). Human-centered AI. Oxford University Press.

Contents

Part I Optimizing Human-Centric AI Partnerships: Preparing Higher Education for Hybrid Human-AI Intelligence & The Next Cognitive Revolution

1 The Fifth Industrial Revolution: A New Cognitive Era & Unprecedented Human-AI Collaboration 3
Unprecedented Change & Transformation in Higher Education 3
The Rapid Evolution of Generative AI & Disruptive Changes in Higher Education 6
The Road to Artificial General Intelligence & Artificial Superintelligence 8
Accelerating Toward Superintelligence: The Dawn of a New Cognitive Era 10
Preparing Higher Education for Advanced AI 12
New Cognitive Revolutions Through Human-AI Collaboration & Synergy 13
Human-AI Co-Creation and Its Effects on Innovation 15
Reconfiguring Higher Education for the Intelligence Age 16
Design Thinking and Change Management for AI Integration 18
Conclusion: Building a Human-Centric AI Collaboration & Synergy Model 21
End of Chapter Workbook Activities: Engaging with GenAI as a Strategic Partner & Stakeholder 22
References 23

2 Hybrid Human-AI Intelligence: A Vision for Innovation & Higher Education's Future 29
Hybrid Human-AI Intelligence: Collaboration and Synergy 30
Human-AI Synergy & Collaboration: Building a Vision for Higher Education 30
Guiding Ethical Principles for Human–GenAI Synergy 32
Conceptualizing GenAI's Influence: A Stakeholder Lens 34
Conclusion: Ethical Human-AI Partnerships in Higher Education 35
References 38

3 Human-AI Collaboration (HAIC) in Higher Education: Defining Roles, Responsibilities, and Oversight with GenAI 41
Human-AI Collaboration (HAIC) in Higher Education 42
Applying a Human-AI Collaboration (HAIC) Framework to GenAI 44
HAIC Factors: Goals, Interaction, and Task Allocation 47
Conceptualizing GenAI's Influence: A Stakeholder Lens 49
Prompting with Clarity and Responsibility 49
Conclusion 50
References 52

4 The Emergence of Generative AI: Navigating Short-Term Viability and Long-Term Sustainability in Human-AI Partnerships 55
The Dual Impact of Technology and Change: Understanding Synergy & Emergence 56
Synergy, Emergence, & Disruptive Technologies 57
Decoding Human-AI Partnerships: Viability vs. Sustainability in Higher Education Synergies 61
The Long-Term View: Is it Healthy for the Future? 62
Building Sustainable Human-AI Collaboration: The Urgent Need for Planning 66
Human-AI Synergy in Higher Education: Cohort Effects of COVID-19 and GenAI 68
Conclusion: Ethically & Effectively Navigating the Challenges & Opportunities with GenAI 72
References 74

5 **Applying & Customizing the Core Frameworks for Optimizing Human-AI Partnerships in Higher Education** 79
Human-Centric AI Synergy (HCAS) Framework Summary 80
HCAS Principles, Components, Goals, and Objectives 80
Regulation, Synergy, & Augmentation (RSA) Framework Summary 83
Limits of Self-Regulation & Co-Regulation 84
Enhancing Human-Centric Skills (EHCS) Framework Summary 85
Enhancing Academic Outcomes (EAO) 86
Authentic High-Impact Practices (AHIPs) Framework Summary 87
Integrating the Five Frameworks 89
Focus on Ongoing Revision 91
Conclusion 92
References 94

Part II Optimizing Human-Centric AI Partnerships for Advancing the Ethical & Effective Implementation, Piloting, Evaluation, & Scaling of Generative AI in Higher Education: A Customizable 4-Phase Framework

6 **Phase 1: Strategic Planning and Stakeholder Engagement** 103
The Synergy of Design Thinking and Change Management 105
Aligning AI Initiatives with Institutional Mission and Values 106
Understanding AI/GenAI Opportunities and Challenges in HE 107
Planning for Diverse Stakeholder Engagement 108
Establishing the Foundation for Ethical AI Governance 110
Strategic Decision: Building In-House AI Systems vs. Integrating Third-Party Solutions 112
Conclusion: Laying the Foundation for Responsible AI Integration 115
References 117

7 **Forming Smart Teams and Stakeholder Mapping (Phase 1)** 121
Forming Diverse "Smart Teams" 123
Defining Roles, Responsibilities, and Team Charters 125
Section 3: Mapping Stakeholders for AI Initiatives 127
Planning for Stakeholder Engagement 131
Section 5: Integrating Key Foundational Concepts 132

User Outcomes and Deliverables (UODs) & Initial Team Actions 137
Conclusion: Looking Ahead & Engaging the Workbook 140
References 141

8 Assessing Institutional Readiness and Capacity (Phase 1) 145
Understanding Stakeholder Needs, Goals, and Context 147
Defining Initial Goals Based on Assessed Needs 149
Assessing Institutional Readiness for AI 152
Evaluating Data Infrastructure and Governance 154
Assessing Technological Capabilities 156
Evaluating Organizational Culture and Change Readiness 158
Evaluating Existing AI Governance and Ethical Frameworks 159
Integrating Findings and Informing Strategy 162
Informing the AI Sourcing Strategy: Proprietary vs. Third-Party/Open-Source 164
User Outcomes and Deliverables (UODs): Concluding Phase 1 Assessment 165
Conclusion: From Assessment to Informed Strategy 166
References 167

9 Establishing AI Vision and Governance (Phase 1) 171
Developing a Shared Vision for Change 173
Formulating a Shared Vision (Vision/Opportunity Statements) 174
Defining Goals, Strategic Planning & Innovation with GenAI 176
Applying Insights: Revisiting the AI Sourcing Strategy 179
Strategic Planning & Roadmap Development 180
User Outcomes & Deliverables (UODs): Finalizing Phase 1 of AI Integration 183
Conclusion: Laying the Foundation for Action 184
Conclusion: Laying the Foundation for Action 185
References 186

10 Designing Context-Specific Pilot Programs (Phase 1) 189
Strategic Roadmap & AI Pilot Design 191
User Outcomes & Deliverables (UODs) for Pilot Design 199
Preparing for Pilot Implementation: Governance, Barriers, and Engagement 199
Conclusion 202
References 203

11 **Phase 2: Implementing Human-Centric AI Initiatives** 207
Pilot Objectives: Empowering Stakeholders and Enabling Continuous Assessment 209
From Strategic Planning to Implementation: Building and Iterating 210
AI Implementation Strategies: Integrating Change Management and Ethical Oversight 213
AI Pilot Testing, User Experience, and Workflow Optimization 216
Operationalizing Implementation: Documentation and Engagement 218
Role-Specific Implementation Strategies 219
Conclusion 221
References 222

12 **Collecting and Synthesizing Pilot Data (Phase 2)** 229
Systematic Pilot Data Collection Methodologies 230
Initial Pilot Performance Evaluation 233
Refining Workflows & Risk Mitigation Based on Pilot Data 235
Documenting Pilot Findings, Lessons Learned, and Preparing for Phase 3 237
Conclusion 239
References 240

13 **Phase 3: Monitoring and Evaluating Ongoing Initiatives** 243
Goals of Ongoing Monitoring & Evaluation 245
AI Performance Evaluation & Benchmarking (Ongoing) 247
AI Ethics & Literacy Audits & Policy Refinement 250
User Outcomes & Deliverables (UODs): Phase 3 Deliverables 258
Conclusion 259
References 261

14 **Phase 4: Scaling and Institutionalizing for Sustainability** 265
The Rationale: Why Scaling and Institutionalizing AI is Essential 267
Establishing Permanent Ethical AI Governance and Leadership Structures 267
Building Sustainable AI Capacity: Competency Frameworks and Lifelong Learning 269
The Process of Scaling AI Integration Across the Institution 273
Expanding AI Applications Across University Functions 274

Strategic Implementation: Alignment and Actionable Steps 276
Key Deliverables for Scaling & Institutionalization 277
Conclusion: Anchoring AI in the Institution's Future 278
References 279

15 Ongoing Adaptation and Refinement (Phase 4) 283
Fostering Adaptive Culture and Managing Change 285
Iterative Policy Development and Adaptive Governance 287
Sustaining Dynamic Stakeholder Coordination 289
User Outcomes & Deliverables (UODs) for Adaptation and Refinement 290
Conclusion 292
References 294

16 Custom AI Ethics and Literacy Frameworks (Phase 4) 297
Workbook Prep 1: Long-Term Ethical Foresight Needs 298
Strategic Imperatives for a 5 to 10-Year Ethical AI Horizon 299
Workbook Prep 2: Planning Sustainable Governance 300
Core Rationale for Utilizing the AI Ethics TLARA Structure for Custom Framework Development 302
The Need for a Unified yet Customizable Approach via the TLARA Structure 302
Workbook Prep 3: Identifying Domain-Specific AI Challenges for Guideline Development 303
Institutionalizing the TLARA-Informed Framework Development Process 306
Workbook Prep 4: Evaluating Governance Structure Readiness 307
TLARA-Informed Guideline Development Across AI Rollout Phases 309
Workbook Prep 5: Aligning TLARA-Informed Development with Current Phase 309
Critical Importance of Developing Customized Multilevel Frameworks: Using the TLARA Structure for HEIs 310
Workbook Prep 6: Communicating the Value of Developing TLARA-Informed Guidelines 311
Strategic Recommendations for Long-Term Governance Based on Custom Frameworks Using the TLARA Structure 312
Workbook Prep 7: Implementing Strategic Recommendations 314

Users Outcomes & Deliverables (UODs) for Long-Term AI Governance Planning 314
References 316

17 **Building an Enduring Culture of AI Literacy (Phase 4)** 321
Workbook Prep 1. Envisioning an AI-Literate Institution 323
Designing Institution-Wide AI Literacy Initiatives 323
Workbook Prep 2. Framing the Institutional Literacy Strategy 326
Implementing Foundational AI Literacy Training 326
Workbook Prep 3. Designing Foundational Training 328
Embedding Applied AI Literacy Across the Institution 329
Developing Role-Specific Applied AI Literacy Modules 329
Workbook Prep 4. Planning Applied Literacy Integration 331
Fostering Communities of Practice and Continuous Learning 331
Establishing AI Learning Hubs and Resources 332
Workbook Prep 5. Planning for Community Building 333
Monitoring, Evaluating, and Sustaining the AI Literacy Culture 334
Linking Literacy Culture to the Institutional AI Maturity Model 334
Workbook Prep 6. Evaluating Literacy Initiatives 335
The Strategic Value of an AI-Literate Institution 336
Workbook Prep 7. Communicating Literacy Culture Value 337
Outcomes & Deliverables (UODs) for Building an AI Literacy Culture 338
Conclusion: Sustaining an Ethical and AI-Literate Future 339
References 341

Appendices 347

Index 373

Abbreviations

4IR	Fourth Industrial Revolution
5IR	Fifth Industrial Revolution
AAC&U	Association of American Colleges and Universities
ACCJC	Accrediting Commission for Community and Junior Colleges
AGI	Artificial General Intelligence
AHIPs	Authentic High-Impact Practices
AI	Artificial Intelligence
AIAS	AI Assessment Scale
AICOS	AI Competency Objective Scale
ALTL	AI Literacy in Teaching and Learning
API	Application Programming Interface
APRU	Association of Pacific Rim Universities
ASI	Artificial Superintelligence
CAIO	Chief AI Officer
CI	Collective Intelligence
CIO	Chief Information Officer
CM	Change Management
CMS	Course Management System
COBIT	[Control Objectives for Information and Related Technologies] *(Inferred Definition)*
CoP	Community of Practice
COPE	Committee on Publication Ethics
CRL	Co-Regulated Learning
CSA	Cloud Security Alliance
CSU	California State University
DEI	Diversity, Equity, and Inclusion
DEIA	Diversity, Equity, Inclusion, and Accessibility

DT	Design Thinking
EAO	Enhancing Academic Outcomes
ECGBL	European Conference on Games Based Learning
EHCS	Enhancing Human-Centric Skills
EPS	Encapsulated PostScript
ERP	Enterprise Resource Planning
ESM	Electronic Supplementary Material
ETHICAL	[Exploration & Evaluation, Transparency & Accountability, Human-Centered Approach, Integrity & Honesty, Continuous Learning & Innovation, Accessibility & Inclusivity, Legal & Ethical Compliance] *(Framework Acronym)*
EU	European Union
FAQ	Frequently Asked Questions
FDC	Faculty Development Center
FERPA	Family Educational Rights and Privacy Act
FT	Full-time
GAN	Generative Adversarial Network
GDPR	General Data Protection Regulation
GenAI	Generative Artificial Intelligence
GLAT	Generative AI Literacy Assessment Test
GPT	Generative Pre-trained Transformer *(Common Understanding)*
GUI	Graphical User Interface
HAIC	Human-AI Collaboration
HAICS	Human-GenAI Collaboration & Synergy
HAII	Human-AI Interaction
HAIT	Human-AI Teaming
H-CAI	Human-Centric AI
HCAI	Human-Centered AI
HCAS	Human-Centric AI Synergy
HCASR	Human-Centric AI Synergy & Regulation
HECVAT	Higher Education Community Vendor Assessment Toolkit *(Inferred Definition)*
HEI	Higher Education Institution
HESA	Higher Education Strategy Associates
HICSS	Hawaii International Conference on System Sciences
HHAIR	Hybrid Human-AI Regulation
HIPs	High-Impact Practices
IA	Intelligence Augmentation
IADIS	International Association for Development of the Information Society
IAPP	International Association of Privacy Professionals
IAP2	International Association for Public Participation
IBM	International Business Machines

ICT	Information and Communication Technology
IDEAS	Instructional Design, Engagement, and Support
IES	Institute of Education Sciences
IoT	Internet of Things
IP	Intellectual Property
IRB	Institutional Review Board
ISACA	[Information Systems Audit and Control Association] *(Inferred Definition)*
ISED	[Innovation, Science and Economic Development Canada] *(Inferred Definition)*
ISO/IEC	International Organization for Standardization / International Electrotechnical Commission
IT	Information Technology
ITcon	[Journal of Information Technology in Construction] *(Inferred Definition)*
JISC	[UK digital, data and technology agency for tertiary education] *(Inferred Definition)*
KPI	Key Performance Indicator
L@S	Learning @ Scale
LaTeX/LaTeX2e	[Document preparation system] *(Common Understanding)*
LEAPS	Leaders in Education: Academy of Presidential Scholars
LLM	Large Language Model
LMS	Learning Management System
LX	Learning Experience
M&E	Monitoring & Evaluation
MAILS	Meta AI Literacy Scale
MCQ	Multiple-Choice Question
MEA	Monitor, Evaluate, and Assess (COBIT Domain)
ML	Machine Learning
MLOps	Machine Learning Operations *(Inferred Definition)*
MMLU-Pro	[Massive Multitask Language Understanding - Professional] *(Inferred Definition)*
MoSCoW	[Must have, Should have, Could have, Won't have] *(Prioritization Method)*
MVP	Minimum Viable Product
NASPA	Student Affairs Administrators in Higher Education
NARA	U.S. National Archives and Records Administration
NCEE	National Center on Education and the Economy
NIST	National Institute of Standards and Technology
NLM	National Library of Medicine
NLP	Natural Language Processing
NPC	Non-Player Character
NTIA	National Telecommunications and Information Administration

OECD	Organisation for Economic Co-operation and Development
OET	Office of Educational Technology (U.S. Department of Education)
OKR	Objectives and Key Results
OSPI	Office of Superintendent of Public Instruction (Washington)
PALS	Personalized Adaptive Learning Session
PD	Participatory Design
PDCA	Plan-Do-Check-Act
PDF	Portable Document Format
PhD	Doctor of Philosophy
PICO	[Population/Problem, Intervention, Comparison, Outcome] *(Framework)*
PII	Personally Identifiable Information
PMBOK	Project Management Body of Knowledge
PPT/PPTX	PowerPoint Presentation
PwC	PricewaterhouseCoopers
R&D	Research and Development
RAG	Retrieval-Augmented Generation
REL	Regional Educational Laboratory
RIT	Rochester Institute of Technology
RMF	Risk Management Framework (NIST AI RMF)
RNL	Ruffalo Noel Levitz
ROI	Return on Investment
RSA	Regulation, Synergy, & Augmentation
SDL	Self-Directed Learning
SEAA	[Euromicro Conference on Software Engineering and Advanced Applications] *(Inferred Definition)*
SHRM	Society for Human Resource Management
SIGITE	[Special Interest Group on Information Technology Education] *(Inferred Definition)*
SIS	Student Information System
SLA	Service Level Agreement
SMR	Small Modular Reactor
SMART	Specific, Measurable, Achievable, Relevant, Time-bound
SMARTIE	Specific, Measurable, Achievable, Relevant, Time-bound, Inclusive, Equitable
SREB	Southern Regional Education Board
SRL	Self-Regulated Learning
SSRN	[Social Science Research Network] *(Common Understanding)*
STEM	Science, Technology, Engineering, and Mathematics
SUNY	State University of New York
TIFF	Tagged Image File Format
TLARA	Teaching, Learning, Assessment, Research, and Administration

TPACK	Technological Pedagogical Content Knowledge
UC	University of California
UDL	Universal Design for Learning
UK	United Kingdom
UMass	University of Massachusetts
UMich	University of Michigan
UNESCO	United Nations Educational, Scientific and Cultural Organization
US	United States
USC	University of Southern California
UTO	University Technology Office (ASU)
UX	User Experience
VAE	Variational Autoencoder
VP	Vice President
VRML	Virtual Reality Modeling Language
WCAG	Web Content Accessibility Guidelines
WEF	World Economic Forum
XAI	Explainable AI
XLS/XLSX	Excel Spreadsheet
XML	Extensible Markup Language

List of Figures

Fig. 3.1 Human-AI collaboration (HAIC). (*Note.* A conceptual diagram illustrating the spectrum of Human-AI collaboration (HAIC), from human-led to AI-driven processes, highlighting the importance of balancing roles for effective synergy) 43

Fig. 3.2 Modes of human-AI collaboration for student discussion posts and grading. (*Note.* This figure depicts different collaborative modes for a common academic task. It shows how the level of human versus AI involvement can shift depending on whether the goal is primarily human-centric (e.g., student-led critical thinking), symbiotic (e.g., AI-assisted feedback with instructor oversight), or AI-centric (e.g., automated initial grading)) 44

Fig. 4.1 Human-AI synergy: emergence and feedback loop diagram. (*Note.* A diagram illustrating the feedback loops within human-AI synergistic systems, showing how emergent properties and outcomes influence future interactions and adaptations) 59

Fig. 4.2 Historical representation of synergies with emergence and feedback loops. (*Note.* A visual model representing how human-technology synergies throughout history have produced emergent outcomes and created feedback loops that shape subsequent development) 65

Fig. 4.3 Synergies and outcomes of historical human-technology collaboration. (*Note.* This diagram illustrates the dynamic interaction between humans (students, educators) and GenAI systems in higher education. Their collaboration produces viable, nonviable, sustainable, and unsustainable synergies, leading to emergent outcomes (e.g., enhanced practices, ethical challenges). Feedback loops demonstrate the iterative nature of this system) 67

Fig. 4.4 Human-GenAI synergy with synergy examples and feedback loops. (*Note.* An illustrative diagram showing specific examples of

human-GenAI synergies and the feedback loops that influence their evolution and impact within an educational system) 69
Fig. 4.5 Human-GenAI synergy across time horizons. (*Note.* This figure visually represents how human-GenAI synergies operate and evolve across different time horizons, from immediate (viable/nonviable) interactions to long-term (sustainable/unsustainable) outcomes) 71
Fig. 4.6 Key takeaways on synergies in higher education. (*Note.* A bubble-and-plate diagram mapping four synergy types (viable, nonviable, sustainable, unsustainable) to three actors (educators, AI, students), illustrating how specific examples (e.g., "Personalized Learning," "Bias Amplification," "Hybrid Roles," "Credential Skill-Gap") cluster under each category) 73
Fig. 5.1 Human-AI synergy in education: collaborative strengths and outcomes. (*Note.* This figure provides a conceptual overview of human-AI synergy in education, illustrating how the collaborative strengths of both humans and AI can lead to enhanced outcomes) 81
Fig. 5.2 Framework integration model for human-centric AI in higher education. (*Note.* Five nested rings in a Venn Diagram depict successive layers of the framework integration model. The outer magenta ring ("Authentic High-Impact Practices") lists core pedagogical and equity-focused activities (e.g., social & emotional learning; diversity, equity & inclusion; student-centered assessment; AI citizenship & ethics). The next blue ring ("Enhancing Academic Outcomes") highlights strategic goal setting, real-time feedback, and accessibility augmentation with AI. The middle magenta ring ("Human-Centric AI Synergy") describes stakeholder empowerment, augmentation of human skills, strength-weakness optimization, and sustainable collaboration. The following blue ring ("Regulation Synergy & Augmentation") covers self-/co-regulation, evidence-based refinement cycles, and iterative goal monitoring. The innermost pale sphere ("Enhancing Human-Centric Skills") focuses on meta-skills (critical thinking, prompt engineering, validation skills) and the cultivation of an AI-literate culture. Each ring is linked by aligned spheres and text blocks to emphasize how each layer builds toward unified, ethical, and effective AI integration) 91
Fig. 6.1 Conceptual diagram: merging change management and design thinking for AI adoption 106
Fig. 6.2 Example structure for institutional AI guidelines 110
Fig. 6.3 NIST AI risk management framework cycle 111
Fig. 7.1 Smart Team charter template—customizable 128
Fig. 7.2 Stakeholder power/interest grid—analysis tool 129
Fig. 7.3 Stakeholder map example—adaptable format 129

Fig. 7.4 AI governance structure example 134
Fig. 7.5 Feedback loop for Smart Team refinement 135
Fig. 8.1 Stakeholder needs assessment summary dashboard 149
Fig. 8.2 SMARTIE goal planning template 151
Fig. 8.3 Sample persona template for AI readiness 153
Fig. 8.4 Competency gap analysis chart 157
Fig. 8.5 Key dimensions of institutional AI readiness 161
Fig. 8.6 Institutional AI readiness dashboard 163
Fig. 8.7 Priority gaps versus quick wins matrix 164
Fig. 9.1 Vision and opportunity statement template 175
Fig. 9.2 Institutional AI maturity roadmap (adapted) 178
Fig. 9.3 Phased transition plan (planning to pilot) 182
Fig. 10.1 AI pilot design template 193
Fig. 10.2 Integrated governance checklist for AI pilots 198
Fig. 11.1 Iterative AI pilot cycle 212
Fig. 11.2 Feedback loop data flow 215
Fig. 12.1 Workflow for essay grading before and after AI pilot integration. (*Note*: Conceptual comparative flowchart illustrating a specific institutional workflow before and after AI pilot integration. The 'After' flow highlights steps involving AI and points where feedback or refinements occurred based on pilot data. (Requires final visual for accuracy)) 235
Fig. 13.1 Continuous quality improvement scorecard 250
Fig. 13.2 Policy refinement feedback loop 255
Fig. 14.1 Permanent AI governance structure chart. (*Note*: An organizational chart showing the relationship between the highest governing body (e.g., board/president), a central AI Council/Advisory Board, dedicated Ethics Committees, key leadership roles (e.g., Chief AI Officer), departmental AI liaisons/ambassadors, and operational units (IT, Academic Affairs, Research Office)) 270
Fig. 14.2 AI Competency framework matrix (advanced). (*Note*: A matrix listing key stakeholder roles (Undergrad, Grad, Faculty, Staff, Admin) as rows and advanced/role-specific AI competencies (e.g., Ethical AI Auditing, AI Pedagogy Design, AI Workflow Automation, Strategic AI Planning, Advanced Prompt Engineering) as columns. Cells could indicate required proficiency levels (e.g., Foundational, Intermediate, Advanced) or link to specific training resources/modules) 272
Fig. 15.1 Iterative AI adaptation cycle. (*Note*: A circular diagram illustrating the AI adaptation and refinement cycle, moving from monitoring/feedback collection through analysis, ideation, implementation, and impact evaluation, with stakeholder engagement inputs) 290

Fig. 15.2 Deliverables integration for continuous improvement. (*Note*: A flowchart depicting how key deliverables integrate: Success Metrics and Risk Reports inform both Updated Training Plans and Governance Refinement Plans, supporting continuous improvement) 293
Fig. 16.1 TLARA-informed implementation process flowchart. (*Note*: A flowchart with three main sequential stages labeled: (1) Establish Permanent AI Ethics Governance. (2) Implement Comprehensive Stakeholder Training. (3) Develop & Scale AI Integration. A feedback loop arrow from Stage 3 points back to Stage 1, indicating continuous improvement) 305
Fig. 16.2 Structuring framework application with TLARA. (*Note*: A diagram illustrating how the TLARA structure organizes the application of foundational institutional frameworks. It shows principles from core frameworks (e.g., HCASR, RSA) flowing into and being organized by the five TLARA domain containers (Teaching, Learning, Assessment, Research, Administration) 306
Fig. 17.1 Core components of AI literacy. (*Note*: A diagram with a central hub labeled "AI Literacy" connected to four segments: Technical Understanding, Critical Evaluation, Effective & Ethical Application, and Equity & Societal Awareness, outlining core components) 322
Fig. 17.2 The institutional AI integration lifecycle: a continuous journey 340

List of Tables

Table 1.1	Challenges facing higher education in the 5IR	5
Table 1.2	Steps or levels to AGI	9
Table 2.1	Holistic mental model of ethical human-GenAI collaboration and synergy	35
Table 3.1	Illustrative scenarios for human-centric collaboration	46
Table 3.2	Example scenarios for symbiotic collaboration	46
Table 3.3	Example scenarios for AI-centric collaboration	47
Table 4.1	Dual impacts of technological revolutions	57
Table 4.2	Timeline of social institutions as they emerged	58
Table 4.3	Key definitions for viable, nonviable, sustainable, and unsustainable synergies	64
Table 4.4	Tasks GenAI can perform in higher education	66
Table 4.5	Key takeaways on synergies in higher education	72
Table 5.1	Comparison of human-centered and human-centric AI synergy	81
Table 5.2	Key principles, components, and goals of human-centric AI synergy (HCAS)	82
Table 5.3	Key principles, components, and goals of regulation, synergy, and augmentation (RSA)	84
Table 5.4	Key principles, components, and goals of enhancing human-centric skills (EHCS)	86
Table 5.5	Key principles, components, and goals of enhancing academic outcomes (EAO)	88
Table 5.6	Key principles, components, and goals of authentic high-impact practices (AHIPs)	90
Table 6.1	Preliminary stakeholder engagement plan outline (template)	109
Table 6.2	Stakeholder engagement strategies by HEI stakeholder group	113
Table 6.3	Comparison of considerations for proprietary versus third-party AI solutions in higher education	114

Table 7.1 Smart Team charter template 125
Table 7.2 Leadership versus management focus in AI integration (Kotter's dual system) 133
Table 7.3 AI council/formal governance committee overview 134
Table 7.4 Comparison of governance considerations: proprietary versus third-party AI 136
Table 7.5 Chapter 7 key deliverables summary 137
Table 8.1 Methods for stakeholder needs assessment in AI initiatives 147
Table 8.2 Defining SMARTIE goals for AI initiatives 150
Table 8.3 SMARTIE goal planning template 150
Table 8.4 Comparison of UX strategies for deepening AI readiness insights 152
Table 8.5 Sample persona template 153
Table 8.6 Key dimensions of data readiness assessment for AI 155
Table 8.7 Key dimensions of technological capability assessment for AI 156
Table 8.8 Components of workforce skills and AI literacy assessment 157
Table 8.9 Key factors in assessing organizational culture and change readiness for AI 159
Table 8.10 Key areas for assessing existing AI governance and ethical frameworks 161
Table 8.11 Key outputs of needs assessment and readiness evaluation synthesis 162
Table 9.1 Characteristics of effective vision and opportunity statements 174
Table 9.2 Asana/anthropic AI maturity model stages (example) 177
Table 9.3 Institutional AI maturity roadmap (adapted) 179
Table 9.4 Key components of an institutional strategic AI plan 181
Table 10.1 AI pilot strategy document template 192
Table 10.2 Categories of pilot evaluation metrics & assessment areas 195
Table 10.3 External ethical & risk frameworks informing AI pilot design 196
Table 10.4 Required ethical & governance protocols in AI pilot design 197
Table 10.5 Alignment of internal (ETHICAL example) and external ethical frameworks for pilot design 197
Table 11.1 Primary objectives of the AI implementation and pilot phase (phase 2) 209
Table 11.2 Minimum viable product (MVP) principles for AI pilots 211
Table 11.3 Leadership versus management roles during AI pilot implementation 214
Table 11.4 Key activities for monitoring ethical guardrails during AI pilots 214
Table 11.5 Role-specific contributions during AI pilot implementation 220
Table 12.1 Activities for optimizing workflows post-pilot implementation 235
Table 12.2 Activities for updating risk mitigation strategies post-pilot implementation 236
Table 13.1 Key areas and example metrics for assessing sustained AI impact in higher education 249

Table 13.2 Key factors in evaluating AI sustainability 250
Table 13.3 Examples of benchmarks and frameworks for AI evaluation in higher education 251
Table 13.4 Types of audits for ongoing AI governance in higher education 253
Table 14.1 Rationale for scaling and institutionalizing AI initiatives 268
Table 14.2 Pillars of permanent ethical AI governance structures 270
Table 14.3 Strategies for building sustainable AI capacity 272
Table 15.1 Shifting focus from scaling phase (Chap. 14) to adaptation phase (Chap. 15) 284
Table 15.2 Components of iterative policy development and adaptive governance for AI 288
Table 15.3 Strategies for sustaining dynamic stakeholder coordination 290
Table 15.4 Key deliverables for ongoing AI adaptation and refinement 292
Table 16.1 Select governance frameworks emphasizing continuous improvement cycles 301
Table 16.2 TLARA structure: domain examples and ethical considerations for institutional guideline development 304
Table 16.3 Process for institutionalizing TLARA-informed framework development 308
Table 17.1 Comparison of selected AI literacy framework dimensions 325
Table A1 Comparison of considerations for proprietary vs. third-party AI solutions in higher education 349

Part I

Optimizing Human-Centric AI Partnerships: Preparing Higher Education for Hybrid Human-AI Intelligence & The Next Cognitive Revolution

1

The Fifth Industrial Revolution: A New Cognitive Era & Unprecedented Human-AI Collaboration

Introduction

This chapter examines the sweeping forces transforming higher education, exploring the social, economic, demographic, and technological factors redefining institutional practices. Learning objectives focus on understanding these forces, recognizing GenAI's role, exploring the 4IR-to-5IR shift, and preparing for ethical, human-centric frameworks discussed later (see the *Human-AI Collaboration & Synergy Guide* information below). While this chapter surveys these disruptions, core frameworks like Human-Centric AI Synergy & Regulation (HCASR) and Regulation Synergy & Augmentation (RSA) are applied throughout the book.

Bridging Concepts and Practice: Detailed explanations and application guides for these HCASR, RSA, and other human-centric frameworks for ethical AI integration are available in the *Human-GenAI Collaboration & Synergy Guide*, downloadable with each chapter.

Unprecedented Change & Transformation in Higher Education

Higher education is undergoing an unprecedented transformation, driven by rapidly converging social, cultural, and technological revolutions. In just two decades (2000–2020), the Fourth Industrial Revolution (4IR) reshaped society

---

Supplementary Information The online version contains supplementary material available at https://doi.org/10.1007/978-3-032-05205-6_1.

D. Woodring, *Advancing Higher Education with Human-Centric AI Partnerships*,
https://doi.org/10.1007/978-3-032-05205-6_1

with hyper-connectivity, artificial intelligence (AI), the Internet of Things (IoT), and big data (Nosta, 2023; Ziatdinov et al., 2024). Now, the Fifth Industrial Revolution (5IR) is accelerating a new era defined by deeper **human-AI collaboration, synergy, and co-evolution**, distinguishing it from the 4IR's focus on connectivity and automation, which will rapidly redefine the concepts of intelligence, work, and education (Nosta, 2023; Ziatdinov et al., 2024).

As the pace of change intensifies, higher education institutions (HEIs) face mounting challenges that threaten their relevance and sustainability (Chaka, 2023). Built on bureaucratic structures from the nineteenth and twentieth centuries—often characterized by siloed departments, slow decision-making processes, and resistance to systemic change—most HEIs struggle to adapt quickly enough to the volatile, uncertain, and complex world of the twenty-first century (Kotter et al., 2021). The convergence of human cognition and AI marks a critical shift: higher education must co-evolve with AI or risk falling behind the impending cognitive and paradigm shifts discussed in the next chapter (Pedreschi et al., 2023).

To thrive within these transformative forces, HEIs must undergo systemic change, transitioning from rigid bureaucracies to agile, innovation-driven ecosystems that ethically and effectively integrate AI (Ahir et al., 2022). This book applies design thinking (Brown, 2019) and structured change management (Kotter et al., 2021) to help institutions adapt quickly and ethically. However, there is no one-size-fits-all model. Institutions must develop context-sensitive, scalable strategies to leverage AI as a strategic partner for advancing teaching, learning, assessment, research, and administration in higher education. However, the external social and technological changes outpacing higher education go beyond AI. The next section briefly discusses other forces transforming HEIs, before delving into those brought by artificial intelligence.

Uncertainty, Complex Changes, & Rapid Technological Development

HEIs worldwide are under immense pressure as multiple forces converge and accelerate. Economic challenges, including declining enrollments and rising operational costs, are leading to budget cuts, staff reductions, and institutional closures (Dewar, 2024) (Table 1.1). Demographic shifts—such as lower birth rates and a shrinking pool of college-age students—have intensified competition among institutions (Clark et al., 2024).

At the same time, political and regulatory uncertainty—from changing accreditation standards to government funding cuts—adds to institutional

Table 1.1 Challenges facing higher education in the 5IR

Challenge	Description	Institutional response	Potential AI solutions
Economic pressures	Declining enrollments, budget constraints, rising operational costs	Financial restructuring, new funding models	AI-driven budgeting, forecasting, resource optimization
Demographic shifts	Lower birth rates, smaller college-age population, competition	Targeted recruitment, online program expansion	Predictive analytics for enrollment, personalized advising
Political/ regulatory uncertainty	Shifting accreditation standards, public funding cuts	Policy vigilance, lobbying, diversified revenue	Adaptive compliance systems, real-time policy monitoring
Technological disruption	Rapid AI, automation, online learning, new skill demands	Curriculum redesign, faculty retraining, flexible delivery	Automated course generation, AI-based teacher training
Social/cultural shifts	Demand for personalized, tech-driven learning; declining public trust	Enhanced student engagement, transparent governance	Chatbots for personalized support, AI-based community outreach
Environmental factors	Need for sustainable infrastructure and resilience to climate issues	Campus sustainability plans, remote/ online strategies	AI-optimized energy management, disaster preparedness tools

Note. This table summarizes the multifaceted challenges confronting higher education institutions, as discussed in Chap. 1. The descriptions are informed by analyses from sources such as Dewar (2024) on economic pressures, Clark et al. (2024) on demographic shifts, and PwC (2025) on technological disruption

instability and complicates long-term planning (Arrojas, 2023). Technological disruptions, particularly in AI, automation, and online learning, are forcing institutions to rethink curricula, retrain faculty, and adapt teaching methods to keep pace with rapid advancements (PwC, 2025).

Social and cultural shifts are further reshaping expectations, with students demanding more personalized, technology-driven learning while public trust in higher education declines (Riley, 2024). Environmental challenges are also driving the need for sustainable infrastructure, requiring institutions to invest in long-term solutions while managing immediate financial constraints (Aver et al., 2021).

Furthermore, the COVID-19 pandemic accelerated the shift to digital learning, disrupted student mobility, and exposed institutional vulnerabilities, making adaptive strategies more urgent than ever (Lund et al., 2021). Moreover, although education is becoming increasingly costly for beneficiaries, its structure, content, and methodologies have not adapted to the magnitude of these shifts (Ali et al., 2022). HEIs must move beyond incremental adjustments and embrace transformational change to align with the demands of the Fifth Industrial Revolution (5IR) and prepare for the next phase of **human-AI collaboration & synergy**.

The Rapid Evolution of Generative AI & Disruptive Changes in Higher Education

The rapid, uncertain, and disruptive evolution of GenAI is transforming higher education, forcing institutions to adapt at an unprecedented pace. The release of ChatGPT-3.5 in November 2022 marked a turning point, as students rapidly adopted AI for generating academic work in 2024 (Rodriguez, 2024; Intelligent.com, 2023; OpenAI, 2022). By February 2023, Google introduced Bard, investing $300 million in Anthropic to develop its Claude AI model (Originality.ai, 2024).

Simultaneously, Microsoft Copilot entered the market following the company's $10 billion investment in OpenAI (Bloomberg, 2023; Warren, 2023). Just a month later, ChatGPT-4 emerged with significant advances in processing text and images (OpenAI, 2023). The pace only accelerated with Google's launch of Gemini in December 2023, replacing Bard and expanding AI's capabilities (Kruppa, 2023).

The rise of autonomous AI agents, such as AutoGPT, extended AI's functionality beyond chatbots. These third-party tools began executing independent tasks like research automation, email management, and meeting coordination, expanding the capabilities and applications of multiple GenAI models as agents (IBM, 2024; Huang et al., 2023). These tools will further disrupt higher education, and other sectors as their capabilities are optimized and are mass adopted in 2025–2026.

The Rise of Advanced Reasoning Models

By late 2024, competition among AI developers intensified, leading to groundbreaking advancements. OpenAI's o1 and o3 models introduced advanced reasoning, handling PhD-level tasks in mathematics, coding, and scientific research (Ouyang et al., 2022). Specifically, OpenAI's o1 model preview demonstrated advanced reasoning capabilities, reportedly handling complex tasks in areas like mathematics and scientific reasoning at a high level (OpenAI, 2024), while the earlier work on instruction following laid groundwork for such developments (Ouyang et al., 2022). Google DeepMind's Gemini 2.0 Flash followed by integrating multimodal processing across text, images, and audio (Morris et al., 2024).

In October 2024, Anthropic unveiled Claude 3.5 Sonnet, introducing the first autonomous AI feature that allowed the model to simulate human-like interactions with computers—performing complex tasks like typing and navigating interfaces independently (Reuters, 2024). OpenAI quickly responded by announcing "Operator" for 2025, an AI system designed for automating coding and travel arrangements (Statt, 2024). These advancements signal a paradigm shift for HEIs: the ability of AI to reason, automate, and operate autonomously means institutions must rethink academic integrity, curriculum design, faculty training, and AI governance. The challenge is no longer if AI will disrupt education but how HEIs will ethically and effectively integrate AI to remain relevant and excel.

Staying Current in the Age of Accelerated AI

AI systems are evolving so rapidly that keeping pace is a significant challenge, even for dedicated experts; relying solely on traditional academic publication cycles is no longer sufficient. Continuous development of AI ethics and literacy skills is essential for all stakeholders in higher education.

To support this, this book incorporates several strategies:

Personalized Adaptive Learning Sessions (PALS): Structured activities throughout this book guide you in partnering with your GenAI model to actively explore chapter concepts, retrieve current research, and practice ethical AI use. PALS leverages the **HCASR (Human-Centric AI Synergy & Regulation)** framework (detailed in the *Human-GenAI Collaboration & Synergy Guide*) to help you augment your learning effectively. **Turn to the Chap. 1 PALS Workbook** for your first guided activity using specific prompts related to the 5IR and AI's impact.

AI Ethics & Literacy Training: Part 3 of the *Human-GenAI Collaboration & Synergy Guide* offers resources conceptually similar to a "Crash Course," using PALS to create customized GenAI ethics and literacy training tailored to user goals.

External Resources: Supplement your learning by following reputable sources that track AI advancements closely. For example, podcasts like *The*

Artificial Intelligence Show provide regular updates and analysis relevant to the topics discussed here, including dedicated coverage of AGI/ASI developments (Roetzer & Kaput, 2021–present).

AI systems are evolving so rapidly that even those who dedicated 100% of their time to keeping up with it have a difficult time. This means that waiting for academic journal articles to come out is not an option. Users must find a way to continually enhance their AI ethics and literacy skills and competencies.

The Road to Artificial General Intelligence & Artificial Superintelligence

The rapid advancement of GenAI is accelerating the progression toward **Artificial General Intelligence (AGI)** and **Artificial Superintelligence (ASI)**—systems capable of surpassing human intelligence across diverse domains. OpenAI and Google have outlined structured roadmaps that define increasing levels of AI capability, moving from specialized conversational models to fully autonomous systems that can reason, innovate, and manage complex operations independently. In 2024, OpenAI introduced a five-level framework for AGI development, beginning with Conversational AI—exemplified by models like ChatGPT, which engage in human-like dialogue (see Table 1.2). The framework advances through **Reasoners**, which solve PhD-level problems, to Autonomous Agents capable of independently completing complex tasks over extended periods. At the highest levels, Innovators generate novel solutions, while Organizational AI functions as independent entities managing large-scale operations. By the end of 2024, OpenAI's o1 and o3 models had demonstrated major advancements in reasoning and problem-solving, marking a significant step toward AGI.

Google DeepMind is also preparing for AGI (see Table 1.2). DeepMind's Gemini 2.0 demonstrated significant advancements in multimodal AI, integrating text, image, and audio processing with high-level reasoning and factual accuracy. It achieved a 76.4% score on the MMLU-Pro reasoning benchmark and a 92.9% performance in the Natural2Code programming test, establishing itself as one of the most advanced AI models available. Leading AI researchers project that AGI could emerge within the next decade, driven by advancements in compute scaling, algorithmic efficiency, and continuous learning models. The transition to ASI is expected to follow rapidly,

Table 1.2 Steps or levels to AGI

Level	OpenAI's 5 steps to AGI	Google's levels of AGI
Level 1	**Conversational AI:** Systems excel at natural conversations (e.g., ChatGPT).	**Emerging Narrow AI:** Simple rule-based systems; performance equal to or better than unskilled humans.
Level 2	**Reasoners:** AI solves complex, Ph.D.-level problems without external tools.	**Competent Narrow AI:** Performs at least at the 50th percentile for a subset of tasks (e.g., Alexa, Siri).
Level 3	**Autonomous agents:** Systems operate autonomously over days, handling intricate tasks with adaptability.	**Expert Narrow AI:** Performs at or above the 90th percentile for skilled tasks in narrow domains (e.g., DALL-E, Grammarly).
Level 4	**Innovators:** AI develops groundbreaking ideas and solutions independently across various fields.	**Virtuoso Narrow AI:** Achieves performance in the 99th percentile for specific tasks (e.g., AlphaGo), surpassing most humans.
Level 5	**Organizational AI:** AI manages entire organizational structures independently, akin to an operational entity.	**Artificial Superintelligence (ASI):** Outperforms 100% of humans in a wide range of general and novel tasks.

Note. This table provides a comparative overview of frameworks developed by leading AI labs to operationalize the path toward Artificial General Intelligence (AGI). The levels for OpenAI are based on its stated AGI development framework, while Google's levels of AGI are adapted from Morris et al. (2024). Definitions of and debates around AGI are highly diverse

introducing both profound benefits and unprecedented challenges. Proponents argue that ASI will accelerate scientific discovery, revolutionizing fields such as medicine, engineering, and theoretical research, while skeptics warn that the transition could lead to serious ethical dilemmas, economic disruption, and geopolitical instability.

The models keep getting smarter. Studies on neural scaling laws (including the so-called Chinchilla scaling) show that as you increase the amount of training data and compute, model performance improves predictably (Hoffmann et al., 2022). These laws explain how and why larger models trained on more tokens tend to achieve lower error rates, effectively "getting smarter" as they scale. However, some of the world's leading AI researchers are planning to create newer architectures for AGI and ASI that may or may not even be built on Generative AI, as discussed next.

Accelerating Toward Superintelligence: The Dawn of a New Cognitive Era

In 2024, Sam Altman, CEO of OpenAI, released the essay The Intelligence Age, outlining the trajectory toward AGI and ASI and predicting that compute power and deep learning breakthroughs would push AI beyond human limitations within the coming years (Altman, 2024, September 23). By 2025, OpenAI publicly asserted confidence in its ability to build AGI, marking a pivotal moment in the advancement of AI agents capable of transforming organizational operations (Altman, 2025). In 2025, Altman further acknowledged both the transformative potential and the uncertainties of AGI and ASI, describing humanity's proximity to superintelligence as "near the singularity; unclear which side" (Altman, 2025). Logan Kilpatrick reinforced the inevitability of ASI by referencing Ilya Sutskever's (former Chief Scientist at Open AI) "Straight Shot to Artificial Superintelligence" strategy, which prioritizes rapid compute scaling over incremental AI advancements (Kilpatrick, 2025).

After announcing the launch of his company *Safe Artificial Superintelligence* (SSI) in June of 2024 (Sutskever, 2024), ten months later SSI was valued at $32 billion, having released no products and operating with a minimalist online presence (PYMNTS, 2025). Shortly after, the main investors of SSI announced the launch of the company Mechanize.work, stating that it is "focused on developing virtual work environments, benchmarks, and training data that will enable the full automation of the economy" (Barnett et al., 2025). Together, this suggests that the emergence of ASI is not a distant possibility but an expected outcome of competitive AI development. The accelerating pace of AI research has shifted the field from theoretical exploration to an engineering challenge, where the risks and rewards are tangible and immediate.

Altman (2025) explains that AI development benefits from scaling laws that not only reduce costs exponentially annually but also drive super-exponential socioeconomic returns—suggesting that, in less than a decade, individual productivity could match the collective capability of today's most accomplished teams (Altman, 2025). This sentiment is echoed by the Marketing AI Institute (2025), which highlighted that the next two years are expected to yield more dramatic breakthroughs than recent progress, with next-generation models such as GPT-5 set to revolutionize fields ranging from scientific research to everyday productivity. Complementing these views, the Financial Times (2025, February 27) reported on innovative ventures like the funding of AI-backed bioscience projects, illustrating how AI's rapid integration is reshaping industries far beyond traditional tech sectors. The rapid evolution of AI will not only redefine technological boundaries but will also reconfigure economic and societal structures (Marketing AI Institute, 2025). However, testing the curve of the scaling laws will require a massive investment in and infrastructure for energy for powering the data centers that power AI, which are akin to "intelligence factors" as discussed below.

Massive Energy Requirements & Funding for Growing Data Centers

To advance the scaling laws as far as possible, AI flagship companies like Google are planning to test these scaling laws power by nuclear energy. Google has already signed a "world first" nuclear power deal with Kairos Power, committing to purchase electricity from small modular nuclear reactors (SMRs) that will provide 500 megawatts of carbon-free energy starting by 2030 (Lawson, 2024). In addition, the "Stargate" project was announced by President Trump alongside leaders from OpenAI, SoftBank, and Oracle, as a joint venture to invest up to $500 billion in U.S. AI infrastructure. The initiative will build data centers (starting in Texas) and related energy generation facilities, with an initial $100 billion commitment that could scale to $500 billion over four years (Reuters, 2025). In fact, Jensen Huang, CEO of NVIDIA, the company that builds these data centers, compares the shift to AI-driven computing to the Industrial Revolution, emphasizing the emergence of factories that produce intelligence—specialized data centers designed to produce AI **Tokens**. These floating-point numbers represent AI-generated outputs such as language, proteins, chemicals, images, and robotic controls. By discovering how to produce AI tokens at an unprecedented scale, Huang suggests that the world will witness an explosion of AI-generated content and solutions (Campolargo, 2024). This mass production of increasingly advanced "intelligence" means there will be large disruptions across all sectors, similar to the advent of printing press, electricity, and/or automobiles.

Huang further explains that just as electricity was a defining factor in previous industrial revolutions, AI tokens will serve as the foundation for a new economic era. He estimates that industries collectively worth $100 trillion will benefit from enhanced productivity through AI-driven services and products. To facilitate financial transactions in this emerging token economy, tools like Stripe will be essential for managing and monetizing AI-generated assets (Campolargo, 2024). But what does this mean for higher education?

Keeping up with AGI/ASI: To keep up with rapid evolution and impact of AI, this book recommends *The Artificial Intelligence Show* podcast with Paul Roetzer and Mike Kaput. Considering starting with *Road to AGI* to keep up with advances in AGI/ASI and what it means for higher education and beyond (Roetzer, 2025).

Critical Thinking Question: How does the role of higher education, faculty, staff, and students change when autonomous agents can answer emails, complete assignments and grading, conduct PhD-Level research, conduct academic research analyzing thousands of variables and studies simultaneously, and form novel research questions and hypotheses?

Preparing Higher Education for Advanced AI

The rapid advance of AI toward AGI/ASI, combined with the massive infrastructure development, has far-reaching implications for higher education. HEIs must proactively address the challenges and opportunities that accompany AI-driven transformation. Traditional pedagogical models must adapt to accommodate AI-assisted learning, faculty training must incorporate AI literacy, and institutions must develop robust governance structures to ensure the ethical deployment of AI. Universities have a responsibility to lead research, policy development, and workforce adaptation strategies so that AGI is not dictated solely by corporate interests. Institutions that fail to integrate AI effectively risk losing their relevance in an era when artificial intelligence is no longer merely a supplementary tool but a primary driver of knowledge creation, problem-solving, and decision-making. Successfully navigating this degree of complexity and disruption necessitates structured, adaptable, and fundamentally human-centric approaches to strategy, integration, and ongoing governance.

However, there is both a silver (and more exciting) lining to these changes, as well as the growing necessity to incorporate AI to deal with rapid change, that have the opportunity to advance teaching, learning, assessment, and research exponentially in higher education as well. This means HEIs and key stakeholders must now focus on principles, theories, frameworks, methods, models, strategies, and approaches from multiple disciplines that optimize human-GenAI collaboration & synergy between stakeholders and GenAI and the emerging systems.

Bridging Concepts to Practice: Applying Human-Centric Frameworks

Given that AI systems are often not designed or deployed in truly human-centric ways (Spencer, 2024), it becomes up to users, stakeholders, and institutions to optimize the way they are used. This book utilizes several core frameworks (see the *Human-GenAI Collaboration & Synergy Guide* and Chap. 5 in Part 1) that offer structured insights and practical strategies for this purpose.

New Cognitive Revolutions Through Human-AI Collaboration & Synergy

Human-AI co-creation highlights how synergy can lead to outputs that neither human nor machine would produce independently (Wu et al., 2021). This synergy manifests in fields such as medicine, where AI-based image analysis supports clinicians' judgments, or in creative industries where generative models offer design suggestions that artists refine (Rane et al., 2024). Within higher education, this approach could augment practices such as group brainstorming, prototyping solutions to engineering problems, or managing interdisciplinary research. AI-based systems like ChatGPT, Gemini, Claude, or specialized co-creator tools can accelerate the brainstorming phase by injecting diverse perspectives, which faculty and students then refine. This means that human-AI partnerships will lead to innovative and groundbreaking scientific methods, models, strategies, and approaches, setting the stage for massive cognitive paradigm shifts.

Fundamental Shifts in Knowledge & Understanding

Consider how moving from the *geocentric* to *heliocentric model*, as well as how transitioning from *Newtonian Physics* to *quantum mechanics* dramatically changed human understanding of reality. These scientific breakthroughs are already emerging from the convergence of human and machine intelligence and collaboration. This is evidenced by several breakthrough showing how **human–AI collaboration** can achieve feats previously deemed unattainable due to the sheer scale of data and resources required. DeepMind's AlphaFold, for instance, has predicted over 200 million protein structures—compressing what would have taken nearly one billion years of traditional research into a single year and saving trillions of dollars in experimental costs (Goldman, 2023; Chiang, 2024; European Bioinformatics Institute, n.d.).

Similarly, Recursion Pharmaceuticals has harnessed AI-driven platforms to revolutionize drug discovery by rapidly screening and identifying potential therapies from vast biological datasets, a process that traditionally spanned decades (Recursion Pharmaceuticals, 2022). In another domain, innovative work by companies like Altana has enabled the mapping of global supply chains in real time; this capability allows businesses to monitor, optimize, and react to disruptions across networks that are too complex for manual analysis (Wilson, 2024). These tools can analyze more data in a day than humans can in millions of years and therefore can see relationships previously hidden from human sight.

Additionally, breakthroughs in AI-enhanced weather forecasting have dramatically reduced prediction times from hours to minutes, empowering more precise and timely responses to climate events and disasters (Chiang, 2024). Collectively, these advancements underscore not only the significant changes ahead, but the transformative potential of integrating human expertise with cutting-edge AI technologies to overcome longstanding challenges across diverse sectors.

Universities can capitalize on these shifts by revising their research and innovation strategies. When AI can independently propose hypotheses or design experiments, human scientists must refine their oversight and creativity, focusing on ethical considerations and conceptual leaps that AI might not handle well (Secundo et al., 2024). The merging of human and machine cognition opens new opportunities for co-evolution, where each partner brings distinct capabilities. AI excels at handling large-scale data and computational intensity, while humans contribute ethical reflection, empathy, and nuanced decision-making (Wilson & Daugherty, 2018).

This interdependence transforms the perceived boundaries of knowledge creation. Students, faculty, and AI systems can engage in a form of collaborative intelligence that surpasses traditional paradigms of teacher-student roles. Instead, classes may structure debates around how to prompt AI effectively and critique or improve its outputs. The emphasis shifts from rote memorization to the capacity to interrogate data, interpret results, and articulate ethical use. However, manifesting this reality in a way that empowers humans (instead of replacing them or replicating their intelligence) requires a focus on **human-centric** principles, goals, and objectives.

Human-Centricity in Design & Use of AI

While human-centered design optimizes the user's interaction with the tool itself, **human-centricity** takes a broader view, ensuring the technology serves larger goals of human flourishing, equity, and empowerment. The *Human-GenAI Collaboration & Synergy Guide* accompanying this book delves deeper into applying human-centric principles (Gooptu, 2024).

This is fundamentally different from "human-centered design" which focuses on immediate usability and satisfaction. Importantly, to ethically and effectively harness the power of AI in ways that empower humans (versus replacing them with AI) and augment their knowledge, skills, and abilities (instead of replicating them with AI), requires a focus on human-centric frameworks, as presented in this book. This is crucial since most AIs are not designed in truly human-centric ways.

Human-AI Co-Creation and Its Effects on Innovation

Despite these possibilities, challenges remain. Over-reliance on AI can reduce direct human practice in critical thinking or creative expression. Some fear that delegating too many tasks to AI might impair students' capacity to learn underlying processes. Institutions must address these challenges by aligning AI usage with broader learning objectives, ensuring that it complements rather than displaces human cognition (Tariq et al., 2024). The responsible integration of GenAI often hinges on contextual factors, including how well the software's limitations are communicated and whether educators develop tasks that require human oversight (Bellemare-Pepin et al., 2024).

Implications for Institutional AI Development and GenAI Integration

Universities seeking to develop proprietary AI models or integrate third-party solutions must consider both ethical and strategic imperatives. For instance, some institutions may invest in specialized large language models (LLMs) with domain-specific expertise, while others may adopt commercial solutions like Microsoft Copilot or Google's Gemini, adapting them for academic purposes. In either approach, organizations face decisions about data governance, transparency, and potential biases. The presence of AI in sensitive processes like admissions or financial aid decisions intensifies concerns about fairness and accountability (Weidinger et al., 2024).

In educational settings, GenAI can streamline tasks such as grading, curriculum design, and personalized learning. Student support services could incorporate AI-based advising or tutoring systems, freeing faculty for more nuanced interactions that require emotional intelligence and deep subject expertise (Shaer et al., 2024). Administrators also benefit from AI in strategic planning, problem-solving, and day-to-day operations. However, these gains require thorough change management strategies to prepare staff, address apprehension, and ensure alignment with academic missions (Kotter et al., 2021).

An expanding role of GenAI as a co-creator raises questions about authorship and intellectual property, especially when entire sections of scholarly work are co-generated by AI tools. Institutions must clarify guidelines on data provenance, attribution, and the permissible scope of AI involvement in assignments (Chaka, 2023). They must also establish governance frameworks that keep administrators, faculty, students, and external stakeholders engaged in shaping AI policies and best practices.

Intro to Smart Teams: Many HEIs are forming "smart teams"—diverse groups of faculty, staff, students, and technologists—to manage AI deployment and continuously monitor evolving needs. This cross-functional collaboration helps harness synergy-based approaches to solve real institutional problems.

In this book, users from a variety of roles are introduced to **User Experience (UX) Strategies** for ethically and effectively implementing, piloting, evaluating, and scaling GenAI. This begins with "UX Warmup Activities" like the one at the end of this chapter. Users can do these activities individually or in teams. But a specific focus of this book is on urging AI leaders in education, no matter what their role, goals, or objectives, to form diverse Smart Teams consisting of undergraduates, graduate students, full-time and part-time faculty, staff, and administrators, to experiment and innovate during the process of GenAI integration.

Reconfiguring Higher Education for the Intelligence Age

For universities, the impending transformations related to AI demand a reconfiguration of multiple layers. Curriculum design must expand to include AI literacy, data analytics, and interdisciplinary perspectives that address both the strengths and weaknesses of intelligent technologies. Professional development for faculty must move beyond mere technical training, fostering comfort with AI tools and designing learning experiences that harness AI effectively (EAB, 2024).

Evaluation methodologies will also evolve. As AI-generated text or code becomes ubiquitous, the focus may shift toward in-class demonstrations of critical thinking, oral defenses of ideas, and collaborative projects that require human insight. Academic integrity policies need to consider how best to incorporate AI usage guidelines, maintain fairness, and preserve creativity. Some programs might require learners to produce "prompt logs" for AI-assisted tasks, while others might treat AI as an extension of the library or reference materials.

Research and publishing also confront new norms. If AI can rapidly generate literature reviews, design research protocols, and interpret data sets, scholars must verify these outputs for accuracy and bias. The interplay between AI-driven knowledge generation and traditional peer review processes remains an open question (Ouyang et al., 2022). Institutions committed to academic rigor will likely invest in protocols that incorporate AI while ensuring thorough human oversight.

Governance structures must incorporate representatives from multiple stakeholder groups—including faculty, students, accreditation bodies, and even AI providers—to develop guidelines that can adapt to shifting technological contexts. Forward-looking leadership can ensure that AI's integration remains consistent with institutional missions and ethical priorities (Kotter et al., 2021).

Ethical, Regulatory, and Governance Considerations

Ethical frameworks for AI usage in higher education typically address transparency, fairness, accountability, and privacy (Weidinger et al., 2024). With generative AI able to compose text or code at advanced levels, intellectual property concerns arise, along with the potential for disinformation if AI models generate inaccurate or misleading content. Institutions must also address possible biases embedded in large language models, especially when the data sets used for training do not reflect diverse perspectives.

Regulatory bodies and governments have begun to develop guidelines for AI usage, but these remain fragmented and can shift quickly. In higher education, accreditation agencies might also adapt standards to include responsible AI practices. While some universities take proactive steps, others risk regulatory censure if they adopt AI hastily without appropriate safeguards (Dotan et al., 2024). A sustainable approach requires vigilant oversight, ongoing risk assessment, and a commitment to refining AI policies in light of new developments.

AI governance is not solely the domain of IT departments or top-level administrators. Shared governance models that involve faculty senates, student councils, and specialized committees can provide more balanced oversight. Including student voices is crucial because they frequently have firsthand insights on AI's day-to-day academic usage (Bozkurt et al., 2024). External partnerships with industry and government agencies can supply resources and perspectives that benefit institutional planning.

The Human-GenAI Collaboration & Synergy Guide

The *Human-GenAI Collaboration & Synergy Guide* is available for download with chapter of this book and contains resources, tools, strategies, and frameworks for realizing the above ethical and effective integration of GenAI through strategic **Human-GenAI Partnerships.** This means strategically implementing AI for some roles and tasks, while keeping others human-led.

Design Thinking and Change Management for AI Integration

Many experts advocate frameworks such as Design Thinking—with its emphasis on empathy and iteration (Brown, 2008)—and structured change management (Kotter, 2012) to ensure AI-driven transformations align with organizational goals and prioritize human-centric values, principles explored further through the frameworks presented in this book. Design Thinking encourages empathy with user needs and iterative prototyping, which can help avoid superficial technology pilots that fail to address deeper challenges. In higher education, these methods can be applied to course redesign, service improvement, or reimagining administrative workflows.

Change management processes offer strategies to minimize resistance and align stakeholder objectives (Kotter, 2012). When introducing advanced AI solutions, leaders may face skepticism from faculty concerned about job security, ethical dilemmas, or the influence of corporate technology within academic settings. Transparent communication, professional development sessions, and pilot projects that involve faculty leaders can ease these transitions. Regular feedback loops, where end-users provide ongoing input, help refine AI usage over time (Brown, 2019; Kotter et al., 2021).

Because AI evolves swiftly, change management cannot be seen as a single-phase initiative. Instead, institutions must cultivate continuous improvement cultures that adapt policies, practices, and training to each new wave of AI features. Partnerships across departments, with external experts, and the formation of Smart Teams composed of diverse stakeholders can help ensure that best practices are shared and that no single unit is left to navigate AI adoption in isolation. However, users, stakeholders, and leaders are encouraged to use the tools provided in this book to experiment and innovate beyond these strategies and approaches, as this is, and will be, uncharted territory for the foreseeable future.

Evidence-Based Planning and Collaborative Problem-Solving

In an environment shaped by accelerated developments, evidence-based planning becomes essential. Research from fields such as social work and education indicates that outcomes improve when interventions are grounded in rigorous data, stakeholder input, and actionable research findings (Head, 2010; Mullen et al., 2015). Higher education institutions can leverage large data sets and advanced analytics provided by AI to optimize resource allocation, identify at-risk students, and evaluate the effectiveness of new teaching strategies.

Collaborative problem-solving frameworks add another dimension by drawing on design thinking or cross-functional teams. This approach encourages shared ownership of institutional challenges (Anderson & Shattuck, 2012). AI systems can process large sets of institutional data, propose possible interventions, and simulate outcomes. Humans then interpret these projections, integrating ethical insights or contextual knowledge. This synergy can reduce guesswork in institutional planning and strengthen accountability, as decisions can be traced back to clear data-driven rationales (Lane et al., 2024). Reviewing this research, it becomes clear that, with the proper frameworks for deployment, use, and integration of these tools, significant benefits can be realized for key stakeholders and HEIs as a whole.

GenAI as an Assistant, Partner, and Co-Creator

Generative AI can operate in different roles, depending on how extensively it is integrated. As an assistant, it handles repetitive tasks or provides quick reference information, freeing academic or administrative staff for more demanding work (Zhai et al., 2024). In a partner capacity, AI becomes a collaborator—offering brainstorming inputs, writing suggestions, or computational analyses that humans refine. The co-creator role extends this synergy further. AI may generate entire drafts, conceptual frameworks, or designs that users critique and adapt. Within higher education, selecting the appropriate level of AI collaboration depends on the complexity of tasks, the institution's ethical guidelines, and the readiness of stakeholders.

However, the depth of AI integration correlates directly with risk. An assistant role typically presents fewer ethical concerns, whereas a co-creator role might raise questions about original authorship, authenticity of work, and the potential overshadowing of student or faculty efforts. A balanced approach that foregrounds clear guidelines and transparency helps institutions leverage AI's strengths without compromising human development or academic values (Rick et al., 2023).

Potential Futures and Ongoing Adaptation

Despite the challenges and uncertainties associated with integrating AI, the higher education sector stands at a moment when bold transformation can yield lasting benefits. Every dimension of the academic enterprise, from finance to faculty workload, is influenced by AI's growing sophistication. As advanced reasoning, automation, and generative capabilities deepen, more responsibilities can shift to AI, potentially expanding the capacity of students and faculty to focus on holistic intellectual growth.

Nonetheless, AI's limitations require careful attention. While **language models** excel at generating text, they do not inherently verify truthfulness. AI's knowledge is shaped by the data used during training, which may contain biases or omissions. For critical tasks in social sciences or human services, these deficits can lead to misrepresentation or discriminatory outcomes if unchecked. The impetus falls on institutions to ensure robust oversight and continuous improvement (Weidinger et al., 2024).

Universities that embrace iterative approaches and frequent evaluations can remain responsive to ongoing changes in AI capabilities. Through cross-departmental committees or specialized AI ethics boards, policies can be updated quickly as new features or challenges arise. This approach echoes principles from software development, where agile methods emphasize real-time user feedback and repeated refinements (Romero, 2023).

Over time, these adaptive strategies can guide institutions to more effective and sustainable AI integrations. The HCASR framework (elaborated in the *Human-GenAI Collaboration & Synergy Guide*, Part 1) offers principles and practical methods for navigating this, addressing challenges like bias, privacy, and over-reliance while fostering ethical collaboration adaptable to unique contexts and goals.

HCASR is a five-component, nonlinear framework that users/HEIs can adapt and customize using their own GenAI models, to optimize ethical and effective human-GenAI collaboration & synergy according to their own unique roles, goals, and contexts.

One of the components is goal setting:

- What goals do you, your team, or your institution need to be setting and achieving consistently to stay commensurate with the rapid advances in technology and changes in society that are impending?

Conclusion: Building a Human-Centric AI Collaboration & Synergy Model

The rapid acceleration of AI marks a crucial phase for higher education. Institutions encounter significant economic and demographic obstacles while simultaneously navigating an exponential rise in AI capacity. Generative AI, once considered a novelty, has become a fundamental force in shaping teaching, learning, research, and administration. As advanced reasoning models inch closer to Artificial General Intelligence, universities must do more than passively watch. They must define ethical frameworks, redesign curricula, and implement governance structures that handle AI's complexities while leveraging its strengths (Pedreschi et al., 2023).

Throughout this chapter, emphasis has been placed on viewing AI not as a replacement for human judgment, but as a partner that can augment creativity, streamline operations, and open novel pathways of inquiry. Such collaboration calls for human-centric approaches, ensuring that oversight, empathy, and ethical considerations remain central (Shneiderman, 2022). By embracing design thinking and structured change management, institutions can integrate advanced AI in a manner consistent with foundational academic values.

The practical implications are wide-ranging. Departments might experiment with AI-assisted lab research, adopt new forms of student evaluation, or employ AI tools for more nuanced administration. As these practices spread, the higher education sector must continuously redefine academic integrity, credentialing, intellectual property, and competency measurement in an environment increasingly shaped by intelligent systems (Chaka, 2023). Having outlined the scale of transformation and the urgency for adaptation, the subsequent chapters, workbook activities, and the accompanying *Human-GenAI Collaboration & Synergy Guide* provide practical frameworks (like HCASR) for ethical integration and synergy, designed to help institutions, teams, and individuals respond strategically and effectively. Part 2, in particular, details a customizable four-phase rollout plan (implementing, piloting, evaluating, scaling) that leverages these frameworks and GenAI's capabilities for personalization, guiding HEIs worldwide toward responsible, ethical, and sustainable AI integration.

End of Chapter Workbook Activities: Engaging with GenAI as a Strategic Partner & Stakeholder

The UX Warmup Activities at the end of each chapter are aligned with the companion workbooks available for download with this chapter. Their goals are to provide training in UX design, ethically and effectively engaging with GenAI, and strategically employing GenAI as a partner for planning, research, UX strategies, living documents, and other outcomes and deliverables, in a human-centric way that empowers users and advances human knowledge, skills, and abilities. All of these activities are based on the core frameworks of this book, available in Part 2 of the *Human-GenAI Collaboration & Synergy Guide* available for download with this chapter.

To optimize these goals, it largely depends on humans doing as much of the work as they can before interacting with their AI models. This helps ensure that AI is used to augment existing human abilities and capabilities, instead of merely simulating or replicating them with AI.

Chapter 1 Workbook Prep

Before engaging with the companion UX and PALS workbooks for Chap. 1, consider the following points to focus your thinking:
Context & Potential Application (Prep for UX Goal Setting/Use Cases):

- Reflecting on the 5IR/GenAI disruptions discussed, what is **one key opportunity or challenge** most relevant to your specific role or context?
- Briefly brainstorm **one potential GenAI use case** that could address this opportunity/challenge.

Stakeholder Perspectives (Prep for UX Empathy Mapping):

- Identify the **primary stakeholders** involved or affected by the opportunity/challenge/use case identified above.
- What are your **initial assumptions** about their perspectives or concerns regarding GenAI in this context? **(Bring these thoughts to the Empathy Mapping activity in the UX Workbook).**

AI-Powered Exploration (Prep for PALS):

- Select **one specific concept or question** from Chap. 1 (e.g., human-AI synergy, AGI implications, 5IR characteristics) that you want to understand more deeply.
- Think about **how you might initially ask a GenAI model** to help you explore this topic. *(This prepares you for the guided learning activities in the PALS Workbook).*

References

Ahir, H., Bloom, N., & Furceri, D. (2022). *The world uncertainty index* (NBER Working Paper No. 29763). National Bureau of Economic Research. https://www.nber.org/papers/w29763

Ali, S. H., Ayad, H., & Al Rubaie, M. T. (2022). Fifth industrial revolution (New perspectives). *International Journal of Business, Management, and Economics, 3*(3), 196–212. https://doi.org/10.47747/ijbme.v3i3.694

Altman, S. (2024, September 23). The intelligence age. *Sam Altman Blog*. https://ia.samaltman.com/

Altman, S. (2025, January 5). Reflections. *Sam Altman Blog*. https://blog.samaltman.com/reflections

Anderson, T., & Shattuck, J. (2012). Design-based research: A decade of progress in education research? *Educational Researcher, 41*(1), 16–25. https://doi.org/10.3102/0013189X11428813

Arrojas, M. (2023, December 12). College accreditation is changing. Here's what students need to know. *BestColleges.com*. https://www.bestcolleges.com/news/college-accreditation-changes-incoming/

Aver, B., Fošner, A., & Alfirević, N. (2021). Higher education challenges: Developing skills to address contemporary economic and sustainability issues. *Sustainability, 13*(22), Article 12567. https://doi.org/10.3390/su132212567

Barnett, M., Besiroglu, T., & Erdil, E. (2025, April 17). Announcing Mechanize, a startup focused on developing virtual work environments. *Mechanize.work*. https://www.mechanize.work/

Bellemare-Pepin, A., Lespinasse, F., Thölke, P., Harel, Y., Mathewson, K. E., Olson, J. A., Bengio, Y., & Jerbi, K. (2024). Divergent creativity in humans and large language models (arXiv:2405.13012). *arXiv*. https://doi.org/10.48550/arXiv.2405.13012

Bloomberg. (2023, January 23). Microsoft makes multibillion-dollar investment in OpenAI. *Bloomberg*. https://www.bloomberg.com/news/articles/2023-01-23/microsoft-makes-multibillion-dollar-investment-in-openai

Bozkurt, A., Xiao, J., Farrow, R., Bai, J. Y. H., Nerantzi, C., Moore, S., Dron, J., Stracke, C. M., Singh, L., & Crompton, H. (2024). The manifesto for teaching and learning in a time of generative AI: A critical collective stance to better navigate the future. *Open Praxis, 16*(4), 487–513. https://doi.org/10.55982/openpraxis.16.4.777

Brown, T. (2008, June). Design thinking. *Harvard Business Review, 86*(6), 84–92.

Brown, T. (2019). *Change by design: How design thinking transforms organizations and inspires innovation* (Rev. and updated ed.). Harper Business.

Campolargo, J. D. (2024, June 13). The AI token revolution by NVIDIA's Jensen Huang. *Medium*. https://medium.com/@johndcampolargo/the-ai-token-revolution-by-nvidias-jensen-huang-e1441a7754d5

Chaka, C. (2023). Fourth industrial revolution—a review of applications, prospects, and challenges for artificial intelligence, robotics and blockchain in higher education. *Research and Practice in Technology Enhanced Learning, 18*, Article e002. https://doi.org/10.58459/rptel.2023.18002

Chiang, Y. S. (2024, March 4). Discover new drugs in months, not years, with AI: DeepMind CEO Demis Hassabis. *GovInsider*. https://www.govinsider.asia/intl-en/article/discover-new-drugs-in-months-not-years-with-ai-deepmind-ceo

Clark, C., Cluver, M., Fishman, T., & Kunkel, D. (2024, April 19). Latest trends in higher education. *Deloitte Insights*. https://www2.deloitte.com/us/en/insights/industry/public-sector/latest-trends-in-higher-education.html

Dewar, J. (2024, March 14). Universities on a financial rollercoaster. *KordaMentha*. https://kordamentha.com/insights/universities-on-a-financial-rollercoaster

Dotan, R., Parker, L. S., & Radzilowicz, J. G. (2024). Responsible adoption of generative AI in higher education: Developing a "points to consider" approach based on faculty perspectives (arXiv:2406.01930). *arXiv*. https://doi.org/10.48550/arXiv.2406.01930

EAB. (2024, March 25). 4 steps academic leaders must take to integrate AI tools into pedagogy. *EAB*. https://eab.com/resources/blog/strategy-blog/integrate-ai-tools-into-pedagogy/

European Bioinformatics Institute. (n.d.). AlphaFold protein structure database. Retrieved May 10, 2025, from https://alphafold.ebi.ac.uk

Financial Times. (2025, February 27). Sam Altman-backed Retro Biosciences to raise $1bn for project to extend human life. *Financial Times*. https://www.ft.com/content/25a473ea-9f87-474a-8729-bc5287df853a

Goldman, S. (2023, July 27). A year ago, DeepMind's AlphaFold AI changed the shape of science — but there is more work to do. *VentureBeat*. https://venturebeat.com/ai/a-year-ago-deepminds-alphafold-ai-changed-the-shape-of-science-but-there-is-more-work-to-do/

Gooptu, A. (2024, July 3). Artificial Intelligence technology should revolutionize humanity making AI human-centric, endurable and supportive of UN SDGs. *AIFOD*. https://aifod.org/artificial-intelligence-technology-should-revolutionize-humanity-making-ai-human-centric-endurable-and-supportive-of-un-sdgs/

Head, B. W. (2010). Reconsidering evidence-based policy: Key issues and challenges. *Policy and Society, 29*(2), 77–94. https://doi.org/10.1016/j.polsoc.2010.03.001

Hoffmann, J., Borgeaud, S., Mensch, A., Buchatskaya, E., Cai, T., Rutherford, E., de las Casas, D., Hendricks, L. A., Welbl, J., Clark, A., Hennigan, T., Noland, E., Millican, K., van den Driessche, G., Damoc, B., Guy, A., Osindero, S., Simonyan, K., Elsen, E., … Sifre, L. (2022). Training compute-optimal large language models. (arXiv:2203.15556). *arXiv*. https://doi.org/10.48550/arXiv.2203.15556

Holland, S. (2025, January 21). Trump announces private-sector $500 billion investment in AI infrastructure. *Reuters*. https://www.reuters.com/technology/artificial-intelligence/trump-announce-private-sector-ai-infrastructure-investment-cbs-reports-2025-01-21/

Huang, Y., Shi, J., Li, Y., Fan, C., Wu, S., Zhang, Q., Liu, Y., Zhou, P., Wan, Y., Gong, N. Z., & Sun, L. (2023). MetaTool benchmark for large language models: Deciding whether to use tools and which to use (arXiv:2310.03128) [cs.SE]. *arXiv*. https://doi.org/10.48550/arXiv.2310.03128

IBM. (2024, October 15). What is AutoGPT? AI's autonomous tool explained. *IBM Think Blog*. https://www.ibm.com/blog/what-is-autogpt/

Intelligent.com. (2023, May 8). One third of college students used ChatGPT for schoolwork during the 2022–23 academic year. *Intelligent.com*. https://www.intelligent.com/one-third-of-college-students-used-chatgpt-for-schoolwork-during-the-2022-23-academic-year/

Kilpatrick, L. [@officiallogank]. (2025, January 4). Ilya's straight shot to artificial superintelligence [Tweet]. *X*. https://x.com/officiallogank/status/1873788158610928013

Kotter, J. P. (2012). *Leading change*. Harvard Business Review Press.

Kotter, J. P., Akhtar, A., & Gupta, R. (2021). *Change: How organizations achieve hard-to-image results in uncertain and volatile times*. Wiley.

Kruppa, M. (2023, December 6). Google announces AI system Gemini after turmoil at rival OpenAI. *The Wall Street Journal.* https://www.wsj.com/tech/ai/google-announces-ai-system-gemini-after-turmoil-at-rival-openai-10835335

Lane, J. N., Boussioux, L., Ayoubi, C., Chen, Y. H., Lin, C., Spens, R., Wagh, P., & Wang, P. (2024). *The narrative AI advantage? A field experiment on generative AI-augmented evaluations of early-stage innovations* (Harvard Business School Technology & Operations Management Unit Working Paper No. 25-001). Social Science Research Network. https://doi.org/10.2139/ssrn.4933189

Lawson, A. (2024, October 15). Google to buy nuclear power for AI data centres in 'world first' deal. *The Guardian.* https://www.theguardian.com/technology/2024/oct/15/google-buy-nuclear-power-ai-datacentres-kairos-power

Lund, S., Madgavkar, A., Manyika, J., Smit, S., Ellingrud, K., & Robinson, O. (2021, February 18). *The future of work after COVID-19.* McKinsey & Company. https://www.mckinsey.com/featured-insights/future-of-work/the-future-of-work-after-covid-19

Marketing AI Institute. (2025, April 23). Sam Altman, GPT-5, and the future of AI (The AI Show E135). Marketing AI Institute. https://www.marketingaiinstitute.com/blog/sam-altman-gpt-5-future-of-ai-the-ai-show-e135

Morris, M. R., Sohl-dickstein, J., Fiedel, N., Warkentin, T., Dafoe, A., Faust, A., Farabet, C., & Legg, S. (2024). Levels of AGI for operationalizing progress on the path to AGI. *Proceedings of the 41st International Conference on Machine Learning (ICML 2024).* (arXiv:2311.02462v4) [cs.AI]. https://doi.org/10.48550/arXiv.2311.02462

Mullen, E., et al. (2015). Evidence based practice in social work. *Social Work Research, 39*(3), 201–210.

Nosta, J. (2023, October 6). The 5th industrial revolution: The dawn of the cognitive age. *Psychology Today.* https://www.psychologytoday.com/us/blog/the-digital-self/202310/the-5th-industrial-revolution-the-dawn-of-the-cognitive-age

OpenAI. (2022, November 30). Introducing ChatGPT. *OpenAI.* https://openai.com/blog/chatgpt/

OpenAI. (2023, March 14). GPT-4. *OpenAI.* https://openai.com/research/gpt-4

OpenAI. (2024, May 13). Introducing OpenAI o1 preview. *OpenAI.* https://openai.com/index/introducing-openai-o1-preview/

Originality.ai. (2024, October 18). Anthropic AI statistics and developments in artificial intelligence. *Originality.ai.* https://originality.ai/blog/anthropic-ai-statistics

Ouyang, L., Wu, J., Jiang, X., Almeida, D., Wainwright, C. L., Mishkin, P., Zhang, C., Agarwal, S., Slama, K., Ray, A., Schulman, J., Hilton, J., Kelton, O., Miller, L., Simens, M., Askell, A., Welinder, P., Christiano, P., Leike, J., & Lowe, R. (2022). Training language models to follow instructions with human feedback (arXiv:2203.02155). *arXiv.* https://doi.org/10.48550/arXiv.2203.02155

Pedreschi, D., Pappalardo, L., Ferragina, E., Baeza-Yates, R., Barabási, A.-L., Dignum, F., Dignum, V., Eliassi-Rad, T., Giannotti, F., Kertész, J., Knott, A., Ioannidis, Y., Lukowicz, P., Passarella, A., Pentland, A. S., Shawe-Taylor, J., & Vespignani, A. (2023). Human-AI coevolution (arXiv:2306.13723v2) [cs.AI]. *arXiv.* https://doi.org/10.48550/arXiv.2306.13723

PwC. (2025). *Enter the future of work with AI*. Retrieved April 24, 2025, from https://www.pwc.com/us/en/services/ai/ai-and-the-future-of-work.html

PYMNTS. (2025, April 13). OpenAI co-founder's firm value jumps sixfold after $6 billion funding round. *PYMNTS.com*. https://www.pymnts.com/artificial-intelligence-2/2025/safe-superintelligences-value-jumps-sixfold-after-6-billion-funding-round/

Rane, S., Bruna, P. J., Sucholutsky, I., Kello, C., & Griffiths, T. L. (2024). Concept alignment (arXiv:2401.08672) [cs.LG]. *arXiv*. https://doi.org/10.48550/arXiv.2401.08672

Recursion Pharmaceuticals, Inc. (2022, March 23). Form 10-K Annual Report for the fiscal year ended December 31, 2021 (Form 10-K). U.S. Securities and Exchange Commission. https://www.sec.gov/Archives/edgar/data/1601830/000160183022000027/rxrx-20211231.htm

Reuters. (2024, October 22). Anthropic releases AI feature to automate mouse clicks for coders. *Reuters*. https://www.reuters.com/technology/artificial-intelligence/anthropic-releases-ai-automate-mouse-clicks-coders-2024-10-22

Rick, S. R., Giacomelli, G., Wen, H., Laubacher, R. J., Taubenslag, N., Heyman, J. L., Knicker, M., Jeddi, Y., Maier, H., Dwyer, S., Ragupathy, P., & Malone, T. W. (2023). Supermind ideator: Exploring generative AI to support creative problem-solving (arXiv:2311.16253). *arXiv*. https://doi.org/10.48550/arXiv.2311.16253

Riley, J. (2024, April 4). What is happening to U.S. higher education? *Vanderbilt Business News*. https://business.vanderbilt.edu/news/2024/04/04/what-is-happening-to-u-s-higher-education/

Roetzer, P. (Host). (2025, April 19). Road to AGI (and beyond) #1 — The AI timeline is accelerating (No. 141) [Audio podcast episode]. *The Artificial Intelligence Show*. Marketing AI Institute. https://podcasts.apple.com/lt/podcast/141-road-to-agi-and-beyond-1-the-ai-timeline-is-accelerating/id1548733275?i=1000701052291

Roetzer, P., & Kaput, M. (Hosts). (2021–present). The artificial intelligence show [Audio podcast]. Marketing AI Institute.

Rodriguez, S. (2024, November 13). Study reveals medical students' use of ChatGPT in education and calls for ethical guidelines. Keck School of Medicine of USC. https://keck.usc.edu/news/study-reveals-medical-students-useof-chatgpt-in-education-and-calls-for-ethical-guidelines/

Romero, M. (2023, July 5–7). Lifelong learning challenges in the era of artificial intelligence: A computational thinking perspective [Paper presentation]. International Research Meeting in Business and Management (IRMBAM) 2023, Nice, France. https://irmbam2023.sciencesconf.org/resource/page/id/19

Secundo, G., Spilotro, C., Gast, J., & Corvello, V. (2024). The transformative power of artificial intelligence within innovation ecosystems: A review and a conceptual framework. *Review of Managerial Science*. https://doi.org/10.1007/s11846-024-00828-z

Shaer, O., Cooper, A., Mokryn, O., Kun, A. L., & Ben Shoshan, H. (2024). AI-augmented brainwriting: Investigating the use of LLMs in group ideation (arXiv:2402.14978v2) [cs.HC]. *arXiv*. https://doi.org/10.48550/arXiv.2402.14978

Shneiderman, B. (2022). *Human-centered AI*. Oxford University Press.

Spencer, P. T. (2024). Human-centric NLP or AI-centric illusion?: A critical investigation (arXiv:2412.10939). *arXiv*. https://doi.org/10.48550/arXiv.2412.10939

Statt, N. (2024, November 13). OpenAI to launch autonomous AI agent called operator in January 2025. *The Verge*. https://www.theverge.com/2024/11/13/24295879/openai-agent-operator-autonomous-ai

Sutskever, I. [@ilyasut]. (2024, June 19). I am starting a new company [Tweet, quoting SSI Inc.]. *X*. https://x.com/ilyasut/status/1803472978753303014

Tariq, S.B., Zhang, J. & Gilal, F.G. (2024). From code to connection: The role of responsible artificial intelligence (RAI) and leaders' RAI symbolization in fueling high-tech employee innovation. *Kybernetes, Vol. ahead-of-print* No. ahead-of-print. https://doi.org/10.1108/K-02-2024-0409

Warren, T. (2023, February 7). Microsoft and Google are about to Open an AI battle. *The Verge*. https://www.theverge.com/2023/2/7/23587767/microsoft-google-open-ai-battle-search-bing

Weidinger, L., Mitchell, M., & Sordoni, A. (2024). The impact of generative artificial intelligence on (mis)information and the information ecosystem (arXiv:2401.05377). *arXiv*. https://doi.org/10.48550/arXiv.2401.05377

Wilson, H. J., & Daugherty, P. R. (2018, July–August). Collaborative intelligence: Humans and AI are joining forces. *Harvard Business Review*. https://hbr.org/2018/07/collaborative-intelligence-humans-and-ai-are-joining-forces

Wilson, J. (Host). (2024, August 21). Transforming global supply chains with AI (No. 341) [Audio podcast episode]. In Everyday AI. Your Everyday AI. https://www.youreverydayai.com/transforming-global-supply-chains-with-ai/

Wu, Z., Ji, D., Yu, K., Zeng, X. C., Wu, D., & Shidujaman, M. (2021). AI creativity and the human-AI co-creation model. In M. Kurosu (Ed.), *Human-computer interaction. Theoretical approaches and design methods* (pp. 171–190). Springer. https://doi.org/10.1007/978-3-030-77

Zhai, X. (2024). Transforming teachers' roles and agencies in the era of generative AI: Perceptions, acceptance, knowledge, and practices (arXiv:2410.03018) [cs.CY]. *arXiv*. https://arxiv.org/abs/2410.03018

Ziatdinov, R., Atteraya, M. S., & Nabiyev, R. (2024). The fifth industrial revolution as a transformative step towards Society 5.0. *Societies, 14*(2), Article 19. https://doi.org/10.3390/soc14020019

2

Hybrid Human-AI Intelligence: A Vision for Innovation & Higher Education's Future

Introduction

As demonstrated in Chap. 1, higher education is grappling with unprecedented disruptions and opportunities fueled by evolving AI capabilities, demographic shifts, and financial pressures. This chapter focuses on building some logic and rationale around concrete frameworks, methods, and guidelines to ethically and effectively integrate Generative AI (GenAI) into teaching, learning, assessment, research, and administration. It builds on the reality that GenAI can be used as a strategic partner and transform academic practices when embraced with strategic planning, diverse multi-stakeholder engagement, a clear vision, empathy, and human-centric design—core elements of design thinking (Brown, 2019), structured change management (Kotter et al., 2021), the core frameworks of this book (see the *Human-AI Collaboration & Synergy* Guide available for download). The sections below explore the logic for pursuing **hybrid human-AI intelligence**—a synergistic combination of human and machine capabilities—while acknowledging critiques regarding the limits of current AI cognition and the complexities of achieving ethical **human-AI partnerships** that advance and empower humans. This exploration suggests that navigating the challenges facing higher education may necessitate sustainable partnerships with AI, engaging with advanced systems like **Large Language Models (LLMs)**, and the models soon to proceed them, as strategic collaborators.

The approaches discussed align with the core frameworks presented throughout this book, such as Human-Centric AI Synergy (HCAS), Regulation Synergy & Augmentation (RSA), and the overarching Human-Centric AI Synergy & Regulation (HCASR) process. These frameworks, detailed in Chap. 5 and the Human-GenAI Collaboration & Synergy Guide, offer strategies for mitigating AI's design-induced challenges and optimizing its ethical use.

Supplementary Information The online version contains supplementary material available at https://doi.org/10.1007/978-3-032-05205-6_2.

D. Woodring, *Advancing Higher Education with Human-Centric AI Partnerships*,
https://doi.org/10.1007/978-3-032-05205-6_2

Hybrid Human-AI Intelligence: Collaboration and Synergy

Hybrid human-AI intelligence is defined as the synergistic integration of human cognitive abilities with artificial intelligence systems to create a collective intelligence that enhances decision-making, creativity, and problem-solving (Grossi et al., 2022). It involves strategically combining human knowledge, skills, and abilities with the capabilities of AI systems to achieve outcomes neither could alone. The rationale is to leverage nuanced human judgment alongside AI's efficiency, creating adaptive, ethical systems for complex challenges. This aligns with this book's foundational HCAS framework, where AI strategically empowers humans and augments their capabilities.

In this text, this concept is primarily expressed through **Human-GenAI Collaboration & Synergy**. "Collaboration" highlights the interactive exchange between humans and AI, a dynamic partnership leveraging complementary strengths: human empathy, context awareness, creativity, and ethical judgment interacting with GenAI's data processing, pattern recognition, and prediction. "Synergy" refers to the emergent outcomes—enhanced decision-making, innovation, problem-solving—that surpass the contributions of either party alone (Punyani, 2024). This dual focus aims for partnerships that are both efficient and creatively valuable. The following sections explore emerging research on these partnerships and their potential in higher education.

Human-AI Synergy & Collaboration: Building a Vision for Higher Education

Scholars are exploring emerging forms of ethical and effective human-AI synergy from multiple vantage points, including goal setting, problem-solving, prompt engineering, and evaluation (Batista et al., 2024; Bisconti et al., 2023; Wang et al., 2024). A central premise is that human initiative and ethical responsibility must guide AI integration. AI can become a collaborative agent refining human creativity, advancing knowledge, and enabling inclusive decision-making. This chapter envisions a cohesive model for adopting, refining, and sustaining effective human–AI collaboration or "Human-AI Partnerships" in HEIs.

Grounding Collaboration in Structured Interdisciplinarity

Effective human–GenAI synergy often arises from structured interdisciplinary teams blending technical expertise with educational principles and ethical oversight (Bisconti et al., 2023). Members from computer science, pedagogy, philosophy, and administration contribute diverse insights. Such teams construct shared vocabulary and use scenario-building to predict outcomes and manage AI ethically, ensuring alignment with academic integrity, data privacy, and fairness (Cañas, 2022). This is crucial, despite the practical hurdles of fostering genuine interdisciplinary work in often-siloed university structures.

Interdisciplinary initiatives can transform how stakeholders balance human oversight and AI tasks. A faculty group might pair pedagogical expertise with an AI specialist's proficiency. For example, if AI flags student disengagement patterns, the teaching team interprets these within the educational context, adjusting their methods. This interplay fosters **reflective hybrid intelligence**, where humans and AI refine each other's contributions through continuous feedback, promoting meaningful human control over the collaborative process (Jonker et al., 2023).

From Tools to Collaborative Partners

A notable shift occurs when institutions view AI not just as a service but as an interactive collaborator (Wang et al., 2024). Research highlights how effective agency negotiation allows educators and AI to exchange iterations of documents or plans. An instructor might draft an outline, request AI suggestions, refine them based on pedagogical goals, and receive further adapted input from the AI. GenAI evolves from a static resource into an adaptive participant learning from user feedback.

This collaboration demands deliberate communication. Humans must shape AI prompts precisely to elicit ethically aligned outputs (Sathvika et al., 2024). Vague prompts yield poor results, while structured instructions enable coherent solutions. This iterative process strengthens the AI's understanding and fosters synergy between human self-regulation (guiding one's own use of AI) and AI adaptation (Batista et al., 2024).

Bridging Concepts and Practice: Concepts like self-regulation and co-regulation (mutual adjustment between human and AI) are operationalized within the Regulation, Synergy, & Augmentation (RSA) framework; specific strategies using RSA to mitigate AI challenges are provided in the Human-GenAI Collaboration & Synergy Guide (Chap. 9).

Guiding Ethical Principles for Human–GenAI Synergy

Ethical frameworks act as guardrails. Transparent institutional policies set boundaries on data use, mandate bias detection, and require disclosure of AI involvement (Hagendorff, 2024), guarding against manipulation or plagiarism and fostering AI literacy.

These guidelines address the complexities around **explainable AI (XAI)**, which promotes clarity on algorithmic reasoning (Hagendorff, 2024). In this context, users benefit from AI providing rationales or flagging uncertainty. Without explainability, educators might adopt suggestions lacking crucial context (e.g., a scheduling tool ignoring time zones). By requiring AI to articulate assumptions, users can correct oversights, reinforcing human-centric collaboration.

Refining Problem-Solving and Decision-Making

AI can enhance decision-making through data-driven insights and creative expansion (Lane et al., 2024; Rick et al., 2023). GenAI can generate alternative research designs or administrative options. However, humans must maintain oversight, evaluating AI outputs against local requirements, ethics, and constraints.

When AI suggests innovations, educators must compare these with established research. If AI proposes consolidating courses online, decision-makers consider faculty readiness, student feedback, and community buy-in. AI's analysis expedites planning but depends on human moral reasoning and contextual understanding for fair outcomes aligned with institutional values (Chan et al., 2023).

Prompting with Clarity and Responsibility

Prompt engineering plays a pivotal role in shaping AI's contributions (Kepel & Valogianni, 2024). Structured, context-rich prompts improve alignment (Sathvika et al., 2024). This involves ethical prompt design to guide AI towards balanced, inclusive perspectives. For instance, a prompt for an equitable grading policy should request scenarios reflecting diverse student backgrounds. If diversity isn't mentioned, the output may lack essential viewpoints (Lin, 2024). By explicitly requesting consideration of multiple perspectives, requiring evidence for claims, or asking the AI to flag uncertainties, structured prompts can steer outputs toward greater fairness and reliability.

Bridging Concepts and Practice: Guidance for practical application across the key components of ethical AI interaction—goal setting, prompt engineering, research augmentation, human-centric implementation, and continuous improvement (corresponding to HCASR Components 1–5)—is detailed in the Human-GenAI Collaboration & Synergy Guide.

Multi-Stakeholder Collaboration and Support

Robust multi-stakeholder engagement is crucial for successful AI integration (Jin et al., 2024). Faculty, students, administrators, IT, and external partners hold distinct responsibilities. While a student might use AI for drafting, an advisor verifies standards. Administrators coordinate pilots, and leadership ensures alignment with goals.

Engagement includes AI developers and ethical review boards. For example, developers can modify AI based on policy concerns, while boards review usage data for discrimination or misrepresentation. This collaboration fosters **distributed moral responsibility**, sharing accountability among creators, implementers, and monitors (Cañas, 2022).

Using AI as a Strategic Partner: Every HEI is situated in a unique socio-historical context, which means there is no one-size-fits-all approach to AI integration in higher education (Carmody-Bubb, 2023). As institutions around the world integrate GenAI, contextual data from institutions can be ethically and effectively used to (1) guide custom tailored tools, resources, and solutions, and (2) mitigate Western (and other) biases that will be inherent within models as they are adopted globally. Part 2 of this book provides a customizable four-phase rollout plan for implementing and scaling AI long-term using AI as a strategic partner.

Continuous Iteration and Reflective Improvement with AI

Continuous evaluation keeps AI a living component, not a static commodity (Dotan et al., 2024). Institutions can embed cyclical processes—collecting feedback, analyzing data, revisiting guidelines—to adapt AI. Termly evaluations of an AI learning analytics platform, for instance, check alignment with student performance trends.

These evaluations can benefit from reflective hybrid intelligence, where user insights shape AI enhancements, and AI results prompt human reflection

(Jonker et al., 2023). This synergy creates a dynamic loop improving workflows and stakeholder capabilities (Kotwel, 2024), reinforcing co-regulation between automation and human judgment.

Conceptualizing GenAI's Influence: A Stakeholder Lens

As AI systems expand their scope—offering strategic advice, influencing academic norms, and shaping policy dialogues—a useful conceptual approach involves viewing AI's influence through a stakeholder lens (Younger, 2024; Socha, 2024). While AI lacks consciousness or inherent stakes, its capabilities allow it to function as if it were a participant with influence, impacting institutional strategies and the behavior of human actors.

Conceptualizing AI this way is crucial for operationalizing human-centric collaboration. As AI increasingly performs tasks at or beyond human levels, understanding its functional role within workflows becomes essential for effective task allocation (a core aspect of the HAIC framework discussed in Chap. 3). Treating AI merely as an inert tool overlooks its capacity to shape outcomes and interactions.

Furthermore, adopting a stakeholder perspective encourages more responsible development and deployment. If AI's potential influence on decisions (e.g., resource allocation based on AI predictions) or its role in academic work (e.g., co-authorship) is not explicitly considered within governance frameworks, its integration may proceed without adequate ethical foresight or mechanisms for accountability. By acknowledging AI's functional role and influence, institutions can proactively design policies and oversight structures. For example, major policy shifts influenced by AI predictions might require vetting by cross-functional committees, ensuring human values guide the interpretation and application of AI-generated insights. This approach harmonizes AI's computational power with institutional obligations, reinforcing human-centric principles (Socha, 2024).

A Synthesis for Inclusive and Adaptive Collaboration

The movement toward ethical human–GenAI synergy culminates in a vision where interdisciplinary teams define, implement, and refine AI practices. GenAI participates as an intelligent collaborator, bridging aspirations with analytics, generating insights while deferring ultimate authority to human expertise.

Table 2.1 Holistic mental model of ethical human-GenAI collaboration and synergy

Dimension	Core aim	Human role	GenAI contribution
Structured interdisciplinarity	Harness multiple fields' insights.	Coordinate scenario-building and hypothesis setting.	Provide automated analysis, bridging domain data sets.
From tools to collaborative partners	Foster AI co-creation.	Manage prompts, refine outputs with context.	Adapt suggestions in response to human feedback.
Ethical principles	Align AI with academic values.	Enforce transparency, fairness, reliability.	Generate traceable outputs that reveal reasoning paths.
Problem-solving and decision-making	Expand creative solutions.	Assess AI outputs for feasibility and ethics.	Supply data-driven alternatives and rationales.
Prompting and responsibility	Improve clarity in requests.	Frame instructions ethically and inclusively.	Translate user input into coherent, context-aware responses.
Multi-stakeholder collaboration	Engage all roles in planning.	Oversee pilot programs, unify departmental feedback.	Integrate technical updates to align with evolving needs.
Continuous iteration	Refine synergy over time.	Conduct regular reviews, adapt policies.	Evolve algorithms and highlight performance shifts.
GenAI as stakeholder	Assign AI an active advisory role.	Integrate AI outputs within governance frameworks.	Provide iterative data insights that shape strategic decisions.

Note. This table presents a holistic mental model for ethical human-GenAI collaboration, outlining eight key dimensions. For each dimension, it defines the core aim and delineates the respective roles and contributions of both humans and GenAI, emphasizing a synergistic partnership

Table 2.1 offers a model for how stakeholders might approach human–AI collaboration dimensions in a cycle of mutual reinforcement. Each dimension flows into the next, fostering a co-evolving system merging AI's analysis with human intuition, ethical vigilance, and contextual sensitivity. These partnerships enable continuous learning for all participants.

Conclusion: Ethical Human-AI Partnerships in Higher Education

Emerging research offers insights into strategic, ethical human-AI partnerships in higher education. These collaborations can encourage deeper thinking, fairer decisions, and flexible resource use, particularly when guided by interdisciplinary teams, clear ethics, shared oversight, and open communication. Viewing GenAI's influence through a stakeholder lens helps protect academic integrity while exploring AI's possibilities.

The above research is just the tip of the iceberg. AI tech is advancing rapidly so that it's hard for AI experts to even keep up. It is recommended that users run the AI ethics and literacy prompts from the **Crash Course in GenAI Ethics & Literacy** in Part 3 of the *Human-AI Collaboration & Synergy Guide* through their preferred models frequently to stay up to date on the rapidly emerging innovations and how they might impact or be harnessed in higher education. Podcasts like the **Artificial Intelligence Show** by Paul Roetzer and Mike Kaput are also great ways to stay up to date (Roetzer & Kaput, 2021–Present).

Chapter 3 will detail approaches for optimizing these partnerships using a human–AI collaboration (HAIC) framework adapted for higher education. Chapter 4 examines the importance of understanding short- and long-term outcomes of human-AI collaboration. Part 2 of the book operationalizes these concepts through a customizable four-phase rollout plan for implementing, piloting, evaluating, and scaling GenAI ethically and effectively in HEIs.

Bridging Theory to Practice in Part 2 of the Book: Implementation, Piloting, Evaluation, and Scaling GenAI: As the book progresses into Human-AI Collaboration (HAIC) in higher education in the next chapter and beyond, the book's frameworks will become operationalized and provide solid foundation for ethical and effective integration of GenAI. As the book continues into Part 2, a customizable and adaptable four-phase rollout plan, using GenAI as a strategic partner and stakeholder, is presented to optimize the ethical and effective implementation, piloting, evaluation, and scaling of GenAI in HEIs.

Chapter 2 Workbook Prep

These questions help you connect the concepts of hybrid human-AI intelligence and collaboration from this chapter to the practical activities in the Chap. 2 UX and PALS Workbooks.

Connecting Concepts to User Needs (Prep for UX Personas/Stories & PALS Goals):

- Reflecting on the ideas of hybrid intelligence and human-AI synergy, identify one specific goal or challenge for a key stakeholder group (e.g., students improving research skills, faculty managing grading workload, staff streamlining communication) in your context.
- How could a collaborative approach with GenAI (rather than just using it as a tool) potentially help address this specific goal or challenge? (This prepares you for creating personas and user stories in the UX Workbook and refining goals in the PALS Workbook.)

Brainstorming Applications & Ethical Considerations (Prep for UX Use Cases & PALS Ethics/Stakeholder Exploration):

- Based on the stakeholder need identified above, brainstorm one concrete way GenAI could be applied as a collaborative partner (e.g., AI co-generating research questions with students, AI summarizing meeting notes for staff review, AI suggesting diverse assessment options for faculty).
- What is one key ethical consideration (e.g., data privacy, potential bias, maintaining human oversight, distributed responsibility) raised in this chapter that would be crucial to address for this specific application? (This prepares you for developing use cases in the UX Workbook and exploring ethical/stakeholder concepts in the PALS Workbook.)

Preparing for Deeper Exploration with AI (Prep for PALS Advanced Prompts):

- Identify one specific concept from Chap. 2 you'd like to explore further using GenAI as a research assistant (e.g., practical examples of "reflective hybrid intelligence," specific techniques for "ethical prompt design," case studies of "interdisciplinary AI teams").
- How might you initially prompt your GenAI model to begin exploring this concept? (This prepares you for the guided research and advanced learning prompts in the PALS Workbook.) The following questions will help prepare you for the Chap. 2 PALS Workbook and the Chap. 2 UX Workbook.

Applying HCASR 2 – Effective Prompting & Prompt Engineering
One of the best ways to harness optimal human-AI collaboration & synergy is to be able to ethically and effectively communicate with your model. Recall that HCASR components and customizable downloads/user guides are available to help you through the process. But AI is advancing so rapidly.

- How will you keep up with prompting more advanced multimodal, vision, and action models as they emerge?
- How can these models be used to advance your role and field beyond what was capable before these advances in AI?
- Update your goals to include any questions you have from the above chapter, any questions about optimizing specific tasks, roles, goals, or objectives with AI, and any questions about prompting or prompt engineering you want to explore on a deeper level.

You are now ready to download and engage with the Chap. 2 UX and PALS Workbooks.

References

Batista, J., Mesquita, A., & Carnaz, G. (2024). Generative AI and higher education: Trends, challenges, and future directions from a systematic literature review. *Information, 15*(11), Article 676. https://doi.org/10.3390/info15110676

Bisconti, P., Orsitto, D., Fedorczyk, F., Brau, F., Capasso, M., De Marinis, L., Eken, H., Merenda, F., Forti, M., Pacini, M., & Schettini, C. (2023). Maximizing team synergy in AI-related interdisciplinary groups: An interdisciplinary-by-design iterative methodology. *AI & Society, 38*, 1443–1452. https://doi.org/10.1007/s00146-022-01518-8

Brown, T. (2019). *Change by design: How design thinking transforms organizations and inspires innovation* (Rev. ed.). Harper Business.

Carmody-Bubb, M. (2023). *Cognition and decision making in complex adaptive systems: The human factor in organizational performance.* Springer.

Cañas, J. J. (2022). AI and ethics when human beings collaborate with AI agents. *Frontiers in Psychology, 13*, Article 836650. https://doi.org/10.3389/fpsyg.2022.836650

Chan, A. J., Huyuk, A., & van der Schaar, M. (2023). Optimizing human-AI collaboration by learning convincing explanations (arXiv:2311.07426). *arXiv.* https://doi.org/10.48550/arXiv.2311.07426

Dotan, R., Parker, L. S., & Radzilowicz, J. G. (2024). Responsible adoption of generative AI in higher education: Developing a "points to consider" approach based on faculty perspectives (arXiv:2406.01930). *arXiv.* https://doi.org/10.48550/arXiv.2406.01930

Grossi, D., Mackay, W., Oliehoek, F., Paiva, A., Perez-Ortiz, M., Schlobach, S., & Tielman, M. (2022). Editorial: Special track on hybrid human-artificial intelligence. *Journal of Artificial Intelligence Research, 74*, 1315–1319. https://doi.org/10.1613/jair.1.13761

Hagendorff, T. (2024). Mapping the ethics of generative AI: A comprehensive scoping review (arXiv:2402.08323). *arXiv.* https://doi.org/10.48550/arXiv.2402.08323

Jin, Y., Yan, L., Echeverria, V., Gašević, D., & Martinez-Maldonado, R. (2024). Generative AI in higher education: A global perspective of institutional adoption policies and guidelines (arXiv:2405.11800). *arXiv.* https://doi.org/10.48550/arXiv.2405.11800

Jonker, C. M., Siebert, L. C., & Murukannaiah, P. K. (2023). Reflective hybrid intelligence for meaningful human control in decision-support systems (arXiv:2307.06159). *arXiv.* https://doi.org/10.48550/arXiv.2307.06159

Kepel, D., & Valogianni, K. (2024). Autonomous prompt engineering in large language models (arXiv:2407.11000). *arXiv.* https://doi.org/10.48550/arXiv.2407.11000

Kotter, J. P., Akhtar, V., & Gupta, G. (2021). *Change: How organizations achieve hard-to-imagine results in uncertain and volatile times*. Wiley.

Kotwel. (2024). Continuous learning: Iterative improvement in AI development. *Kotwel*. Retrieved April 24, 2025, from https://kotwel.com/continuous-learning-iterative-improvement-in-ai-development/

Lane, J. N., Boussioux, L., Ayoubi, C., Chen, Y. H., Lin, C., Spens, R., Wagh, P., & Wang, P. H. (2024). The narrative AI advantage? A field experiment on generative AI-augmented evaluations of early-stage innovations (Harvard Business School Working Paper No. 25-001). Harvard Business School. https://www.hbs.edu/ris/Publication%20Files/25-001_8a7a50a6-b136-44b8-9c03-aa687ac65a9e.pdf

Lin, Z. (2024). Beyond principlism: Practical strategies for ethical AI use in research practices (arXiv:2401.15284). *arXiv*. https://doi.org/10.48550/arXiv.2401.15284

Punyani, A. (2024, March 11). Exploring human-AI collaboration: Synergy between humans and machines. *Convin Blog*. https://convin.ai/blog/human-ai-collaboration-insights

Rick, S. R., Giacomelli, G., Wen, H., Laubacher, R. J., Taubenslag, N., Heyman, J. L., Knicker, M. S., Jeddi, Y., Maier, H., Dwyer, S., Ragupathy, P., & Malone, T. W. (2023). Supermind Ideator: Exploring generative AI to support creative problem-solving (arXiv:2311.01937). *arXiv*. https://doi.org/10.48550/arXiv.2311.01937

Roetzer, P. (Host), & Kaput, M. (Host). (2021–present). *The artificial intelligence show* [Audio podcast]. Marketing AI Institute. https://www.marketingaiinstitute.com/podcast-showcase

Sathvika, A. L., Purnima, A., Harshitha, S., Anusha, P., & Nafisa, S. (2024). Prompt engineering for improved AI responsiveness: Methods and case studies. *International Journal of Creative Research Thoughts, 12*(11), 468–471. https://www.ijcrt.org/papers/IJCRT2411056.pdf

Socha, B. (2024). There is no AI without alignment: How stakeholder collaboration ensures AI success. *Strategy+Business*. Retrieved April 24, 2025, from https://www.strategy-business.com/blog/There-is-no-AI-without-alignment

Wang, X., Chen, Y., & Zhang, L. (2024). Tools or teammates? Examining agency negotiation in human-GenAI collaboration (SSRN Scholarly Paper No. 4956265). Social Science Research Network. https://ssrn.com/abstract=4956265

Younger, R. (2024). AI is not just a technology. It has become a stakeholder. Oxford Answers. Retrieved April 24, 2025, from https://www.sbs.ox.ac.uk/oxford-answers/ai-not-just-technology-it-has-become-stakeholder

3

Human-AI Collaboration (HAIC) in Higher Education: Defining Roles, Responsibilities, and Oversight with GenAI

Introduction

This chapter applies a proposed framework for **Human-AI Collaboration (HAIC)** created by Fragiadakis et al. (2024) to applications in higher education, focusing on how GenAI can be ethically and effectively integrated into teaching, learning, assessment, research, and administration. By examining the three modes of HAIC—Human-Centric, Symbiotic, and AI-Centric—readers will learn how to assign roles, responsibilities, and oversight to AI systems so they function as productive partners rather than impersonal tools. This chapter focuses on the roles and tasks of users, key stakeholders, and GenAI, considering it a stakeholder, as discussed in Chap. 2.

This allows users to begin to see what types of outcomes can emerge (e.g., desired or undesired) from different types of relationships between humans and AI systems. To help users build a foundational mental model of ethical and effective collaboration with AI, this chapter uses **Design Thinking** strategies for visualizing **human-AI partnerships**. This helps set the foundation for optimizing **Human-Centric AI Synergy** in Part 2, where AI systems are used in ways that empower humans and augment their knowledge, skills, and abilities, rather than merely replacing them or replicating their intelligence with AI.

The chapter also discusses **Human-Centric AI (H-CAI),** an approach emphasizing human control and augmentation, and its alignment with ethical principles, ensuring that GenAI deployments advance equity, institutional integrity, and student success. With the understanding that humans and technology are co-evolving together and can produce unwanted consequences. **User Experience (UX)** strategies, which focus on optimizing the interaction between humans and technology, help users build new competencies for optimizing GenAI in higher education. Through stakeholder mapping and reflective exercises, higher education leaders, faculty, staff, and students can assess their current AI usage and plan for viable and sustainable human-AI partnerships.

Supplementary Information The online version contains supplementary material available at https://doi.org/10.1007/978-3-032-05205-6_3.

D. Woodring, *Advancing Higher Education with Human-Centric AI Partnerships*,
https://doi.org/10.1007/978-3-032-05205-6_3

Human-AI Collaboration (HAIC) in Higher Education

Collaboration has always defined higher education—whether between students, faculty, administrators, or industry partners. Now, a new era is unfolding with unprecedented human-AI collaboration, driven by AI's growing ability to replicate knowledge work and creative processes. When used strategically, AI is not just another tool but can function as an active collaborator (Atchley et al., 2024) and stakeholder (see Chap. 2), augmenting human decision-making, research, and administrative workflows. However, achieving effective and ethical collaboration is complex, facing critiques regarding AI's limitations and the practical challenges of deep human-machine integration. This chapter explores how AI fits into higher education ecosystems using a proposed framework, examining how institutions can foster productive, ethical, and sustainable human-AI partnerships while ensuring academic integrity and positive learning outcomes.

By examining Human-Centric, Symbiotic, and AI-Centric modes of collaboration, based on an emerging framework (Fragiadakis et al., 2024), HEIs can conceptualize how to tailor AI's role to specific tasks. The chapter addresses ethical implications and the importance of stakeholder engagement. Ultimately, effective HAIC rests on strategically combining computational power and human creativity, balancing efficiency with the core mission of fostering human knowledge and intellectual growth (Fig. 3.1).

By examining **Human-Centric**, **Symbiotic**, and **AI-Centric modes** of collaboration, higher education institutions (HEIs) can tailor AI's role to specific tasks—ranging from routine grading to strategic institutional planning. The chapter also addresses the ethical implications of delegating authority to AI systems and underscores the importance of stakeholder engagement in designing AI interactions. Ultimately, HAIC rests on the principle that computational power and human creativity function best in tandem, balancing efficiency gains with the core mission of higher education: fostering human knowledge and intellectual growth.

Applying Three Modes of HAIC to Online Discussion Posts

This section briefly, and at a basic level, considers the three modes of AI in relationship to the various ways both students and instructors have been exploring different ways to use AI in everyday tasks, such as writing and

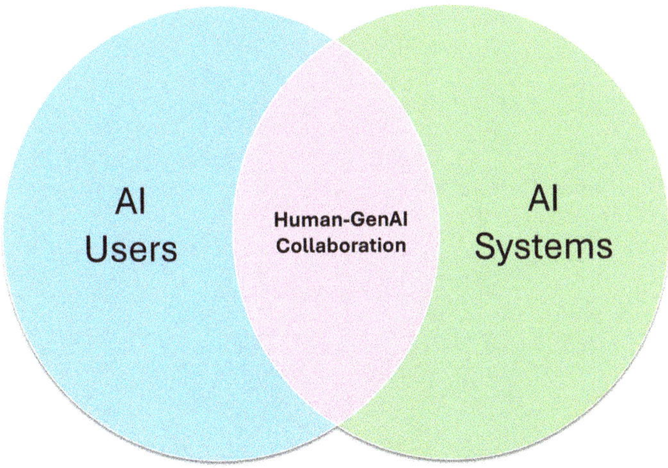

Fig. 3.1 Human-AI collaboration (HAIC). (*Note.* A conceptual diagram illustrating the spectrum of Human-AI collaboration (HAIC), from human-led to AI-driven processes, highlighting the importance of balancing roles for effective synergy)

grading discussion posts (see Fig. 3.2). In a Human-Centric approach, students write posts entirely on their own, drawing on course materials and personal reflection. Instructors read and evaluate these posts using rubrics, their professional judgment, and written feedback—relying very little on AI tools. This mode supports deep learning, critical thinking, and ethical decision-making.

The Symbiotic mode represents a collaborative relationship between humans and AI. Students may use AI to help generate ideas, clarify writing, or organize their thoughts—but they remain responsible for revising and shaping the final post. Instructors might use AI to help summarize trends in student responses, provide draft feedback, or streamline repetitive parts of the grading process, while still applying their expertise and judgment. This mode can enhance both learning and efficiency when used responsibly.

In the AI-Centric mode, AI plays the leading role. For students, this could mean using AI to generate most or all of a post with minimal editing. For instructors, it might involve relying on automated grading tools or AI-generated summaries to assess student work. While this mode can save time, it also raises questions about originality, fairness, and the quality of learning and feedback. In practice, many real-world situations blend elements from all three modes. The goal is to find the right balance for each task, ensuring that learning remains meaningful and that the use of AI supports rather than replaces human insight.

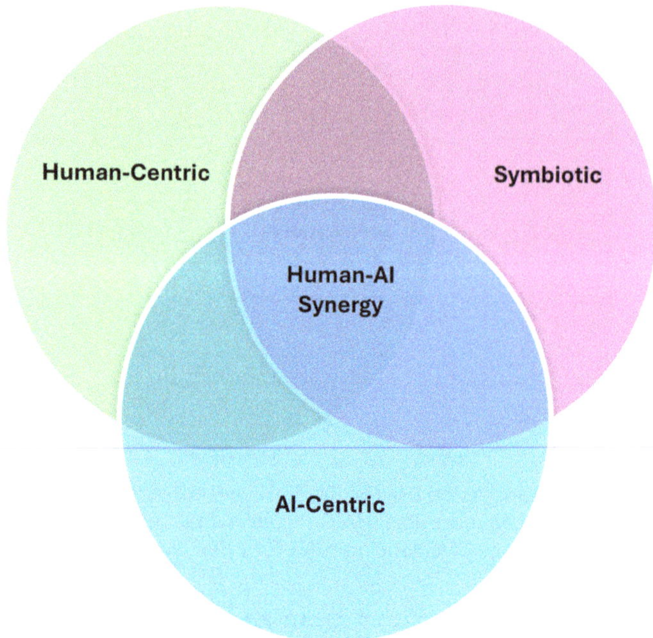

Fig. 3.2 Modes of human-AI collaboration for student discussion posts and grading. (*Note.* This figure depicts different collaborative modes for a common academic task. It shows how the level of human versus AI involvement can shift depending on whether the goal is primarily human-centric (e.g., student-led critical thinking), symbiotic (e.g., AI-assisted feedback with instructor oversight), or AI-centric (e.g., automated initial grading))

Applying a Human-AI Collaboration (HAIC) Framework to GenAI

Human-AI Collaboration (HAIC) involves dynamic partnerships where humans and AI combine complementary skills to achieve shared goals (Fragiadakis et al., 2024). Understanding HAIC is rapidly becoming a core competency for higher education leaders and practitioners navigating the complexities of GenAI integration, especially given current widespread uses, such as instructors using AI to grade potentially AI-generated student work (Katsamakas et al., 2024). This section applies a useful, recently proposed framework from Fragiadakis et al. (2024) to human-GenAI partnerships, demonstrating how its principles can help conceptualize GenAI implementation and governance in HEIs.

According to this framework, HAIC operates in three illustrative modes—Human-Centric, Symbiotic, and AI-Centric—each defining distinct roles and interactions. In higher education, these modes can be strategically considered across roles (students, instructors, staff, administrators) to foster integrity, critical thinking, and ethical AI use.

Introduction to HAIC

According to Fragiadakis et al. (2024), HAIC operates in three modes—Human-Centric, Symbiotic, and AI-Centric—each defining distinct roles and interactions between humans and AI. In higher education, these modes should be strategically tailored across roles (students, instructors, faculty, staff, and administrators) to foster academic integrity, promote critical thinking, and ensure ethical AI use. The following sections detail each mode.

Human-Centric AI Collaboration: AI as a Supportive Tool

The Human-Centric mode prioritizes human decision-making, with AI serving as a supportive tool. This mode is conceptually ideal where human expertise and ethical judgment are paramount. Illustrative applications include students using AI for guidance while retaining critical judgment, or instructors employing AI for administrative support while controlling pedagogy. This mode conceptually ensures academic integrity by preventing overreliance on AI and prioritizing human oversight (University of Rochester, Office of the Provost, n.d.; Wynants et al., 2025) (Table 3.1).

Symbiotic AI Collaboration: Humans & AI as Partners

The Symbiotic mode represents a balanced partnership where humans and AI actively contribute (Atchley et al., 2024). Illustrative applications include administrators using AI for operational efficiency while retaining strategic control, or adaptive learning platforms tailoring content delivery while instructors provide deeper support. This mode emphasizes mutual support and adaptability, potentially offering versatility for institutional growth (Holstein & Aleven, 2022) (Table 3.2).

AI-Centric AI Collaboration: AI Takes the Lead

The AI-Centric mode assigns the primary role to AI. While potentially effective for highly routine tasks (e.g., bulk data processing), its application in core

Table 3.1 Illustrative scenarios for human-centric collaboration

Scenario	Task	AI role	Human role	Benefits
Student learning	Homework support, summarizing notes	AI suggests resources but final understanding requires active student reflection	Students exercise critical judgment to validate AI's suggestions	Reinforces deep learning, fosters reflection
Faculty teaching	Lesson plan development	AI provides data insights or quick brainstorming	Instructor tailors final content, ensures alignment with learning outcomes	Saves time on data tasks, preserves pedagogical values
Administrative	Budget forecasting	AI generates predictive models	Administrators interpret results and weigh intangible factors	AI efficiency with a human ethical lens

Note. This table presents three scenarios illustrating the human-centric mode of collaboration, where human judgment, critical thinking, and contextual decision-making remain paramount, and AI serves a supportive or augmentative function

Table 3.2 Example scenarios for symbiotic collaboration

Scenario	Task	AI role	Human role	Benefits
Student-instructor	Adaptive learning (real-time feedback)	AI monitors performance, suggests resources	Instructor provides context, emotional support, and deeper interventions	Highly personalized, fosters engaged learning
Research collaboration	Data analysis in large-scale studies	AI clusters data sets, identifies trends	Researchers interpret findings, form hypotheses, and ensure ethical compliance	Speeds discovery while preserving human insight
Admin and staff	Scheduling and resource optimization	AI automates routine tasks, real-time updates	Staff handle exceptions, strategic decisions	Efficiency gains, high-level human oversight

Note. This table provides examples of symbiotic collaboration, where human and AI partners work interdependently, each contributing distinct strengths to achieve a shared goal more effectively than either could alone

Table 3.3 Example scenarios for AI-centric collaboration

Scenario	Task	AI role	Human role	Concerns
Administrative	Bulk data processing, standard emails	AI executes nearly all steps	Occasional auditing by staff	Potential errors or bias may go unnoticed
Exam proctoring	Automated monitoring for cheating	AI flags anomalies in real time	Staff seldom intervene unless alerted	Over-reliance on AI can affect fairness, privacy

Note. This table details AI-centric collaboration scenarios, where the AI performs the majority of the task with minimal human intervention. It also highlights the primary concerns associated with this mode, such as the potential for unmonitored errors and ethical issues

educational functions is limited due to risks of diminished human agency, reduced feedback literacy, and ethical concerns like bias or lack of transparency. Illustrative examples might include automated exam proctoring or standard report generation, but extending this to tasks requiring nuanced judgment is generally discouraged within this framework (Fragiadakis et al., 2024).

The AI-Centric mode assigns the primary decision-making role to AI, with minimal human input (see Table 3.3). While effective for routine and repetitive tasks, such as scheduling or automated report generation, this mode has limited applicability in higher education due to its potential to diminish human agency and exacerbate ethical concerns. For example, over-reliance on AI for grading or assessments could undermine academic trust, diminish feedback literacy for educators and students, and introduce biases that compromise fairness. This mode is most appropriate for backend administrative functions but should not be extended to roles requiring human judgment and creativity (Fragiadakis et al., 2024). The next section applies the three primary factors outlined by the framework.

HAIC Factors: Goals, Interaction, and Task Allocation

Goals

Clearly defined goals are essential. This includes individual goals (e.g., a student aiming to improve writing clarity using an AI assistant; a researcher using AI to accelerate literature review) and collective goals (e.g., a department improving feedback timeliness using AI grading support; an institution enhancing accessibility via AI tools). Aligning individual efforts with broader institutional objectives ensures coherence and directs the collaboration effectively.

Interaction: Effective Interaction Hinges on Several Subfactors

1. **Communication**: This involves more than just the prompt; it includes how the AI presents information back to the user. For example, an AI tool providing feedback on student writing needs an intuitive interface that clearly distinguishes suggestions from definitive corrections.
2. **Feedback Mechanisms**: Human feedback improves AI, and AI feedback can inform humans. An instructor correcting an AI's inaccurate summary refines the AI's future performance for that context; AI identifying common student misconceptions can prompt the instructor to adjust their teaching.
3. **Adaptability**: AI systems should adjust to context. An AI tutor should adapt its difficulty level based on student progress; an administrative AI should adapt its scheduling suggestions based on updated university calendars or priorities.
4. **Trust and Safety**: Transparency is key. An AI tool used in admissions must operate with clear guidelines regarding data privacy (FERPA) and undergo audits for potential bias to build trust among applicants and administrators.

Task Allocation

Strategic allocation leverages complementary strengths. This means assigning tasks based on suitability. For instance, AI might excel at processing large datasets to identify potential research correlations (data-heavy task), while the human researcher handles the ethical considerations of the research design, interprets the context of the correlations, and exercises creative judgment in forming hypotheses. Evaluating the efficiency (e.g., time saved) and appropriateness (e.g., ethical implications, quality of outcome) of this allocation is crucial for success.

Conceptualizing GenAI's Influence: A Stakeholder Lens

As AI systems expand their scope—offering strategic advice, influencing academic norms, and shaping policy dialogues—a useful conceptual approach involves viewing AI's influence through a stakeholder lens (Younger, 2024; Socha, 2024). While AI lacks consciousness or inherent stakes, its capabilities allow it to function as if it were a participant with influence, impacting institutional strategies and the behavior of human actors.

Conceptualizing AI this way is crucial for operationalizing human-centric collaboration. As AI increasingly performs tasks at or beyond human levels, understanding its functional role within workflows becomes essential for effective task allocation—deciding which mode (Human-Centric, Symbiotic, AI-Centric) is appropriate and how responsibilities are shared. Treating AI merely as an inert tool overlooks its capacity to shape outcomes and interactions, potentially leading to inefficient or ethically problematic implementations (e.g., the unsustainable synergy of AI grading AI-generated work without critical human oversight).

Furthermore, adopting a stakeholder perspective encourages more responsible development and deployment. If AI's potential influence on decisions (e.g., resource allocation based on AI predictions) or its role in academic work (e.g., co-authorship) is not explicitly considered within governance frameworks, its integration may proceed without adequate ethical foresight or mechanisms for accountability. By acknowledging AI's functional role and influence, institutions can proactively design policies and oversight structures. For example, major policy shifts influenced by AI predictions might require vetting by cross-functional committees, ensuring human values guide the interpretation and application of AI-generated insights. This approach helps integrate AI's power ethically, aligning it with institutional missions and human-centric principles (Socha, 2024).

Prompting with Clarity and Responsibility

Effective prompt engineering is pivotal (Kepel & Valogianni, 2024). Structured, context-rich prompts improve alignment (Sathvika et al., 2024). This involves ethical prompt design to guide AI towards balanced, inclusive perspectives. For instance, a prompt for an equitable grading policy should request scenarios reflecting diverse student backgrounds. If diversity isn't mentioned, the output may lack essential viewpoints (Lin, 2024). Structured prompts can enhance fairness by explicitly asking the AI to consider multiple viewpoints, justify its reasoning, cite supporting evidence (if applicable), or flag areas of uncertainty, thereby mitigating the risk of biased or unsubstantiated outputs.

Conclusion

Human-AI Collaboration (HAIC) offers a structured way to conceptualize GenAI integration across higher education. By considering tasks through Human-Centric, Symbiotic, or AI-Centric modes, institutions can aim to optimize synergy while respecting human oversight, creativity, and ethics, potentially leveraging AI as an effective collaborator (Atchley et al., 2024). Applying the HAIC framework helps lay the groundwork for the practical implementation strategies discussed later in this book.

Looking Forward to Part 2: This role–task mapping and HAIC approach will be further piloted and scaled in Part 2's four-phase rollout, where you will evaluate these human-GenAI collaboration and synergy outcomes, refine them, and extend them across the institution.

Applying Core Frameworks: The application of concepts and activities from this chapter and the PALS/UX workbooks can be enhanced with the core frameworks for this book. **Human-Centric AI Synergy (HCAS)** and **Regulation, Synergy & Augmentation (RSA)** can be accessed in **Chaps. 8 and 9** of the *Human-GenAI Collaboration & Synergy Guide.*

- HCAS offers a concrete conceptual foundation for both human-centric AI and symbiotic AI in higher education. However, instead of focusing on design, AI design principles are translated to focus on how AI is strategically used through human-AI partnerships.
- RSA operationalizes HCAS with evidence-based self-regulated and co-regulated learning approaches that help mitigate the design induced challenges—threats to academic integrity, human-centric skills, and the quality of teaching and learning—brought to higher education by GenAI.

Whether you are just reading this chapter to enhance your AI literacy skills, or whether you are employing this book and GenAI as a strategic partner for implementation and scaling in Part 2, this chapter and these frameworks provide a powerful way to conceptualize ethical and effective human-GenAI collaboration.

Applying HCASR 3 & HAIC to Academic Research – Rapid Research & Review

One of the most crucial areas of ethical and effective integration of GenAI is academic research. There are several issues: Models may hallucinate research or insights. They might not be able to read and summarize certain documents like PDFs accurately. Models might only read abstracts and not have access to deep insights.

But there are several benefits: Models can expose you to interdisciplinary research and insights you would never encounter during a traditional academic database search. While some models have a hard time reading PDFs, listening to the PDF and having it summarize, review, and organize key insights from articles can save you a lot of time. After you have all your research insights together, you can use the capabilities of your model to create custom frameworks or otherwise synthesize the text into various forms that would traditionally take months to do on your own (see Chap. 5).

It's all about leveraging the strengths of yourself or your team and AI while mitigating their weaknesses through strategic, evidence-based practices and policies that advance and empower you, your team, and your stakeholders. Answer the below questions to enhance your GenAI research competencies and prepare for the workbook activities.

Chapter 3 Workbook Prep

These questions help you connect the HAIC framework from this chapter to the practical activities in the Chap. 3 UX and PALS Workbooks.

Applying HAIC Modes to Your Context (Prep for UX Activity 1 & PALS Task Classification):

- Identify one specific task in your role (teaching, research, admin, student work) where GenAI could potentially assist.
- Considering the Human-Centric, Symbiotic, and AI-Centric descriptions, which HAIC mode feels like the most appropriate starting point for this task, and why?

Considering Stakeholder Interactions (Prep for UX Activity 2):

- For the task and potential AI use identified above, which other roles or groups (students, colleagues, supervisors, support staff) might be directly or indirectly affected by this human-AI collaboration?
- Briefly note one potential synergy (e.g., shared efficiency gain) or one potential conflict (e.g., differing expectations about AI's role) that might arise from this interaction.

Ethical Foresight (Prep for UX Ethics Check & PALS Reflection):

- What is the primary ethical consideration (e.g., accuracy, bias, privacy, academic integrity, human oversight) you would need to address when implementing the AI assistance for the task you identified?
- What is one practical safeguard or check you could put in place?

Preparing for Deeper AI Exploration (Prep for PALS Advanced Prompts):

- Select one HAIC concept from this chapter (e.g., Symbiotic mode characteristics, Interaction factors like trust, Task Allocation principles) that you want to understand better.
- How might you initially prompt a GenAI model to provide more detailed explanations or examples related to this specific concept?

References

Atchley, P., Pannell, H., Wofford, K., Hopkins, M., & Atchley, R. A. (2024). Human and AI collaboration in the higher education environment: opportunities and concerns. *Cognitive Research: Principles and Implications, 9*(1), Article 20. https://doi.org/10.1186/s41235-024-00547-9

Fragiadakis, G., Smith, J., & Chen, L. (2024). Evaluating human-AI collaboration: A review and methodological framework (arXiv:2407.19098). *arXiv.*

Holstein, K., & Aleven, V. (2022). Designing for human–AI complementarity in K–12 education. *AI Magazine, 43*(2), 239–248. https://doi.org/10.1002/aaai.12058

Katsamakas, E., Pavlov, O. V., & Saklad, R. (2024). Artificial intelligence and the transformation of higher education institutions: A systems approach. *Sustainability, 16*(14), Article 6118. https://doi.org/10.3390/su16146118

Kepel, D., & Valogianni, K. (2024). Autonomous prompt engineering in large language models (arXiv:2407.11000). *arXiv.* https://doi.org/10.48550/arXiv.2407.11000

Lin, Z. (2024). Beyond principlism: Practical strategies for ethical AI use in research practices (arXiv:2401.15284). *arXiv.* https://doi.org/10.48550/arXiv.2401.15284

Sathvika, A. L., Purnima, A., Harshitha, S., Anusha, P., & Nafisa, S. (2024). Prompt engineering for improved AI responsiveness: Methods and case studies. *International Journal of Creative Research Thoughts, 12*(11), 468–471. https://www.ijcrt.org/papers/IJCRT2411056.pdf

Socha, B. (2024). There is no AI without alignment: How stakeholder collaboration ensures AI success. *Strategy+Business*. Retrieved April 24, 2025, from https://www.strategy-business.com/blog/There-is-no-AI-without-alignment

University of Rochester, Office of the Provost. (n.d.). *Generative AI use in education*. Retrieved May 10, 2025, from https://www.rochester.edu/provost/gen-ai-education/

Wynants, S., Childers, G., De La Torre Roman, Y., Budar-Turner, D., & Vasquez, P. (2025). Ethical principles AI framework for higher education. CSU AI Commons, California State University, Fullerton. https://genai.calstate.edu/communities/faculty/ethical-and-responsible-use-ai/ethical-principles-ai-framework-higher-education

Younger, R. (2024). AI is not just a technology. It has become a stakeholder. *Oxford Answers*. Retrieved April 24, 2025, from https://www.sbs.ox.ac.uk/oxford-answers/ai-not-just-technology-it-has-become-stakeholder

4

The Emergence of Generative AI: Navigating Short-Term Viability and Long-Term Sustainability in Human-AI Partnerships

Introduction: This Significance of Humanity's Most Powerful Technology & AI

Human Language is the most powerful technology humans have ever created—it forms the foundation of human civilization (Harari, 2014). Language began as a tool for survival and communication in small groups. It allowed early humans to share information, solve problems, and build trust. Over time, language became the foundation for mass cooperation, as it enabled people to create and share stories and myths. These narratives allowed humans to organize themselves into larger groups by establishing shared values, beliefs, and norms, forming the beginnings of culture (Tomasello, 2019). Through this, humans were able to create **information networks** that preserved and passed down knowledge over generations, making cooperation on a massive scale possible (Harari, 2014).

All major social, cultural, and technological revolutions resulted from human language (Harari, 2014). For example, the Scientific Revolution was driven by the ability to share and refine knowledge through written communication, as evidenced by Isaac Newton's *Principia Mathematica*, which laid the foundation for modern physics (Westfall, 1980). All complex social structures emerged from the mass cooperation that language afforded groups of humans (Tomasello, 2018). This ability to attach meaning to shared symbols allowed humans to collaborate in large numbers toward common goals (Deacon, 1997). For instance, the concept of money only exists because people collectively agree on the value assigned to physical or digital symbols like coins, bills, and cryptocurrencies (Searle, 1995). Without this shared agreement, the system of trade and global commerce would not exist.

Supplementary Information The online version contains supplementary material available at https://doi.org/10.1007/978-3-032-05205-6_4.

D. Woodring, *Advancing Higher Education with Human-Centric AI Partnerships*,
https://doi.org/10.1007/978-3-032-05205-6_4

It is rather significant then, that for the past 70 years, humans have been teaching their language to another powerful technology: Artificial Intelligence (AI)—The simulation and replication of human intelligence by machines, particularly computers (Russell & Norvig, 2020). The development of Natural Language Processing (NLP) dates back to Alan Turing's seminal 1950 paper, Computing Machinery and Intelligence. Turing proposed a framework for natural language understanding, marking the foundation of teaching machines to process human language (Turing, 1950). Early advancements focused on syntax and word-based models, such as machine translation projects in the 1950s and ELIZA, an early NLP chatbot developed in the 1960s (Weizenbaum, 1966). This chapter expands on the significance of advances in AI, lessons from history, and looks into the future with illustrative examples of both desired and undesired outcomes emerging from variations in human-AI collaboration and synergy in higher education.

Chapter 2 and 3 Recap: Building on the three modes of human-AI collaboration in Chap. 2, this chapter expands on the synergy between human-AI collaboration that can produce both desirable and undesirable outcomes—representing the dual-nature impact of technologies on society throughout history. Understanding these potential outcomes in terms of short-term vs. long-term viability and sustainability provides insights for optimizing human-GenAI collaboration and synergy in the multi-phase rollout in Part 2.

The Dual Impact of Technology and Change: Understanding Synergy & Emergence

Every major technological revolution has been a double-edged sword (Bostrom, 2014). From the innovations in farming tools and techniques to the scientific advancements in physics, engineering, and programming knowledge that brought the digital revolution, each revolution advanced societies in many ways but also came with a cost (see Table 4.1) (Harari, 2014). For example, the printing press democratized knowledge but also facilitated the spread of misinformation (Eisenstein, 1982). Similarly, AI systems enhanced problem-solving and innovation, enabling breakthroughs in fields as diverse as healthcare, finance, and engineering (Russell & Norvig, 2020; Topol, 2019; Agrawal et al., 2018). Yet, AI systems trained on biased data have reproduced societal inequities, while reliance on machine-driven automation has displaced workers in traditional industries (Obermeyer et al., 2019).

Design Thinking & Change Management Insights: In design thinking terms (Brown, 2019), these dual effects illustrate the importance of empathizing with users and prototyping incremental solutions. Likewise, Kotter's adaptive leadership model (Kotter et al., 2021) warns that harnessing "the

Table 4.1 Dual impacts of technological revolutions

Revolution and timeline	Description
Agricultural Revolution (~10,000 BCE)	Expanded food supply and human settlements but caused environmental harm and created social inequalities (Diamond, 1999).
Bronze and Iron Ages (~3300 BCE–500 BCE)	Enhanced farming and weapons, but also fueled wars and resource exploitation (McAnany & Yoffee, 2009).
Printing Press (fifteenth century CE)	Boosted literacy and knowledge, but also spread false information and propaganda (Eisenstein, 1982).
Scientific Revolution (~sixteenth–eighteenth century CE)	Advanced understanding and technology but raised ethical concerns about human experiments and dangerous weapons (Shapin, 2018).
Industrial Revolution (~eighteenth–nineteenth century CE)	New technology boosted productivity and modernized society, but caused pollution, unsafe working conditions, and overcrowded cities (Stearns, 2018).
Electricity Revolution (nineteenth century CE)	Improved work and home life through electrification, but increased dependency on finite resources and environmental problems (Hughes, 1993).
Digital Revolution (late twentieth–early twenty-first century CE)	Improved communication, education, and business, but spread misinformation, violated privacy, and increased social isolation (Allcott et al., 2019; Twenge et al., 2018).
Artificial Intelligence (twenty-first century CE)	Automates tasks and solves complex problems, but raises concerns about bias, data privacy, transparency, and potential job displacement (Russell & Norvig, 2020).

Note. This table provides a historical overview of major technological revolutions, summarizing their dual impacts by describing both the advancements and the concurrent societal or ethical challenges they introduced

good edge" of technology requires transparency, shared vision, and continuous iteration.

Synergy, Emergence, & Disruptive Technologies

The above examples demonstrate two fundamental concepts that are important for understanding the impacts of disruptive technologies on social institutions: **synergy** and **emergence**. Synergy refers to the idea that the combined effects of a group are greater than the sum of individual contributions (Corning, 2018). When humans began cooperating in small groups, shared language emerged as an outcome. It evolved organically, not from a single individual, but from the collective needs and activities of a group—the sum of the whole was greater than its parts. Once it emerged, language itself became a foundation for other **emergent properties:** religion, government, education, economy, law, science, media, and technology, to name a few (see Table 4.2).

Table 4.2 Timeline of social institutions as they emerged

Institution and timeline	Description
Family (~2 million BCE–present)	The first social institution, based on biological relationships and survival cooperation (Chapais, 2010).
Communities (~10,000 BCE–present)	Communities formed around farming, introducing property ownership and stewardship (Bogucki, 1999).
Government (~3500 BCE–present)	Emergence of complex governance systems in early city-states (McAnany & Yoffee, 2009).
Religion (prehistoric–present)	Organized religions predating written history, evolving into complex belief systems (Harari, 2014).
Legal Systems (~2000 BCE–present)	Formalized laws and legal systems, starting with early codes like Hammurabi's (Westbrook, 2003).
Military (prehistoric–present)	Organized warfare and defense systems evolving with societal complexity (Gat, 2006).
Education (ancient civilizations–present)	Formal education institutions from ancient civilizations to modern systems (Marrou, 1982).
Labor Economy (~1760–present)	Industrial factories and worker unions redefining labor and economic exchange (Stearns, 2013).
Media (~1450–present)	Evolution of mass communication from printing press to digital age (Briggs & Burke, 2005).
Healthcare (ancient civilizations–present)	Development of medical practices and public health systems across cultures (Porter, 1999).
Digital Technology (~1950–present)	From early computers to ubiquitous digital technologies shaping society (Ceruzzi, 2003).

Note. This table presents a chronological overview of the emergence of key social institutions, providing a brief description and approximate timeline for each

In other words, language emerged from synergy created from human collaboration. Emerging language then created a **feedback loop** where enhanced communication enabled larger-scale cooperation. Together, mass cooperation and language enhanced creativity, innovation, problem solving, and decision-making within and across groups. Enhanced cooperation and organization, in turn, led to the emergence of complex social systems, governance, and technologies. These **nonlinear feedback loops** continue to shape human progress, as each innovation builds on and amplifies the previous ones (Tomasello, 2018; Harari, 2014).

These patterns of emergence and synergy are not only visible in the evolution of language and institutions, but also in the design and integration of technology—particularly in the case of artificial intelligence. As humans increasingly collaborate with GenAI systems, new emergent properties continue to arise from this interaction, enabling capabilities neither humans nor machines could achieve alone. Figure 4.1 illustrates this dynamic process by modeling **Human–AI Synergy** as a system of layered emergence and feedback.

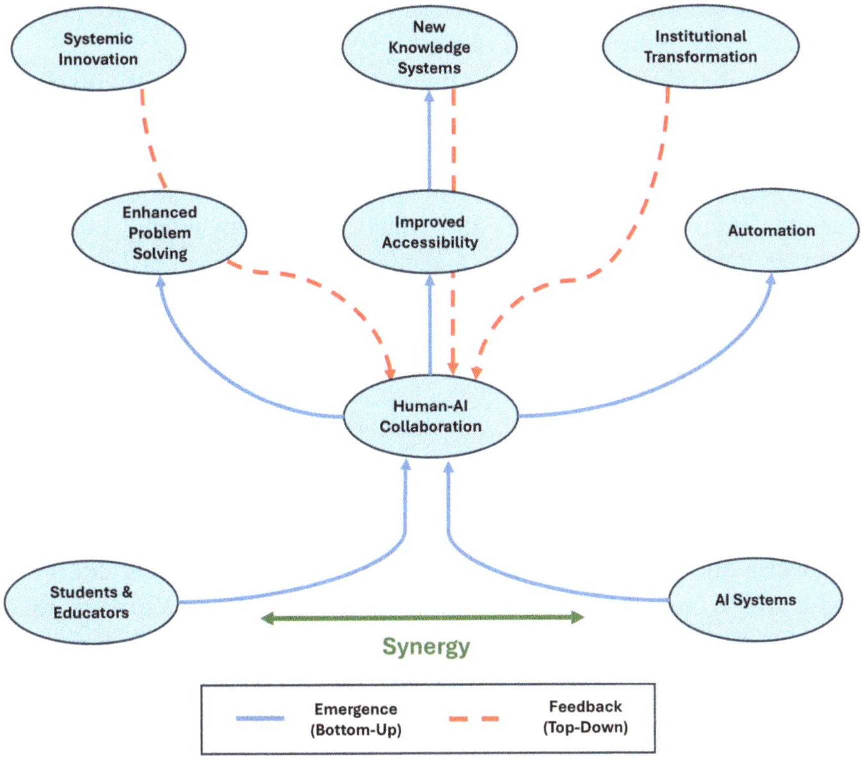

Fig. 4.1 Human-AI synergy: emergence and feedback loop diagram. (*Note.* A diagram illustrating the feedback loops within human-AI synergistic systems, showing how emergent properties and outcomes influence future interactions and adaptations)

Human-AI Synergy & Emergence

Similarly, early AI systems created feedback loops by automating tasks and improving problem-solving, which amplified human innovation in fields like healthcare, finance, and engineering, setting the foundation for even more complex technological advancements (Russell & Norvig, 2020).

These processes often culminate in **meta-emergence**, where new properties and relationships emerge from the interaction of previously emergent systems. For example, the invention of the printing press (an emergent property) revolutionized the way information was shared, leading to the spread of literacy and education. Over time, this facilitated the *Scientific Revolution*, which created a new system of knowledge generation and validation. The interaction between the printing press and scientific inquiry eventually gave rise to modern universities, which institutionalized the pursuit and transmission of knowledge. This cycle of meta-emergence demonstrates how innovations build upon one another, often leading to outcomes that are compounding, interactive, and unpredictable (Eisenstein, 1982).

Just as the printing press revolutionized the dissemination of knowledge while posing challenges such as misinformation and loss of traditional authority structures (Eisenstein, 1982), AI brings both opportunities and risks. Integrating AI into higher education offers the potential for personalized learning, improved accessibility, and enhanced problem-solving capacities. However, it also necessitates addressing ethical concerns, such as data privacy, algorithmic bias, and the risk of over-reliance on AI tools that could undermine critical thinking and creativity (Russell & Norvig, 2020). It is likely the impact of GenAI on higher education will have the same effect as the printing press, if not more pronounced.

Visualizing Nonlinear Human-AI Synergy & Emergence

To further understand how the integration of GenAI in higher education might evolve into systems-level transformation, it is helpful to view these developments through the lens of layered emergence and feedback. Figure 4.1 visualizes this process by modeling Human–AI Synergy as a recursive system in which each level of collaboration gives rise to new emergent and meta-emergent outcomes.

At the core of this system is Human–AI Collaboration, where human capabilities—such as creativity, ethical reasoning, and contextual knowledge—interact with AI's speed, automation, and pattern recognition. This collaboration generates a series of first-order emergent properties, including Enhanced Problem-Solving, Improved Accessibility, and Automation. These outcomes enable efficiencies and affordances that would not be possible through human or AI efforts alone.

However, the model shows that emergence does not stop at this initial layer. As illustrated in Fig. 4.1, these capabilities compound and give rise to meta-emergent outcomes, including Systemic Innovation, New Knowledge Systems, and Institutional Transformation. These are not simply additive gains—they represent shifts in the way knowledge is created, systems are governed, and institutions function. For example, just as the printing press eventually enabled the rise of the modern university, AI-driven collaboration could lead to new forms of educational infrastructure, scholarly knowledge generation, and learner-centered models.

Crucially, the diagram also highlights feedback loops, shown as top-down arrows from meta-emergent systems back to the foundational layer. These feedback mechanisms demonstrate how higher-order changes—such as institutional reform or shifts in epistemological frameworks—can influence the nature of future Human–AI interactions. The result is a dynamic system in which emergence and recursion are always in play, reinforcing the idea that GenAI's impact will not be linear but compounding, interactive, and continually evolving.

Decoding Human-AI Partnerships: Viability vs. Sustainability in Higher Education Synergies

History reveals that mass cooperation around significant social, cultural, and technological shifts inevitably leads to far-reaching, long-term, and often unpredictable consequences (Carmody-Bubb, 2023). This lesson is especially pertinent today as humans and Artificial Intelligence (AI) begin to collaborate across numerous domains, including the vital sector of higher education. This Human-AI Collaboration is poised to create novel synergies and emergent properties with potentially unprecedented effects. To navigate this new landscape responsibly, we must differentiate between interactions that offer short-term functionality and those that foster long-term health and ethical alignment within educational ecosystems around the world.

Therefore, we introduce two crucial dimensions for evaluating these collaborations: viability (short-term effectiveness) and sustainability (long-term health and consequences). Understanding these concepts is central to the principles and goals discussed throughout this book.

The Short-Term View: Does it Work Now?

In the immediate term, human-AI interactions can be classified based on their operational success or failure.

First, let's consider viable synergies. A viable synergy is essentially a collaboration between humans and AI that produces a positive, functional outcome in the short term; it achieves an immediate goal, solves a current problem, or makes a process work effectively right now (Barile et al., 2014). The focus here is squarely on immediate functionality, efficiency, problem-solving, and achieving current objectives. Think of it like building a temporary bridge across a small stream using readily available logs: it's viable because it gets you across today and serves the immediate purpose. A practical example in higher education involves a department overwhelmed by common prospective student email inquiries implementing a simple GenAI chatbot trained on their FAQ documents. When the chatbot successfully answers 80% of basic questions almost instantly, significantly freeing up staff time, it represents a viable synergy. It solves an immediate operational problem (staff overload) and achieves a short-term goal (faster response times).

Conversely, nonviable synergies can be identified. These are interactions or collaborations that fail to produce the intended beneficial result or simply cannot function effectively even in the short term (Barile et al., 2014). The interaction breaks down, fails to achieve the goal, or creates immediate problems that outweigh any benefits. Such situations are characterized by malfunction, immediate failure, an inability to meet current needs, and the creation of more problems than solutions right away.

Using the bridge analogy, this would be like trying to use those same logs to span a wide river; the attempt fails immediately as the structure collapses, rendering it nonviable for the intended purpose from the start. In a higher education context, imagine an instructor attempting to use a new, poorly designed GenAI tool to automatically grade complex, nuanced essays in a philosophy course. If the AI consistently misunderstands arguments, gives irrelevant feedback, and assigns nonsensical grades, confusing students and forcing the instructor to spend more time correcting errors than grading manually, this represents a nonviable synergy. The collaboration fails to achieve the desired outcome (accurate, helpful grading) and creates immediate negative consequences.

It is crucial to recognize, however, that viability only indicates if something works now. It doesn't indicate if it's a good idea in the long run. This critical distinction leads to the concept of sustainability.

The Long-Term View: Is it Healthy for the Future?

Beyond immediate functionality, consideration must be given to the long-term impacts of human-AI interactions on individuals, institutions, and society.

This leads to the consideration of sustainable synergies. A sustainable synergy is a collaboration that is not only viable (it works now) but is also healthy and beneficial over the long term (Pedercini et al., 2019). It fundamentally considers the broader, extended impacts on people, the institution, society, and potentially the environment. Its focus is on fostering long-term health, ensuring ethical alignment, promoting equity, supporting human development, encouraging responsible resource use, generating positive future impacts, and maintaining adaptability. Returning to the bridge analogy, a sustainable synergy is like building a sturdy, well-engineered stone bridge: it works now, is built to last using durable materials responsibly, and serves the community for generations without causing unintended harm downstream.

Consider a university implementing an AI-powered adaptive learning platform in introductory math courses. This is initially viable by providing immediate, personalized practice. It becomes sustainable when it is integrated thoughtfully into the curriculum, complementing rather than replacing instructor interaction and focusing on building foundational understanding. Sustainability is further achieved when instructors receive training to interpret AI data and enhance their teaching, when student data is handled ethically with clear policies and consent, when the system is regularly audited for bias to ensure equity, and when feedback loops allow ongoing improvement based on user input. Such a synergy supports student learning, empowers instructors, upholds ethical standards, and contributes positively to the educational mission over time.

Finally, it is important to be wary of unsustainable synergies. These collaborations might appear successful or viable in the short term but carry hidden costs or negative consequences that emerge over time (Pedercini et al., 2019). They tend to deplete crucial resources—such as human skills, trust, or environmental resources—create inequities, violate ethical principles, or cause long-term harm to individuals, the institution, or society. The focus here is on recognizing potential long-term harm, resource depletion (skills, trust, environment), ethical erosion, inequity, and negative future consequences that might arise despite short-term gains. The analogy is building that temporary log bridge (viable now) but doing so by cutting down all the trees along the bank, leading to massive soil erosion and flooding next rainy season. The short-term fix created a long-term disaster.

In higher education, an example of an unsustainable synergy could be a university encouraging widespread, unchecked use of a powerful GenAI writing tool across all courses solely to quickly "improve" student writing scores (a viable short-term goal). This approach becomes unsustainable as students grow overly reliant on the tool, hindering the development of their own critical thinking, research synthesis, and authentic writing voice (depleting human skills). Faculty may struggle to assess genuine understanding versus polished AI output, potentially leading to grade inflation and eroding assessment value (damaging academic integrity). Inequities can arise based on access or prompting skills, and the institution's reputation might suffer if graduates lack expected communication competencies. Furthermore, unclear ethical guidelines can cause widespread confusion about plagiarism. This synergy, by focusing only on a short-term metric, ultimately undermines core educational goals and creates significant long-term problems. These four synergy types, which build on established systems theory and sustainability research, are summarized in Table 4.3.

These concepts are not entirely new. Throughout history, collaborations between humans and technological systems have generated a spectrum of outcomes. For example, the printing press fostered viable synergies between printers, authors, and educators, leading to increased literacy (Krebs & Bossel, 1997). Over time, this evolved into sustainable synergies driving advancements in education and science. However, resistance from institutions fearing

Table 4.3 Key definitions for viable, nonviable, sustainable, and unsustainable synergies

Type of synergy	Definition
Viable synergies	Collaborative interactions that produce immediate positive outcomes and function successfully in the short term (Barile et al., 2014).
Nonviable synergies	Interactions that fail to generate beneficial results or cannot sustain themselves in the present context (Barile et al., 2014).
Sustainable synergies	Efforts that maintain positive impacts over time, considering long-term economic, social, and environmental factors (Pedercini et al., 2019).
Unsustainable synergies	Interactions that may be beneficial in the short term but cannot maintain positive effects without depleting resources or causing harm (Pedercini et al., 2019; UNESCO, 2021).

Note. This table defines four types of synergies, distinguishing between short-term functionality (viable/nonviable) and long-term endurance and impact (sustainable/unsustainable)

loss of control created nonviable synergies hindering adoption, while practices like over-harvesting trees for paper led to unsustainable synergies causing deforestation (Gomez da Silva et al., 2020).

Similarly, the Industrial Revolution saw viable synergies in factory production boosting efficiency and wealth, but also nonviable synergies through worker exploitation (Humphries, 2011). Long-term, sustainable synergies emerged through technological advancements improving living standards, yet unsustainable synergies manifested as severe environmental degradation and social inequality (Hopkins, 1982).

As Fig. 4.2 illustrates, these historical interactions between humans and technology generated the four synergy types, leading to broader emergent social outcomes—both positive (like modern education) and negative (like environmental harm). Crucially, feedback loops exist: positive outcomes reinforce innovation, while negative ones can disrupt collaboration. This demonstrates that synergy and emergence are dynamic, interconnected processes shaping history.

The Emergence of Generative AI (GenAI) in Higher Education

The latest powerful technological system demanding attention is Generative AI (GenAI)—AI capable of creating novel text, audio, image, and video content from existing data (Shi et al., 2024). GenAI is particularly disruptive in higher education because, for the first time, AI can replicate many core academic tasks previously exclusive to students and educators (see Table 4.4). This highlights GenAI's practical utility but also raises profound questions

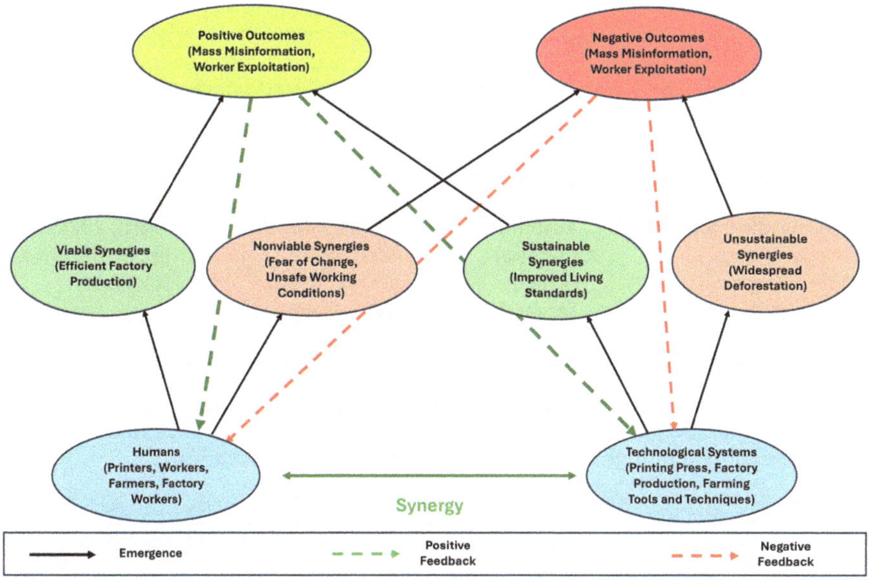

Fig. 4.2 Historical representation of synergies with emergence and feedback loops. (*Note*. A visual model representing how human-technology synergies throughout history have produced emergent outcomes and created feedback loops that shape subsequent development)

about what is considered knowledge (ontology) and how it is obtained and understood (epistemology) (Eke, 2023).

Optimizing Human-GenAI Collaboration: Synergies and Emergence Today

The advent of GenAI necessitates new forms of human-AI collaboration. How these systems are designed and how students and educators choose to use them will inevitably create viable, nonviable, sustainable, and unsustainable synergies within higher education ecosystems globally. Since 2023, this has been unfolding rapidly, yielding unforeseen short- and long-term impacts (see Fig. 4.3).

Figure 4.3 depicts this dynamic interaction. Humans (students, educators) bring context and goals, while GenAI systems offer efficiency and content generation. This interaction produces the four synergy types: viable synergies offering short-term wins like automated grading; nonviable synergies resulting from problematic uses leading to bias or mistrust; sustainable synergies involving ethical, strategic integrations like well-designed adaptive learning; and unsustainable synergies where short-term convenience leads to long-term harms like over-reliance. These synergies, in turn, lead to emergent outcomes

Table 4.4 Tasks GenAI can perform in higher education

Category	Description
Academic Writing and Communication	Curates knowledge and generates papers, speeches, scripts, presentations, and literature reviews (Li, 2024).
STEM and Technical Work	Writes, debugs, and explains code; performs simulations, modeling, and data analysis (Garrido-Merchan, 2023).
Teaching and Pedagogical Support	Provides personalized tutoring, designs curricula, and automates test creation and grading (Reuters, 2024).
Creativity and Interdisciplinary Work	Creates art, music, and literature; facilitates collaborative problem-solving; and synthesizes interdisciplinary research (Ding, 2024).
Language and Translation Work	Translates languages in real-time, assists in language learning, and proofreads and edits text (Kim et al., 2024).

Note. This table categorizes and describes various tasks that Generative AI can perform within the context of higher education, supported by recent citations

such as enhanced educational practices and innovation (from viable/sustainable interactions) or ethical challenges and systemic risks (from nonviable/unsustainable interactions). Crucially, feedback loops (blue dashed arrows in Fig. 4.3) show that these outcomes influence future human behavior and AI development, highlighting the iterative nature of emergence, feedback, and adaptation in the **Human-AI system**.

Building Sustainable Human-AI Collaboration: The Urgent Need for Planning

As stakeholders and institutions integrate GenAI, careful planning and research are paramount. We must adapt to a technology whose inner workings are not fully understood, even by its creators (Anthropic, 2024). This "black box" nature presents opportunities but also significant risks. Viable synergies (e.g., automated grading) might streamline tasks, but if implemented without considering long-term impacts, they can easily become nonviable (amplifying bias) or unsustainable (reducing human oversight and critical thinking) (Anthropic, 2024).

Why This Distinction Matters Urgently for GenAI in Higher Education

Understanding these four synergy types is not merely academic; it's essential for ethical leadership and navigating the AI era (Shneiderman, 2022). The temptation to focus solely on viability—Does it save time now? Does it solve

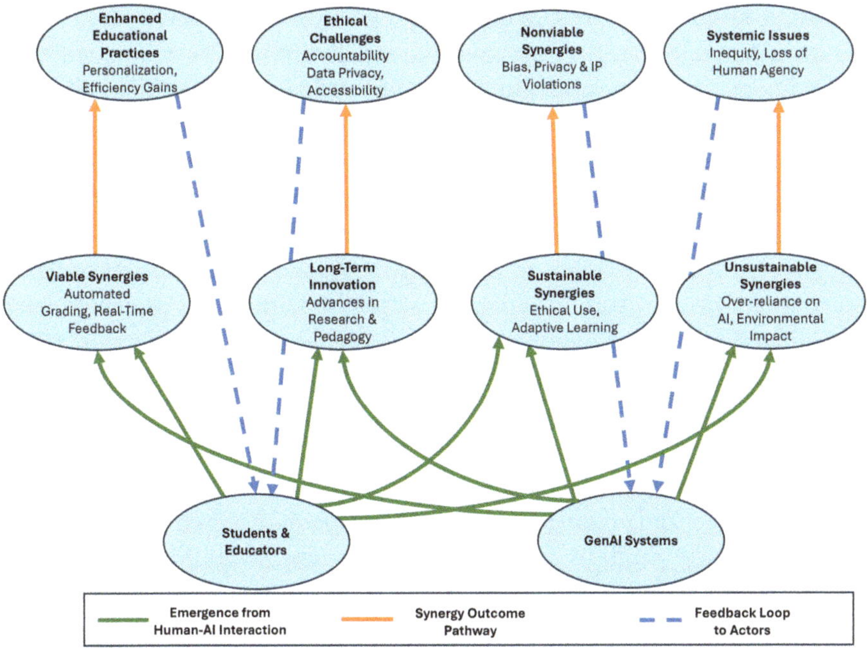

Fig. 4.3 Synergies and outcomes of historical human-technology collaboration. (*Note.* This diagram illustrates the dynamic interaction between humans (students, educators) and GenAI systems in higher education. Their collaboration produces viable, nonviable, sustainable, and unsustainable synergies, leading to emergent outcomes (e.g., enhanced practices, ethical challenges). Feedback loops demonstrate the iterative nature of this system)

this immediate problem?—is strong. However, pursuing short-term fixes without evaluating long-term sustainability is perilous with GenAI.

The risk of unsustainable viability is significant. Many GenAI applications offer tempting immediate benefits. But if these lead to deskilling, academic integrity erosion, bias amplification, or privacy violations, they become unsustainable, ultimately betraying the core mission of education. The superficial "efficiency" of AI generating work and AI grading it, without deep human engagement, exemplifies a potentially viable but profoundly unsustainable loop.

Conversely, the need for human-centric sustainability is paramount. This book advocates for prioritizing sustainable synergies. This demands a human-centric approach (Shneiderman, 2022), asking not just "Can AI do this?" but "How can AI empower humans, augment skills, uphold ethics, and benefit the long-term health of educational ecosystem?" We must design AI integrations that are both effective today and ethically sound for the future.

Furthermore, emergence amplifies the stakes. The outcomes of these synergies are not isolated; they trigger emergence—potentially large-scale shifts in institutional culture, learning value, and the definition of knowledge itself.

Fostering sustainable synergies encourages positive emergence, while tolerating unsustainable ones risks negative, potentially irreversible consequences.

Therefore, as users, teams, and institutions evaluate any GenAI application, they must consciously apply both the viability and sustainability lenses. The objective isn't simply AI adoption, but the cultivation of human-AI partnerships that are effective today and contribute to a thriving, ethical, and human-centered future for higher education. Understanding these dynamics from a human-centric perspective, as encouraged by the User Experience (UX) warmup activities in Part 1, is the first step. The frameworks presented later, particularly the four-phase rollout in Part 2, are designed to guide this critical journey toward conceptualizing, mapping, and implementing truly viable and sustainable human-AI strategies.

As stakeholders and institutions implement and scale GenAI, planning and research become critical. All stakeholders need to gain understanding and adapt to the challenges and opportunities posed by a system that even its own creators don't fully understand. While large language models (LLMs) demonstrate remarkable capabilities, their inner workings remain poorly understood, even by the experts who design them (Anthropic, 2024). This creates immediate opportunities for innovation but also significant risks. For example, viable synergies might emerge through applications like automated grading, curriculum design, or real-time student feedback, streamlining administrative burdens and enhancing personalization. However, nonviable synergies could arise if reliance on these tools diminishes human oversight or amplifies biases, leading to decisions that lack transparency and accountability (see Fig. 4.4) (Anthropic, 2024).

Importance of UX Warmup Activities

This chapter demonstrates why understanding human-AI collaboration and synergies from a human-centric perspective is so important. These warmup activities in Part 1 will help prepare you for thinking like a User Experience (UX) Designer, capable of conceptualizing, mapping, and implementing viable and sustainable human-AI strategies and practices.

Human-AI Synergy in Higher Education: Cohort Effects of COVID-19 and GenAI

The release of ChatGPT-3.5 in 2023 coincided with a critical period for higher education (OpenAI, 2023). Students entering college carried academic, social, and emotional deficits from the disruptions caused by the

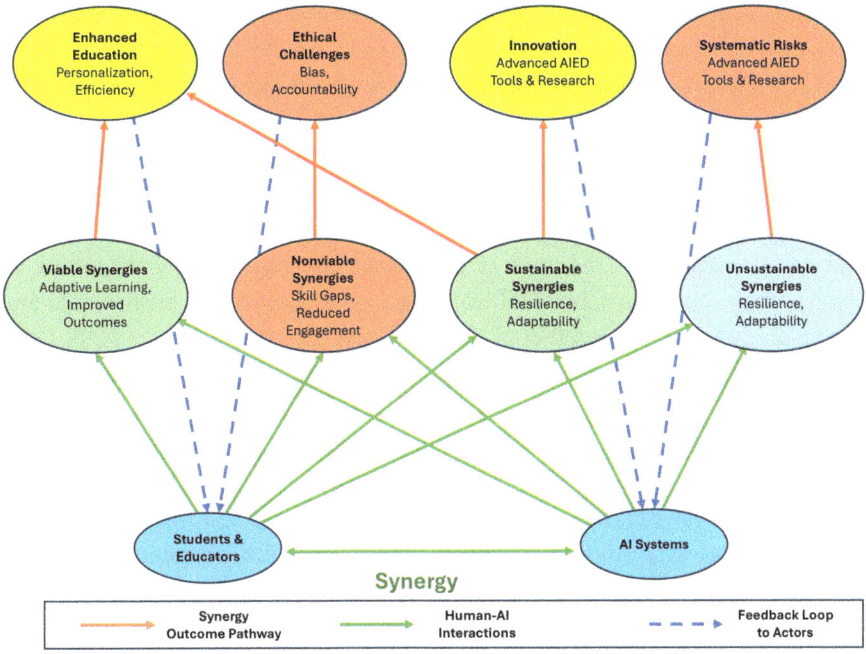

Fig. 4.4 Human-GenAI synergy with synergy examples and feedback loops. (*Note.* An illustrative diagram showing specific examples of human-GenAI synergies and the feedback loops that influence their evolution and impact within an educational system)

COVID-19 pandemic (2020), including prolonged remote learning and limited peer interaction (Engzell et al., 2021; Racine et al., 2021). This enabled both viable and nonviable synergies between students, educators, institutions, and AI in higher education. This section explores the evolving relationship between human intelligence and AI across short-term, medium-term, and long-term horizons, emphasizing their impact on teaching, learning assessment, and other outcomes.

Methodological Caveat: Readers should note that the foregoing timeline (2023–2033) provides a conceptual illustration of potential synergistic effects, not an empirical analysis establishing causality. The discussion did not utilize longitudinal or quantitative data. Its primary aim is to underscore the potential for unpredictable, nonlinear consequences when significant societal events (like the COVID-19 pandemic) coincide with disruptive technological shifts (like GenAI adoption) in complex systems such as higher education. Although this treatment is introductory and simplified, the plausible interactions discussed highlight critical areas where empirical research, exemplified by studies like Katsamakas et al. (2024), is essential to fully understand the likely significant impacts.

2023–2025: Short-Term Viable and Nonviable Synergies

The COVID-19 pandemic (2020) disrupted education systems globally, affecting the educational, social, emotional, and psychological outcomes of students aged 5–24 (Engzell et al., 2021; Racine et al., 2021; Fegert et al., 2020). These students, now aged 8–27, have entered or completed college, bringing with them challenges caused by prolonged remote learning and limited social interaction (Morgül et al., 2020). As students began college in 2023, many relied on ChatGPT for homework, leading to both viable and nonviable synergies (Engzell et al., 2021; Dorn et al., 2020).

Viable synergies include the immediate academic support GenAI can offer, such as aiding students in brainstorming ideas, editing drafts, or solving problems more efficiently, potentially enhancing academic performance (Chan & Hu, 2023).

However, over-reliance on AI (nonviable synergy) may bypass essential learning processes, resulting in gaps in critical skills like independent research or analytical reasoning. This can also undermine trust in the system as instructors struggle to differentiate between student-generated work and AI outputs (Chan & Hu, 2023).

2026–2028: Medium-Term Sustainable and Unsustainable Synergies

As GenAI becomes more entrenched in education, skills gaps may become evident if graduates' competencies do not match their academic credentials. Employers could question the value of degrees if there are mismatches between qualifications and real skills (Katsamakas et al., 2024).

Sustainable synergies emerge when institutions teach students how to collaborate responsibly with AI, producing graduates who excel in hybrid roles that require human creativity and AI proficiency (Atchley et al., 2024). Conversely, unsustainable synergies arise when institutions fail to adapt, perpetuating a disconnect between academic outputs and workforce demands. If grading or academic work relies heavily on AI-generated content, this work may reflect the unexplainable inner workings of an AI rather than understanding, reducing learning quality and originality (Zhai et al., 2024).

Higher education ecosystems may face challenges as they adapt curricula and assessments. For example, a graduate who relied on AI tools like ChatGPT might lack practical experience in market research or data analysis, despite strong grades. Such uncertainties may generate unpredictable ripple effects, including reduced trust in academic credentials and overall trust in higher education (Eaton, 2020; Gao et al., 2022).

2029–2033: Long-Term Impacts on Institutional Integrity & Resilience

By 2033, higher education may face complex emergent outcomes from these synergies. Institutions aligning curricula with **human-centric skills** and **AI ethics** could foster graduates who thrive in an AI-integrated world (sustainable synergies) (see Fig. 4.5). Human-AI collaboration between educators and AI could lead to educational models emphasizing creativity, critical thinking, and ethical reasoning. AI-powered tutors, like **Khanmigo** from **Khan Academy**, exemplify efforts to promote critical thinking (Khan Academy, 2024; Lowe, 2024, pp. 19–56).

However, overuse of AI for learning and teaching could blur the distinction between human and AI-generated work, leading to a potential crisis of credibility. Employers and students may question the relevance of higher education, highlighting the need for careful integration of AI to preserve academic integrity (Xie et al., 2023). Table 4.5 summarizes these key takeaways, providing examples for each type of synergy in the context of higher education.

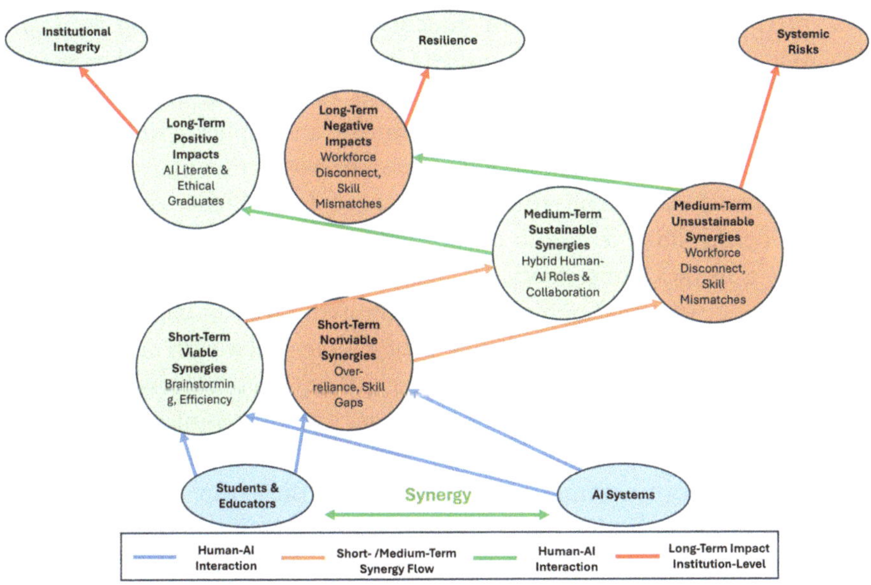

Fig. 4.5 Human-GenAI synergy across time horizons. (*Note.* This figure visually represents how human-GenAI synergies operate and evolve across different time horizons, from immediate (viable/nonviable) interactions to long-term (sustainable/unsustainable) outcomes)

Table 4.5 Key takeaways on synergies in higher education

Type of synergy	Description with citation
Viable synergies	When students and educators use Generative AI (GenAI) as a complementary tool to enhance learning while preserving essential human skills. Adaptive AI systems can personalize educational experiences (Chan & Hu, 2023).
Nonviable synergies	Over-reliance on AI for assignments or grading can reduce student engagement, create skills gaps, and erode academic integrity. Balanced AI integration is crucial to avoid undermining learning processes (Chan & Hu, 2023).
Sustainable synergies	Strategic AI integration in education fosters resilience and adaptability. Developing AI literacy among students prepares them for future challenges and hybrid roles requiring creativity and AI proficiency (Jin & Kim, 2024).
Unsustainable synergies	Overuse or misuse of AI could lead to diminished innovation, critical thinking, and institutional credibility. Policies guiding AI use are necessary to prevent negative feedback loops in education systems (Mallet, 2024).

Note. This table summarizes the key characteristics and implications of viable, nonviable, sustainable, and unsustainable synergies in the context of GenAI use in higher education

Conclusion: Ethically & Effectively Navigating the Challenges & Opportunities with GenAI

The interplay between cohorts of students social, emotional, and academic deficits entering higher education when ChatGPT-3.5 was released presents a complex set of challenges and opportunities for all stakeholders involved. Given that AI systems are designed to simulate and replicate human intelligence, it is easy to understand the problems created by the mass adoption of GenAI. GenAI easily replicates academic work for students and educators. While OpenAI didn't set out to disrupt higher education per se, the negative impact can be best understood as a design-induced challenge. The complex bureaucratic structure(s) of higher education itself means it has its own institutionally induced challenges, making it difficult to evolve with rapidly changing technologies (Christensen & Eyring, 2011). These factors are further complicated by the numerous other challenges and opportunities higher education institutions are facing.

While the above examples are oversimplistic, they help demonstrate the importance of strategically using AI tools in ways that foster viable and sustainable synergies for higher education and society (see Fig. 4.6). Harnessing this synergy is one of the primary goals in the book. Students, educators, and institutions in the process of integrating or planning to integrate generative GenAI into teaching, learning, assessment, or research practices must consider the transformative potential of GenAI while also mitigating potential risks.

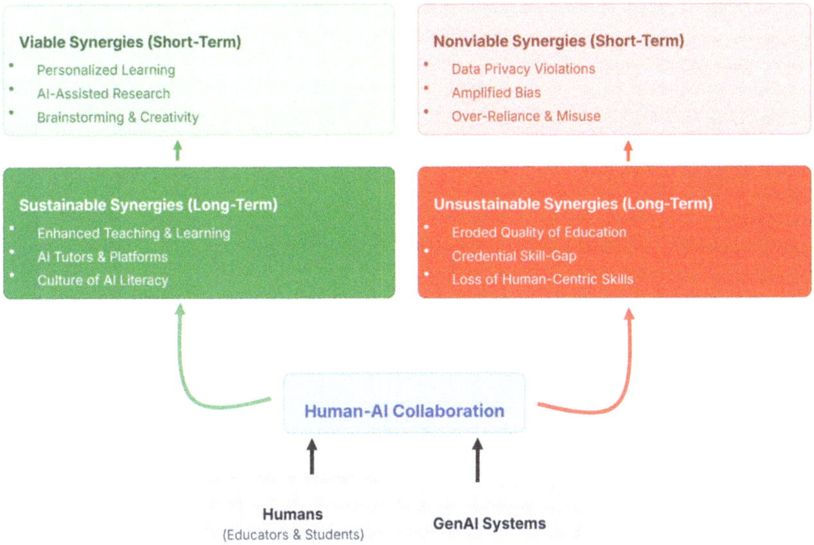

Fig. 4.6 Key takeaways on synergies in higher education. (*Note*. A bubble-and-plate diagram mapping four synergy types (viable, nonviable, sustainable, unsustainable) to three actors (educators, AI, students), illustrating how specific examples (e.g., "Personalized Learning," "Bias Amplification," "Hybrid Roles," "Credential Skill-Gap") cluster under each category)

Chapter 4 Workbook Prep

These questions help you connect the concepts of synergy, emergence, and the dual nature of technology from this chapter to the practical evaluation activities in the Chap. 4 UX and PALS Workbooks.

Analyzing Short-Term vs. Long-Term Outcomes (Prep for UX/PALS Synergy Mapping):

- Think about one specific GenAI application currently used or considered in your context (e.g., AI writing assistance, automated scheduling, AI research tool).
- Based on the chapter's definitions, what is one potential viable (positive short-term) outcome and one potential unsustainable (negative long-term) outcome of this application?

Considering Emergence & Stakeholders (Prep for UX Stakeholder Mapping & PALS Emergence Exploration):

- How might the short-term outcome you identified above emerge into a larger, perhaps unexpected, institutional change over time (positive or negative)?
- Which stakeholder group might be most significantly impacted (positively or negatively) by this potential emergent change?

Preparing for Deeper AI Exploration (Prep for PALS Advanced Prompts):

- Select one core concept from this chapter (e.g., historical parallels of tech disruption, viable vs. sustainable synergy, emergence theory, COVID cohort effects) you want to examine more closely using AI.
- How might you initially prompt a GenAI model to help you find case studies or contrasting viewpoints related to this concept in higher education?

References

Agrawal, A., Gans, J., & Goldfarb, A. (2018). *Prediction machines: The simple economics of artificial intelligence*. Harvard Business Review Press.

Allcott, H., Gentzkow, M., & Yu, C. (2019). Trends in the diffusion of misinformation on social media. *Research in Politics, 6*(2), 2053168019848554. https://doi.org/10.1177/2053168019848554

Anthropic. (2024). Decomposing language models into understandable components. *Anthropic*. https://www.anthropic.com/news/decomposing-language-models-into-understandable-components

Atchley, P., Pannell, H., Wofford, K., Hopkins, M., & Atchley, R. A. (2024). Human and AI collaboration in the higher education environment: Opportunities and concerns. *Cognitive Research: Principles and Implications, 9*, Article 20. https://doi.org/10.1186/s41235-024-00547-9

Barile, S., Iandolo, F., & Calabrese, M. (2014, October). *The viable systems approach (vSa) and its contribution to sustainable business behaviours* [Paper presentation]. 2nd International Symposium on Systems Thinking for a Sustainable Economy, Sapienza University of Rome, Italy.

Bogucki, P. I. (1999). *The origins of human society*. Blackwell Publishers.

Bostrom, N. (2014). *Superintelligence: Paths, dangers, strategies*. Oxford University Press.

Briggs, A., & Burke, P. (2005). *A social history of the media: From Gutenberg to the Internet* (2nd ed.). Polity Press.

Brown, T. (2019). *Change by design: How design thinking transforms organizations and inspires innovation* (Revised and Updated ed.). Harper Business.

Carmody-Bubb, M. (2023). *Cognition and decision making in complex adaptive systems: The human factor in organizational performance* (1st ed.). Springer Cham. https://doi.org/10.1007/978-3-031-31929-7

Ceruzzi, P. E. (2003). *A history of modern computing* (2nd ed.). MIT Press.

Chapais, B. (2010). *Primeval kinship: How pair-bonding gave birth to human society.* Harvard University Press.

Chan, C. K. Y., & Hu, W. (2023). Students' voices on generative AI: Perceptions, benefits, and challenges in higher education (arXiv:2305.00290). *arXiv*. https://arxiv.org/abs/2305.00290

Christensen, C. M., & Eyring, H. J. (2011). *The innovative university: Changing the DNA of higher education from the inside out.* Jossey-Bass.

Corning, P. A. (2018). *Synergistic selection: How cooperation has shaped evolution and the rise of humankind.* World Scientific Publishing Co. Pte Ltd.

Deacon, T. W. (1997). *The symbolic species: The co-evolution of language and the human brain.* W. W. Norton & Company.

Diamond, J. (1999). *Guns, germs, and steel: The fates of human societies.* W. W. Norton & Company.

Ding, Z. (2024). Opportunities and challenges of generative AI in art and design education. *Journal of Education, Humanities and Social Sciences, 25*, 126–132. https://doi.org/10.54097/ehss.v25i.13781

Dorn, E., Hancock, B., Sarakatsannis, J., & Viruleg, E. (2020). *COVID-19 and student learning in the United States: The hurt could last a lifetime.* McKinsey & Company. https://www.mckinsey.com/industries/education/our-insights/covid-19-and-student-learning-in-the-united-states-the-hurt-could-last-a-lifetime

Eaton, S. E. (2020). Academic integrity during COVID-19: Reflections. *Journal of Educational Thought, 53*(3), 277–281.

Eisenstein, E. L. (1982). *The printing press as an agent of change: Communications and cultural transformations in early-modern Europe.* Cambridge University Press.

Eke, D. O. (2023). ChatGPT and the rise of generative AI: Threat to academic integrity? *Journal of Responsible Technology, 13*, Article 100060. https://doi.org/10.1016/j.jrt.2023.100060

Engzell, P., Frey, A., & Verhagen, M. D. (2021). Learning loss due to school closures during the COVID-19 pandemic. *Proceedings of the National Academy of Sciences, 118*(17), Article e2022376118. https://doi.org/10.1073/pnas.2022376118

Fegert, J. M., Vitiello, B., Plener, P. L., & Clemens, V. (2020). Challenges and burden of the Coronavirus 2019 (COVID-19) pandemic for child and adolescent mental health: A narrative review to highlight clinical and research needs. *Child and Adolescent Psychiatry and Mental Health, 14*, Article 20. https://doi.org/10.1186/s13034-020-00329-3

Gao, J., Liu, X., & Wu, H. (2022). The impact of COVID-19 pandemic on higher education students' learning engagement and the role of academic resilience. *Frontiers in Psychology, 13*, Article 906469. https://doi.org/10.3389/fpsyg.2022.906469

Garrido-Merchán, E. C., & Gozalo-Brizuela, R. (2023). Generative AI for software development: A survey. *arXiv preprint arXiv:2310.15830*. https://doi.org/10.48550/arXiv.2310.15830
Gat, A. (2006). *War in human civilization*. Oxford University Press.
Gomez da Silva, G. C., Gibbs, H. K., & Rausch, L. L. (2020). Deforestation dynamics and the role of public policies in conserving the Amazon rainforest. *Current Opinion in Environmental Sustainability, 44*, 28–37. https://doi.org/10.1016/j.cosust.2020.06.005
Harari, Y. N. (2014). *Sapiens: A brief history of humankind*. Harper.
Hopkins, E. (1982). Working hours and conditions during the Industrial Revolution: A re-appraisal. *The Economic History Review, 35*(1), 52–66. https://doi.org/10.1111/j.1468-0289.1982.tb01186.x
Hughes, T. P. (1993). *Networks of power: Electrification in Western society, 1880–1930*. Johns Hopkins University Press.
Humphries, J. (2011). *Childhood and child labour in the British industrial revolution*. Cambridge University Press. https://doi.org/10.1017/CBO9780511973901
Jin, Y., & Kim, J. (2024). Fostering AI literacy in higher education: A review of competencies and pedagogical strategies. *Computers and Education: Artificial Intelligence, 6*, 100213. https://doi.org/10.1016/j.caeai.2024.100213
Katsamakas, E., Pavlov, O. V., & Saklad, R. (2024). Artificial intelligence and the transformation of higher education institutions: A systems approach. *Sustainability, 16*(14), Article 6118. https://doi.org/10.3390/su16146118
Khan Academy. (2024). *Khanmigo: AI-powered guide for every student*. https://www.khanacademy.org/khan-labs
Kim, H., Kim, S., & Park, J. (2024). A comparative study of machine translation and large language models for Korean-English translation. *Journal of Digital Contents Society, 25*(1), 145–154.
Kotter, J. P., Akhtar, V., & Gupta, G. (2021). *Change: How organizations achieve hard-to-imagine results in uncertain and volatile times*. Wiley.
Krebs, F., & Bossel, H. (1997). Emergent value orientation in self-organization of an animat. *Ecological Modelling, 99*(2–3), 143–164. https://doi.org/10.1016/S0304-3800(96)00050-6
Li, Y., Sha, L., Yan, L., Lin, J., Raković, M., Galbraith, K., & Chen, G. (2024). Can large language models write a good literature review? A comprehensive evaluation. *arXiv preprint arXiv:2402.01986*. https://doi.org/10.48550/arXiv.2402.01986
Lowe, A. (2024). The more things change: The ethical impacts of artificial intelligence in higher education. *Research Issues in Contemporary Education, 9*(2), 19–56. https://files.eric.ed.gov/fulltext/EJ1435770.pdf
Mallet, R. (2024). Navigating the ethical landscape of generative AI in higher education. *Journal of Academic Ethics*. https://doi.org/10.1007/s10805-024-09512-x
Marrou, H. I. (1982). *A history of education in antiquity* (G. Lamb, Trans.). University of Wisconsin Press. (Original work published 1948)

McAnany, P. A., & Yoffee, N. (Eds.). (2009). *Questioning collapse: Human resilience, ecological vulnerability, and the aftermath of empire*. Cambridge University Press.

Morgül, E., Kallitsoglou, A., & Essau, C. A. (2020). Psychological effects of the COVID-19 lockdown on children and families in the UK. *Revista de Psicología Clínica con Niños y Adolescentes, 7*(3), 42–48. https://doi.org/10.21134/rpcna.2020.mon.2049

Obermeyer, Z., Powers, B., Vogeli, C., & Mullainathan, S. (2019). Dissecting racial bias in an algorithm used to manage the health of populations. *Science, 366*(6464), 447–453. https://doi.org/10.1126/science.aax2342

OpenAI. (2023). GPT-4 technical report. *arXiv preprint arXiv:2303.08774*. https://doi.org/10.48550/arXiv.2303.08774

Pedercini, M., Arquitt, S., Collste, D., & Herren, H. (2019). Harvesting synergy from sustainable development goal interactions. *Proceedings of the National Academy of Sciences, 116*(46), 23021–23028. https://doi.org/10.1073/pnas.1817276116

Porter, R. (1999). *The greatest benefit to mankind: A medical history of humanity*. W. W. Norton & Company.

Racine, N., McArthur, B. A., Cooke, J. E., Eirich, R., Zhu, J., & Madigan, S. (2021). Global prevalence of depressive and anxiety symptoms in children and adolescents during COVID-19: A meta-analysis. *JAMA Pediatrics, 175*(11), 1142–1150. https://doi.org/10.1001/jamapediatrics.2021.2482

Reuters. (2024, January 18). *Arizona State University becomes first university to partner with OpenAI*. Reuters. https://www.reuters.com/technology/arizona-state-university-becomes-first-university-partner-with-openai-2024-01-18/

Russell, S., & Norvig, P. (2020). *Artificial intelligence: A modern approach* (4th ed.). Pearson.

Searle, J. R. (1995). *The construction of social reality*. Free Press.

Shapin, S. (2018). *The scientific revolution* (2nd ed.). University of Chicago Press.

Shi, J., Jain, R., Doh, H., Suzuki, R., & Ramani, K. (2024). An HCI-centric survey and taxonomy of human-generative-AI interactions (arXiv:2310.07127). *arXiv*. https://arxiv.org/abs/2310.07127

Shneiderman, B. (2022). *Human-centered AI*. Oxford University Press.

Stearns, P. N. (2013). *The industrial revolution in world history* (4th ed.). Westview Press.

Stearns, P. N. (2018). *The industrial revolution in world history* (4th ed.). Routledge.

Tomasello, M. (2018). *A natural history of human thinking*. Harvard University Press.

Tomasello, M. (2019). *Becoming human: A theory of ontogeny*. The Belknap Press of Harvard University Press.

Topol, E. J. (2019). *Deep medicine: How artificial intelligence can make healthcare human again*. Basic Books.

Turing, A. M. (1950). Computing machinery and intelligence. *Mind, 59*(236), 433–460. https://doi.org/10.1093/mind/LIX.236.433

Twenge, J. M., Joiner, T. E., Rogers, M. L., & Martin, G. N. (2018). Increases in depressive symptoms, suicide-related outcomes, and suicide rates among U.S. adolescents after 2010 and links to increased new media screen time. *Clinical Psychological Science, 6*(1), 3–17. https://doi.org/10.1177/2167702617723376

UNESCO. (2021). *Recommendation on the ethics of artificial intelligence.* UNESCO. https://unesdoc.unesco.org/ark:/48223/pf0000381137

Weizenbaum, J. (1966). ELIZA—a computer program for the study of natural language communication between man and machine. *Communications of the ACM, 9*(1), 36–45. https://doi.org/10.1145/365153.365168

Westbrook, R. (Ed.). (2003). *A history of ancient Near Eastern law* (Vol. 1). Brill.

Westfall, R. S. (1980). *Never at rest: A biography of Isaac Newton.* Cambridge University Press.

Xie, Y., Wu, S., & Chakravarty, S. (2023). AI meets AI: Artificial intelligence and academic integrity: A survey on mitigating AI-assisted cheating in computing education. In *Proceedings of the 24th Annual Conference on Information Technology Education (SIGITE '23)* (pp. 79–83). ACM. https://doi.org/10.1145/3585059.3611449

Zhai, C., Wibowo, S., & Li, L. D. (2024). The effects of over-reliance on AI dialogue systems on students' cognitive abilities: A systematic review. *Smart Learning Environments, 11,* Article 28. https://doi.org/10.1186/s41239-024-01138-6

5

Applying & Customizing the Core Frameworks for Optimizing Human-AI Partnerships in Higher Education

Introduction to 5 Human-Centric Frameworks for Ethical AI Integration

Higher Education Institutions (HEIs) worldwide are currently navigating significant changes driven by the capabilities of Generative Artificial Intelligence (GenAI). Research indicates that tools like ChatGPT can substantially alter traditional approaches to teaching, learning, assessment, research, and administration (Chang & Kidman, 2023; Cotton et al., 2023). This situation requires clear frameworks that integrate ethical principles with practical guidance to help stakeholders optimize GenAI's benefits responsibly (Guo & Lee, 2023).

This chapter introduces five core frameworks developed to meet this need:

1. Human-Centric AI Synergy (HCAS)
2. Regulation, Synergy, & Augmentation (RSA)
3. Enhancing Human-Centric Skills (EHCS)
4. Enhancing Academic Outcomes (EAO)
5. Authentic High-Impact Practices (AHIPs)

These frameworks, detailed comprehensively in Part 2 of the accompanying *Human-GenAI Collaboration & Synergy Guide*, provide the conceptual underpinnings and strategic approaches for ethical and effective GenAI integration. This chapter serves to summarize these frameworks, explain their interrelations, and prepare users to apply and customize them within their specific roles and institutional contexts. By utilizing the detailed guidance and practical steps available in the downloadable Guide, users can adapt these frameworks to align with diverse needs, goals, objectives, and desired outcomes across higher education.

Supplementary Information The online version contains supplementary material available at https://doi.org/10.1007/978-3-032-05205-6_5.

D. Woodring, *Advancing Higher Education with Human-Centric AI Partnerships*,
https://doi.org/10.1007/978-3-032-05205-6_5

Human-Centric AI Synergy (HCAS) Framework Summary

HCAS serves as the foundational conceptual framework for this guide. Its primary purpose is to translate principles from AI design research into practical, use-centric applications. This approach helps higher education stakeholders mitigate potential design-induced challenges or unintended consequences associated with using general-purpose GenAI tools. The core logic involves understanding AI limitations through continuous ethics and literacy training, strategically leveraging the complementary strengths of humans and AI in collaborative activities, and ensuring human-centric values (prioritizing well-being, equity, fairness, transparency, sustainability) guide GenAI use and policy development (Berretta et al., 2023; Shneiderman, 2022; Sigfrids et al., 2023).

HCAS emphasizes a human-centric perspective, which extends beyond the focus of human-centered design (often centered on immediate usability) to include long-term well-being, ethical alignment, and societal impact (Shneiderman, 2022; Sigfrids et al., 2023).

Access the Complete HCAS Framework

For a comprehensive guide to HCAS principles, components, goals, objectives, practical examples, and customization prompts, please refer to Part 2, Section 1 of the downloadable *Human-AI Collaboration & Synergy Guide* (Table 5.1 and Fig. 5.1).

HCAS Principles, Components, Goals, and Objectives

HCAS adapts principles from fields like Human-AI Interaction (HAII), Human-AI Collaboration (HAIC), Human-AI Teaming (HAIT), and Intelligence Augmentation (IA) into practical usage strategies (Bao et al., 2023; Berretta et al., 2023; McCaffrey, 2018; Sundar, 2020; Xia & Maes, 2013; Xu et al., 2022; Zhao et al., 2022). Key components derived from these principles include developing personalized learning paths, utilizing collaborative problem-solving tools, establishing clear roles in human-AI teams, implementing continuous feedback loops, and adhering to ethical usage guidelines.

Table 5.1 Comparison of human-centered and human-centric AI synergy

Aspect	Human-centered AI synergy	Human-centric AI synergy
Definition	Collaboration between humans and AI, optimized for individual usability, adaptivity, and seamless interaction.	Collaboration between humans and AI, aligned with societal values, ethical considerations, and collective well-being.
Focus	Enhancing user experience, productivity, and task-specific outcomes.	Ensuring fairness, transparency, and shared accountability in broader societal applications of AI.
Viability and sustainability	Prioritizes viability for specific domains, such as higher education, through practical, intuitive system designs.	Focuses on ensuring long-term sustainability for societal growth, avoiding harm to equity or ethics.
Application	AI-powered customer service tools that adapt to individual user queries and preferences.	AI systems in criminal justice designed to ensure equitable treatment and avoid discriminatory outcomes.
Example	A human-centered design for an AI assistant that intuitively schedules meetings based on a user's habits and needs.	Human-centric synergy in urban planning where AI analyzes data to recommend policies that balance equity, sustainability, and growth.

Note. This table distinguishes between human-centered and human-centric approaches to AI synergy, comparing their definitions, focus areas, and application examples to clarify their distinct goals

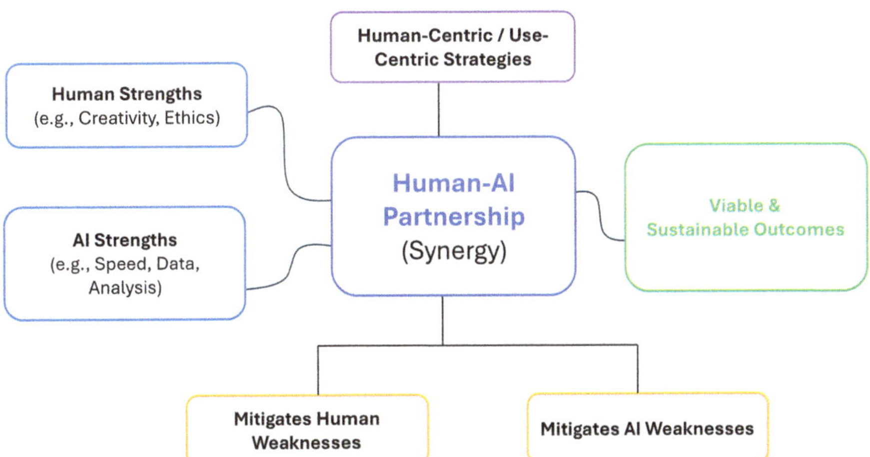

Fig. 5.1 Human-AI synergy in education: collaborative strengths and outcomes. (*Note.* This figure provides a conceptual overview of human-AI synergy in education, illustrating how the collaborative strengths of both humans and AI can lead to enhanced outcomes)

Table 5.2 Key principles, components, and goals of human-centric AI synergy (HCAS)

Core principles	Key components	Overarching goals
• Human-centric values and ethics • Synergy (human + AI strengths) • Intelligence augmentation focus • Ethical alignment and transparency • Meaningful human control/oversight	• Personalized learning paths • Collaborative problem-solving tools • Clear roles and responsibilities in human-AI teams • Continuous feedback loops • Ethical usage guidelines and mitigation of design-induced challenges	• Enhance academic integrity • Develop human-centric skills • Improve academic outcomes • Foster pedagogical innovation • Promote societal well-being and sustainability through responsible AI use

Note. A conceptual model illustrating how the five core frameworks interlock. Each framework is represented by a labeled component: HCAS (The "Why"), RSA (The "How"), EHCS (The "Skills"), EAO (The "Results"), and AHIPs (The "Where"). Together, these integrated components work synergistically to build a culture of AI-empowered stakeholders

The overarching goals of HCAS are to enhance academic integrity, develop human-centric skills, improve academic outcomes, foster pedagogical innovation, and promote societal sustainability through responsible AI integration (Handayani, 2019; Lodge et al., 2023; McArthur, 2023; Nekipelova & Nekipelov, 2022; Sigfrids et al., 2023) (Table 5.2).

HCAS Summary & Download

Given the evolving legal and intellectual landscape in the early days of GenAI integration (and the design and capabilities of GenAI systems), "Framework Summaries" were created for each of the core frameworks in this book and are available in Part 2 of the *Human-GenAI Collaboration & Synergy Guide*. These are available for download and can be used as personal or team guides for optimizing ethical and effective human-GenAI collaboration. Likewise, these summaries can be downloaded and uploaded to GenAI systems for various applications in human-centric AI in higher education. In this chapter and corresponding workbook, users will engage with the framework summaries and downloads to create custom frameworks for optimizing human-centric AI that are directly aligned with needs, goals, objectives, and desired outcomes of their educational context.

Regulation, Synergy, & Augmentation (RSA) Framework Summary

While HCAS provides the conceptual foundation, the Regulation, Synergy, & Augmentation (RSA) framework offers the operational strategies needed to achieve ethical and effective GenAI integration in practice. RSA translates HCAS principles into evidence-based actions for teaching, learning, assessment, research, and administration. It specifically addresses how users can interact productively and responsibly with GenAI systems, particularly those not designed explicitly for educational augmentation, mitigating risks related to academic integrity and skill erosion (Cotton et al., 2023; Michel-Villarreal et al., 2023).

RSA operationalizes HCAS by integrating principles from Self-Regulated Learning (SRL), empowering individuals to manage their AI interactions (Greene & Azevedo, 2007; Panadero, 2017; Zimmerman, 1994), and Co-Regulated Learning (CRL), fostering collaborative oversight and shared responsibility (Berretta et al., 2023; Lodge et al., 2023). It guides users to employ GenAI for Intelligence Augmentation (IA), enhancing human capabilities rather than replacing them (Sadiku et al., 2021; Shneiderman, 2022), and implements Hybrid Human-AI Regulation (HHAIR) models where human judgment actively oversees AI contributions (Molenaar, 2022).

The core principles of RSA emphasize ethical AI use, SRL and CRL integration, disciplinary adaptation, an IA focus, leveraging Collective Intelligence (CI), promoting lifelong learning, and implementing HHAIR. Key components include structured goal setting and prompt design, strategic implementation for augmentation, adaptive feedback mechanisms, monitoring and self-reflection processes, clear ethical guidelines and integrity checks, and structures supporting collaboration (Boekaerts, 1999; Chang et al., 2023; Johri et al., 2023; Katsamakas et al., 2024; Lase & Nkosi, 2023; Tsai et al., 2020; Winne & Hadwin, 1998). The goals of RSA align with HCAS but focus on the operational strategies needed to enhance learning, promote integrity, develop skills, foster innovation, and align with workforce needs through regulated, synergistic, and augmentative practices (Cotton et al., 2023; Panadero, 2017; Zielinski, 2024).

Table 5.3 Key principles, components, and goals of regulation, synergy, and augmentation (RSA)

Core principles	Key components	Overarching goals
• Ethical AI use and responsibility	• Structured goal setting and prompt design	• Enhance learning effectiveness and depth
• Self-regulated learning (SRL)	• Strategic implementation for augmentation	• Promote academic integrity
• Co-regulated learning (CRL)	• Adaptive feedback mechanisms (AI and human)	• Develop essential human-centric skills
• Intelligence augmentation (IA) focus	• Monitoring and self-reflection processes	• Foster pedagogical innovation
• Leverage collective intelligence (CI)	• Clear ethical guidelines and integrity checks	• Align education with workforce needs
• Promote lifelong learning habits	• Structures supporting collaboration (human-human and human-AI)	• Ensure sustainable and ethical AI integration
• Hybrid human-AI regulation (HHAIR)	• Continuous training in AI ethics and literacy	

Note. This table outlines the operational elements of the RSA framework, detailing the principles, components, and goals that guide the practical regulation of human-AI interactions to achieve effective and ethical synergy

Access the Complete RSA Framework

For a comprehensive guide to RSA principles, components, goals, objectives, practical examples, and customization prompts, please refer to Part 2, Section 2 of the downloadable *Human-AI Collaboration & Synergy Guide* (Table 5.3).

Limits of Self-Regulation & Co-Regulation

RSA acknowledges the limitations of self- and co-regulation with opaque AI systems (Hadwin et al., 2011; Lodge et al., 2023) and stresses the need for supportive institutional policies, robust training, and hybrid regulatory approaches where human oversight remains central.

Preparing to Customize Frameworks

In this chapter, users are preparing to apply the advanced capabilities of LLMs/GenAI systems to customize the core frameworks of this book to advance ethical and effective GenAI integration.

Users should have their goals and objectives, contextual data, and other living documents they've prepared through this book to optimize the activities in the workbook.

Enhancing Human-Centric Skills (EHCS) Framework Summary

The Enhancing Human-Centric Skills (EHCS) framework provides an evidence-based pedagogical approach for strategically using GenAI to actively cultivate essential human competencies, rather than allowing technology use to inadvertently erode them. It focuses on fostering skills like critical thinking, complex problem-solving, creativity, communication, collaboration, ethical reasoning, and adaptability—skills crucial for success in an AI-integrated world (Fadel & Trilling, 2009; World Economic Forum, 2023).

EHCS draws directly from empirical research investigating GenAI's impact on skills in educational settings. While early studies showed promise under specific conditions (Essel et al., 2024; Guo & Lee, 2023; Michalon & Camacho-Zuñiga, 2023; Yilmaz, 2023), recent findings emphasize that benefits are highly conditional and misuse can hinder learning (Lee et al., 2024; Liu et al., 2024; Wecks et al., 2024). Therefore, EHCS emphasizes pedagogical designs that require active human engagement, critical evaluation, ethical reflection, and regulation of the AI interaction process (aligning with RSA).

Core principles of EHCS include integrating GenAI within High-Impact Practices (HIPs), promoting personalized and active learning, fostering self-directed learning with regulation, ensuring continuous feedback loops, implementing scaffolded learning designs, embedding ethical AI use and literacy, and emphasizing critical engagement with and verification of AI outputs (Barana et al., 2023; Essel et al., 2024; Guo & Lee, 2023; Kuh, 2008; Lee et al., 2024). Key components involve defining skill-focused objectives, providing detailed prompt engineering guidelines for skill development, integrating scaffolded activities, implementing interactive multi-source feedback, conducting verification exercises, facilitating ethics discussions, promoting reflective practice, ensuring continuous instructional improvement, and designing assessments that measure skill development directly (Essel et al., 2024; Guo & Lee, 2023; Michalon & Camacho-Zuñiga, 2023). The goals of EHCS are to enhance critical thinking, foster creativity, improve communication/collaboration, promote ethical reasoning, and cultivate lifelong learning skills.

Table 5.4 Key principles, components, and goals of enhancing human-centric skills (EHCS)

Core principles	Key components	Overarching goals
• Integration within high-impact practices (HIPs)	• Defining skill-focused learning objectives	• Enhance critical thinking and complex problem-solving
• Personalized and active learning	• Detailed prompt Engineering guidelines for skill development	• Foster creativity and innovation
• Fostering self-directed learning and regulation	• Integrating scaffolded learning activities	• Improve communication and collaboration skills
• Continuous, multi-source feedback loops	• Implementing interactive feedback (peer, AI, instructor)	• Promote ethical reasoning and responsible AI use
• Scaffolded learning design	• Conducting verification exercises (checking AI outputs)	• Cultivate adaptability and lifelong learning mindsets
• Embedding ethical AI use and literacy	• Facilitating ethics discussions and reflection	• Ensure development over AI-induced skill erosion
• Critical engagement with AI outputs	• Designing assessments measuring skill development directly	
	• Promoting reflective practice on learning process	
	• Ensuring continuous instructional improvement	

Note. This table summarizes the pedagogical elements of the EHCS framework, which is designed to strategically use GenAI to cultivate essential human skills within educational settings

Access the Complete EHCS Framework

For a comprehensive guide to EHCS principles, components, goals, objectives, practical examples, and customization prompts, please refer to Part 2, Section 3 of the downloadable *Human-AI Collaboration & Synergy Guide* (Table 5.4).

Enhancing Academic Outcomes (EAO)

The Enhancing Academic Outcomes (EAO) framework offers a targeted approach for leveraging GenAI technologies to directly improve measurable student success metrics, such as grades, retention rates, assessment scores, and mastery of specific competencies. While related to EHCS (which focuses on underlying skills), EAO concentrates on the application of skills and AI support towards achieving tangible academic performance improvements (Elkhodr et al., 2023; Toktosunova et al., 2023).

EAO is grounded in empirical research, acknowledging mixed findings: while some studies show GenAI can enhance outcomes through personalization and feedback (Alneyadi & Wardat, 2023; Elkhodr et al., 2023; Luo et al., 2024; Toktosunova et al., 2023; Zhang et al., 2024), others demonstrate negative impacts when AI is used as a shortcut, hindering deep learning (Abbas et al., 2024; Liu et al., 2024; Wecks et al., 2024). Consequently, EAO emphasizes strategic, regulated integration.

Core principles include customization of AI tools to specific contexts, ensuring equity of access, leveraging personalization for individual needs, supporting critical thinking and self-assessment in service of achieving outcomes, facilitating collaboration, and integrating responsible AI practices (Abbas et al., 2024; Elkhodr et al., 2023; RIT Center for Teaching and Learning, n.d.; Toktosunova et al., 2023). Key components involve outcome-oriented goal setting and prompting, implementing immediate and actionable feedback mechanisms, designing interactive learning activities, providing support for diverse learning needs, ensuring balanced integration with traditional pedagogy, embedding AI literacy and ethical training, and robustly assessing the impact on outcomes (Abbas et al., 2024; Elkhodr et al., 2023; Pane et al., 2015; Sun et al., 2023; Toktosunova et al., 2023). The goals of EAO focus on enhancing measurable learning outcomes, fostering critical application of knowledge, supporting diverse students equitably, promoting autonomous learning for success, facilitating collaboration for better group results, and boosting academic self-efficacy.

Access the Complete EAO Framework

For a comprehensive guide to EAO principles, components, goals, objectives, practical examples, and customization prompts, please refer to Part 2, Section 4 of the downloadable *Human-AI Collaboration & Synergy Guide* (Table 5.5).

Authentic High-Impact Practices (AHIPs) Framework Summary

The Authentic High-Impact Practices (AHIPs) framework provides a pedagogical and andragogical response designed for the GenAI era. It addresses the challenge of AI replication by integrating established High-Impact Practices

Table 5.5 Key principles, components, and goals of enhancing academic outcomes (EAO)

Core principles	Key components	Overarching goals
• Customization to specific contexts	• Outcome-oriented goal setting and prompting	• Enhance measurable learning outcomes (grades, retention)
• Equity of access and use	• Implementing immediate and actionable feedback mechanisms	• Foster critical application of knowledge and skills
• Personalization for individual needs	• Designing interactive learning activities	• Support diverse student populations equitably
• Supporting critical thinking and self-assessment	• Providing support for diverse learning needs	• Promote autonomous learning for academic success
• Facilitating collaboration for outcomes	• Ensuring balanced integration with traditional pedagogy	• Facilitate Collaboration for Better Group Results
• Integrating responsible AI practices	• Embedding AI literacy and ethical training relevant to outcomes	• Boost academic self-efficacy and confidence
	• Robustly assessing impact on specific academic outcomes	

Note. This table outlines the core elements of the EAO framework, focusing on targeted strategies for using GenAI to demonstrably and responsibly improve student success metrics

(HIPs) with principles of authentic teaching, learning, and assessment. AHIPs emphasize creating meaningful, complex, real-world educational experiences that inherently require higher-order thinking and skills that GenAI cannot easily replicate (Gulikers et al., 2004; Kuh, 2008; McArthur, 2023).

AHIPs provide the ideal context for applying HCAS and RSA principles, enabling users to leverage GenAI ethically and effectively as an augmentative tool within engaging, meaningful tasks. The framework is guided by twelve core principles emphasizing student-centeredness, well-being, global perspectives, DEI, digital literacy, high-impact design, authenticity, feedback literacy, sustainability, ethical leadership, professional development, and lifelong learning.

Key components fall into two categories: established HIPs (e.g., first-year seminars, collaborative projects, undergraduate research, global learning, service-learning, internships, capstones) and authentic strategies (e.g., real-world problem solving, simulations, reflective portfolios, community-engaged projects, case studies, performance-based assessments). Within these components, GenAI can be strategically integrated—guided by HCAS/RSA—to support aspects like research, brainstorming, data analysis, or simulation, while ensuring human effort remains central to critical thinking, ethical judgment, creativity, and collaboration (Astin & Sax, 1998; Berretta et al., 2023; Carroll, 2014; Challis, 2005; Darling-Hammond & Snyder, 2000; Lanning & Brown, 2019; Maxwell, 2012; Perkins et al., 2024; Seifert et al., 2014; Smith & Baik, 2021; Sokhanvar et al., 2021; Sotiriadou et al., 2020; Tukibayeva & Gonyea, 2014; Varela, 2019; White, 2018).

The goals of AHIPs include enhancing learning authenticity and impact, promoting academic integrity through engagement, developing holistic human-centric skills, facilitating lifelong learning, and optimizing ethical human-AI synergy within meaningful contexts (Blaschke, 2021; Cotton et al., 2023; Farrelly & Baker, 2023; Lee et al., 2024; Lodge et al., 2023; McArthur, 2023).

Access the Complete AHIPs Framework

For a comprehensive guide to AHIPs principles, components, goals, objectives, practical examples, and customization prompts, please refer to Part 2, Section 5 of the downloadable *Human-AI Collaboration & Synergy Guide* (Table 5.6).

Integrating the Five Frameworks

The five frameworks presented—HCAS, RSA, EHCS, EAO, and AHIPs—are designed to be complementary, addressing different but interconnected facets of GenAI integration in higher education. While each can be applied individually, their true strength lies in their potential for synergistic integration.

- HCAS provides the overarching ethical and philosophical compass.
- RSA offers the practical strategies for managing the human-AI interaction itself (regulation, synergy, augmentation).
- EHCS focuses specifically on pedagogical designs to cultivate human skills using AI augmentatively.
- EAO targets strategies for leveraging AI to improve measurable academic outcomes responsibly.
- AHIPs create the authentic, engaging learning contexts where the other frameworks can be most effectively applied.

Table 5.6 Key principles, components, and goals of authentic high-impact practices (AHIPs)

Core principles	Key components	Overarching goals
• Student-centeredness and well-being • Global perspectives and interculturalism	**Established HIPs examples:** • First-year seminars/ experiences	• Enhance learning authenticity and impact • Promote academic integrity through meaningful engagement
• Diversity, equity, and inclusion (DEI)	• Collaborative assignments and projects	• Develop holistic human-centric skills
• Digital literacy and fluency • High-impact pedagogical design	• Undergraduate research • Global learning/study abroad	• Facilitate deep, lifelong learning habits • Optimize ethical human-AI synergy in meaningful contexts
• Authenticity in learning and assessment • Feedback literacy and assessment • Sustainability • Ethical leadership and policy • Professional development • Lifelong learning	• Service-learning/ community-based learning • Internships/field experiences • Learning communities • Writing-intensive courses • Capstone courses and projects **Authentic strategies examples:** • Real-world problem-solving • Simulations and role-playing • Reflective portfolios (ePortfolios) • Community-engaged projects • Case studies • Performance-based assessments	• Prepare students for complex real-world challenges

Note. This table summarizes the AHIPs framework, which integrates established high-impact practices (HIPs) with authentic pedagogical strategies to create meaningful learning experiences suitable for ethical GenAI augmentation

In practice, institutions or educators can select and combine elements from these frameworks to create a customized approach tailored to specific goals, disciplines, student populations, or institutional priorities. For example, an institution aiming to improve writing outcomes (EAO goal) might implement an AHIP approach using portfolio assessments. Within this context,

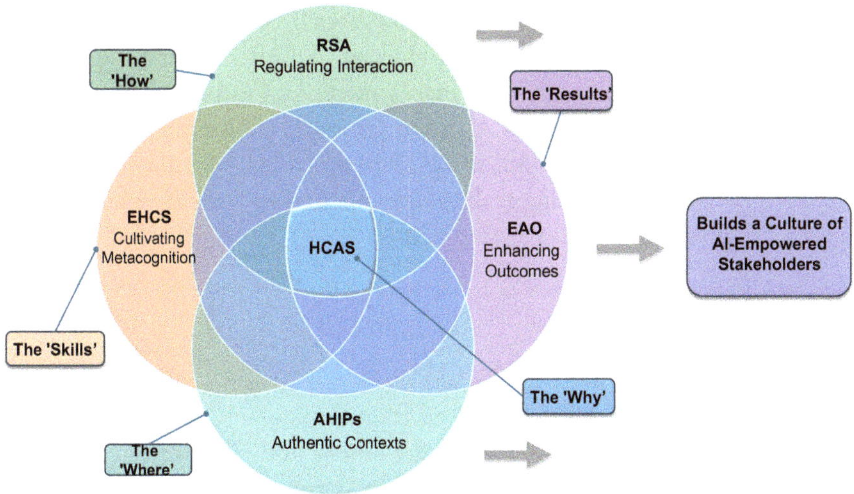

Fig. 5.2 Framework integration model for human-centric AI in higher education. (*Note*. Five nested rings in a Venn Diagram depict successive layers of the framework integration model. The outer magenta ring ("Authentic High-Impact Practices") lists core pedagogical and equity-focused activities (e.g., social & emotional learning; diversity, equity & inclusion; student-centered assessment; AI citizenship & ethics). The next blue ring ("Enhancing Academic Outcomes") highlights strategic goal setting, real-time feedback, and accessibility augmentation with AI. The middle magenta ring ("Human-Centric AI Synergy") describes stakeholder empowerment, augmentation of human skills, strength-weakness optimization, and sustainable collaboration. The following blue ring ("Regulation Synergy & Augmentation") covers self-/co-regulation, evidence-based refinement cycles, and iterative goal monitoring. The innermost pale sphere ("Enhancing Human-Centric Skills") focuses on meta-skills (critical thinking, prompt engineering, validation skills) and the cultivation of an AI-literate culture. Each ring is linked by aligned spheres and text blocks to emphasize how each layer builds toward unified, ethical, and effective AI integration)

they could use RSA principles to guide students in self-regulating their use of AI writing assistants for feedback (not generation), apply EHCS strategies to ensure this process enhances critical revision skills, and ensure the entire initiative aligns with HCAS values regarding academic integrity and human authorship (Fig. 5.2).

Focus on Ongoing Revision

Given the rapid evolution of GenAI capabilities and the emerging research base, it is essential to treat these frameworks not as static documents but as living guides. The academic publishing cycle often lags behind technological breakthroughs. Therefore, institutions and educators must foster environments that encourage near-real-time collaboration, reflection, and adaptation of these frameworks. Sharing experiences through communities of practice, monitoring ongoing research (using processes like HCASR Component 3), and maintaining flexible institutional guidelines (addressed in HCASR Component 5) are crucial. This adaptive approach allows the higher education community to respond effectively to new challenges and opportunities presented by GenAI, ensuring technology integration remains ethical, effective, and aligned with educational missions.

Conclusion

This chapter has introduced five core frameworks—HCAS, RSA, EHCS, EAO, and AHIPs—designed to guide the ethical and effective integration of Generative AI across higher education. Each framework offers a distinct lens and set of strategies, yet they work synergistically to provide a comprehensive approach. From the foundational human-centric values of HCAS and the interaction management strategies of RSA, to the skill-focused pedagogy of EHCS, the outcome-driven applications of EAO, and the authentic contexts provided by AHIPs, these frameworks equip stakeholders with the tools needed to navigate the complexities of GenAI. They are designed to be adaptable, allowing institutions and individuals to select, combine, and customize approaches using GenAI itself as a partner in the process. By applying these frameworks thoughtfully and committing to continuous learning and adaptation, the higher education community can harness the potential of GenAI to augment human capabilities, enhance educational experiences, and advance its core mission responsibly.

Long-Term Scaling, Sustainability & AI Governance: Custom TLARA Frameworks with AI

Ethical and effective AI integration over the long-term requires holistic frameworks and approaches that consider the multiple **levels** and systems within and around HEIs across **time** and **space** (Carmody-Bubb, 2023; Katsamakas et al., 2024). This means examining:

1. Human-AI collaboration and synergy across and between students, faculty, staff, administration, and the institution as a whole.
2. How those emerging or established Human-AI Partnerships and practices impact stakeholders individually and in groups across **time** and in different **sociocultural systems**.

As readers progress into Part 1 of this book, they will encounter a simplified way to understand and address how the above complex factors and relationships impact the values, processes, and outcomes related to **Teaching, Learning, Assessment, Research, and Administration (TLARA)** in HEIs around the world.

The **TLARA** framework is a simple way to organize this. Think of it as five main areas or **levels** where AI can be used:

- **Teaching**: How professors use AI to teach.
- **Learning**: How students use AI to learn.
- **Assessment**: How AI is used for grading or giving feedback.
- **Research**: How AI helps in making new discoveries.
- **Administration**: How the university's staff uses AI for things like admissions or scheduling.

As you progress into Part 2 of this book, you will encounter a dedicated series of activities and exercise in both the main chapter texts and the Personalized Adaptive Learning Sessions (PALS) Workbook in sections called: **"Long-Term Scaling, Sustainability & AI Governance: Custom TLARA Frameworks with AI."** These exercises and activities are designed to facilitate users, teams, HEIs, etc., in employing AI in ethical ways for designing frameworks, policies, tools, resources, and other solutions in ways that 1) are strategically aligned with their institution's and their stakeholders' needs, roles, goals, objectives, outcomes, and sociocultural contexts, and that 2) create and maintain practices, change, innovation, adaptation, and human-AI partnerships that foster long-term sustainability and AI Governance.

Chapter 5 Workbook Prep

These questions help you connect the five core frameworks (HCAS, RSA, EHCS, EAO, AHIPs) from this chapter to the practical application and customization activities in the Chap. 5 UX and PALS Workbooks.

Framework Relevance & Goals (Prep for PALS Framework Selection):

- Which of the five core frameworks (or a specific principle within one) most directly addresses a key goal or challenge you face in your specific role (teaching, learning, research, administration)? Briefly explain the connection.

Application & Measurement (Prep for UX Metrics Activity):

- How might you apply the framework principle identified above to a specific task?
- What is one potential way you could measure the success or impact of applying this principle (e.g., improved efficiency, enhanced student skill, better alignment with ethical guidelines)?

Interaction & Ethics (Prep for UX Prompting & PALS Ethics):

- If using GenAI to support this task/principle, what is the most critical ethical consideration (e.g., ensuring fairness, maintaining human oversight, protecting privacy)?
- How might you prompt the AI assistant to help with the task while upholding this ethical consideration?

Customization & Integration (Prep for PALS Customization):

- Based on your specific context (e.g., student population, resources, institutional policies), do you anticipate needing to combine elements from different frameworks or significantly adapt one to make it work effectively? Why?

References

Abbas, M., Jam, F. A., & Khan, T. I. (2024). Is it harmful or helpful? Examining the causes and consequences of generative AI usage among university students. *International Journal of Educational Technology in Higher Education, 21*, Article 10. https://doi.org/10.1186/s41239-023-00430-5

Alneyadi, F., & Wardat, Y. (2023). Enhancing student achievement in electromagnetism through AI integration. *International Journal of STEM Education, 10*(1), Article 56. https://doi.org/10.1186/s40594-023-00446-y

Astin, A. W., & Sax, L. J. (1998). How undergraduates are affected by service participation. *Journal of College Student Development, 39*(3), 251–263.

Bao, J., Chen, Y., Zhang, H., Zhao, L., & Zhang, Y. (2023). Human-AI synergy in decision-making: A comprehensive review. *Artificial Intelligence Review, 56*(Suppl 1), 145–181. https://doi.org/10.1007/s10462-023-10484-8

Barana, A., Marchisio, M., & Roman, F. (2023). Fostering problem-solving and critical thinking in mathematics through generative artificial intelligence. In *Proceedings of the 20th International Conference on Cognition and Exploratory Learning in the Digital Age (CELDA 2023)* (pp. 186–195). IADIS Press.

Berretta, S., Tausch, A., Ontrup, G., Gilles, B., Peifer, C., & Kluge, A. (2023). Defining human-AI teaming the human-centered way: A scoping review and network analysis. *Frontiers in Artificial Intelligence, 6*, Article 1250725. https://doi.org/10.3389/frai.2023.1250725

Blaschke, L. M. (2021). The dynamic mix of heutagogy and technology: Preparing learners for lifelong learning. *British Journal of Educational Technology, 52*(4), 1629–1645. https://doi.org/10.1111/bjet.13105

Boekaerts, M. (1999). Self-regulated learning: Where we are today. *International Journal of Educational Research, 31*(6), 445–457. https://doi.org/10.1016/S0883-0355(99)00014-2

Carmody-Bubb, M. (2023). *Cognition and decision making in complex adaptive systems: The human factor in organizational performance*. Springer.

Carroll, J. M. (Ed.). (2014). *Innovative practices in teaching information sciences and technology: Experience reports and reflections*. Springer. https://doi.org/10.1007/978-3-319-03656-4

Challis, D. (2005). Committing to quality learning through adaptive online assessment. *Assessment & Evaluation in Higher Education, 30*(5), 519–527. https://doi.org/10.1080/02602930500188991

Chang, C.-H., & Kidman, G. (2023). The rise of generative artificial intelligence (AI) language models: Challenges and opportunities for geographical and environmental education. *International Research in Geographical and Environmental Education, 32*(2), 85–89. https://doi.org/10.1080/10382046.2023.2194036

Chang, J. W., Chang, J. C., & Chang, C. S. (2023). Design principles for using a chatbot that supports self-regulated learning in online learning environments. *Computers & Education: Artificial Intelligence, 5*, Article 100078. https://doi.org/10.1016/j.caeai.2023.100078

Cotton, D. R. E., Cotton, P. A., & Shipway, J. R. (2023). Chatting and cheating: Ensuring academic integrity in the era of ChatGPT. *Innovations in Education and Teaching International, 61*(2), 228–239. https://doi.org/10.1080/14703297.2023.2190148

Darling-Hammond, L., & Snyder, J. (2000). Authentic assessment of teaching in context. *Teaching and Teacher Education, 16*(5–6), 523–545. https://doi.org/10.1016/S0742-051X(00)00015-9

Elkhodr, M., Gide, E., Wu, R. M. X., & Darwish, O. (2023). ICT students' perceptions towards ChatGPT: An experimental reflective lab analysis. *STEM Education, 3*(2), 70–88. https://doi.org/10.3934/steme.2023006

Essel, H. B., Nunoo, F. N. A., Asigbey, J., Amankwah, R., Owusu, A., & Adjei, E. (2024). Development of critical, creative, and reflective thinking skills through AI-assisted collaborative learning: A case of research methodology course. *Educational Technology Research and Development, 72*(3), 1149–1170. https://doi.org/10.1007/s11423-023-10331-6

Fadel, C., & Trilling, B. (2009). *21st century skills: Learning for life in our times*. Jossey-Bass.

Farrelly, T., & Baker, N. (2023). Generative artificial intelligence: Implications and considerations for higher education practice. *Education Sciences, 13*(1), Article 9. https://doi.org/10.3390/educsci13010009

Greene, J. A., & Azevedo, R. (2007). A theoretical review of Winne and Hadwin's model of self-regulated learning: New perspectives and directions. *Review of Educational Research, 77*(3), 334–372. https://doi.org/10.3102/003465430303953

Gulikers, J. T. M., Bastiaens, T. J., & Kirschner, P. A. (2004). A five-dimensional framework for authentic assessment. *Educational Technology Research and Development, 52*(3), 67–85. https://doi.org/10.1007/BF02504676

Guo, Y., & Lee, J. (2023). Leveraging ChatGPT for enhancing critical thinking skills. *Journal of Chemical Education, 100*(12), 4876–4883. https://doi.org/10.1021/acs.jchemed.3c00505

Hadwin, A. F., Järvelä, S., & Miller, M. (2011). Self-regulated, co-regulated, and socially shared regulation of learning. In B. J. Zimmerman & D. H. Schunk (Eds.), *Handbook of self-regulation of learning and performance* (pp. 65–84). Routledge.

Handayani, M. (2019). Integration of sustainable development into higher education curriculum. *Journal of Sustainable Development Education and Research, 3*(1), 43–50. https://doi.org/10.17509/jsder.v3i1.15716

Johri, A., Katz, A. S., Qadir, J., & Hingle, A. (2023). Generative artificial intelligence and engineering education. *Journal of Engineering Education, 112*(3), 572–577. https://doi.org/10.1002/jee.20537

Katsamakas, E., Pavlov, O. V., & Saklad, R. (2024). Artificial intelligence and the transformation of higher education institutions: A systems approach. *Sustainability, 16*(14), Article 6118. https://doi.org/10.3390/su16146118

Kuh, G. D. (2008). *High-impact educational practices: What they are, who has access to them, and why they matter*. Association of American Colleges and Universities.

Lanning, S. L., & Brown, M. (2019). Undergraduate research as a high-impact practice in higher education. *Education Sciences, 9*(3), Article 199. https://doi.org/10.3390/educsci9030199

Lase, E. M., & Nkosi, F. (2023). Human-centric AI: Understanding and enhancing collaboration between humans and intelligent systems. *Algorithm Asynchronous, 1*(1), 33–40. https://doi.org/10.61963/jaa.v1i1.49

Lee, J. S., Suh, J., & Suri, G. (2024). Large language models may reduce perceived cognitive effort but harm task performance. *Scientific Reports, 14*, Article 22777. https://doi.org/10.1038/s41598-024-73552-1

Liu, C.-Y., Chen, S.-Y., & Chen, M.-P. (2024). The effectiveness of ChatGPT in assisting high school students in programming learning: Evidence from a quasi-experimental research. *Interactive Learning Environments*. Advance online publication. https://doi.org/10.1080/10494820.2024.2420659

Lodge, J. M., de Barba, P., & Broadbent, J. (2023). Learning with generative artificial intelligence within a network of co-regulation. *Journal of University Teaching and Learning Practice, 20*(7). https://doi.org/10.53761/1.20.7.02

Luo, H., Cheng, X., & Yang, T. (2024). Exploring the impact of generative artificial intelligence on students' learning outcomes: A meta-analysis. *Education and Information Technologies*. Advance online publication. https://doi.org/10.1007/s10639-024-13033-w

Maxwell, T. W. (2012). Assessment in higher education in the professions: Action research as an authentic assessment task. *Teaching in Higher Education, 17*(6), 686–696. https://doi.org/10.1080/13562517.2012.678298

McArthur, J. (2023). Rethinking authentic assessment: Work, well-being, and society. *Higher Education*. Advance online publication. https://doi.org/10.1007/s10734-022-00822-y

McCaffrey, T. (2018). Human-AI synergy in creativity and innovation. In *Artificial intelligence - emerging trends and applications*. IntechOpen. https://doi.org/10.5772/intechopen.75310

Michalon, B., & Camacho-Zuñiga, C. (2023). ChatGPT, a brand-new tool to strengthen timeless competencies. *Frontiers in Education, 8*, Article 1251163. https://doi.org/10.3389/feduc.2023.1251163

Michel-Villarreal, R., Vilalta-Perdomo, E., Salinas-Navarro, D. E., Thierry-Aguilera, R., & Gerardou, F. S. (2023). Challenges and opportunities of generative AI for higher education as explained by ChatGPT. *Education Sciences, 13*(9), Article 956. https://doi.org/10.3390/educsci13090956

Molenaar, I. (2022). The concept of hybrid human-AI regulation: Exemplifying how to support young learners' self-regulated learning. *Computers and Education: Artificial Intelligence, 3*, Article 100070. https://doi.org/10.1016/j.caeai.2022.100070

Nekipelova, I. M., & Nekipelov, E. S. (2022). The human capital: Vital, soft, hard, meta and mental skills. *Sotsial'no-Ekonomicheskoe Upravlenie: Teoriia i Praktika [Social and Economic Management: Theory and Practice], 18*(1), 60–75.

Panadero, E. (2017). A review of self-regulated learning: Six models and four directions for research. *Frontiers in Psychology, 8*, Article 422. https://doi.org/10.3389/fpsyg.2017.00422

Pane, J. F., Steiner, E. D., Baird, M. D., & Hamilton, L. S. (2015). *Continued progress: Promising evidence on personalized learning*. RAND Corporation. https://www.rand.org/pubs/research_reports/RR1365.html

Perkins, M., Furze, L., Roe, J., & MacVaugh, J. (2024). The Artificial Intelligence Assessment Scale (AIAS): A framework for ethical integration of generative AI in educational assessment. *Journal of University Teaching and Learning Practice, 21*(6). https://doi.org/10.53761/1.21.6.07

RIT Center for Teaching and Learning. (n.d.). *Equity, privacy, and academic integrity in generative AI*. Rochester Institute of Technology. Retrieved April 20, 2025, from https://www.rit.edu/teaching/ai-concerns

Sadiku, M. N. O., Ashaolu, T. J., Ajayi-Majebi, A., & Musa, S. M. (2021). Augmented intelligence. *International Journal of Scientific Advances, 2*(5), 648–650. https://doi.org/10.51542/ijscia.v2i5.17

Seifert, T. A., Gillig, B., Hanson, J. M., Pascarella, E. T., & Blaich, C. F. (2014). The effects of community service participation on student learning and development: A meta-analysis. *Review of Educational Research, 84*(2), 261–288. https://doi.org/10.3102/0034654313502087

Shneiderman, B. (2022). *Human-centered AI*. Oxford University Press.

Sigfrids, A., Leikas, J., Salo-Pöntinen, J., & Koskimies, K. (2023). Exploring human-centric AI: Ethical considerations in system design. *Technology in Society, 74*, Article 102305. https://doi.org/10.1016/j.techsoc.2023.102305

Smith, C., & Baik, C. (2021). High-impact teaching practices in higher education: A best evidence review. *Studies in Higher Education, 46*(5), 908–924. https://doi.org/10.1080/03075079.2019.1684741

Sokhanvar, Z., Salehi, K., & Sokhanvar, F. (2021). Advantages of authentic assessment for improving the learning experience and employability skills of higher education students: A systematic literature review. *Studies in Educational Evaluation, 70,* Article 101030. https://doi.org/10.1016/j.stueduc.2021.101030

Sotiriadou, P., Logan, D., Daly, A., & Guest, R. (2020). The role of authentic assessment to preserve academic integrity and promote skill development and employability. *Studies in Higher Education, 45*(11), 2132–2148. https://doi.org/10.1080/03075079.2019.1582015

Sun, Z., Wu, Y., & Li, K. C. (2023). Investigating the effectiveness of ChatGPT in programming education. *Journal of Educational Technology Development and Exchange, 16*(2), 1–16. https://doi.org/10.18785/jetde.1602.01

Sundar, S. S. (2020). Rise of machine agency: A framework for studying the psychology of human–AI interaction (HAII). *Journal of Computer-Mediated Communication, 25*(1), 74–88. https://doi.org/10.1093/jcmc/zmz026

Toktosunova, A., Fallah, H., Toutouchi, Z., Tadzhibaeva, Z., & Tick, A. (2023). The impact of ChatGPT on learning in higher education – Results of a pilot study. *Practice and Theory in Systems of Education, 18*(1), 31–50. https://doi.org/10.1515/ptse-2023-0003

Tsai, Y.-S., Perrotta, C., & Gašević, D. (2020). Empowering learners with personalised learning approaches? Agency, equity, and transparency in the context of learning analytics. *Assessment & Evaluation in Higher Education, 45*(4), 554–567. https://doi.org/10.1080/03075079.2019.1636076

Tukibayeva, M., & Gonyea, R. M. (2014). High-impact practices and the first-year student. *New Directions for Institutional Research, 2014*(161), 19–35. https://doi.org/10.1002/ir.20068

Varela, O. E. (2019). Multicultural competence: An empirical comparison between intercultural sensitivity and cultural intelligence. *European Journal of International Management, 13*(6), 836–854. https://doi.org/10.1504/EJIM.2019.101819

Wecks, J. O., Voshaar, J., Plate, B. J., & Zimmermann, J. (2024). *Generative AI usage and exam performance* (Bochum Accounting Research Working Paper No. 2). Ruhr University Bochum. https://doi.org/10.2139/ssrn.4973430

White, A. L. (2018). Understanding the university and faculty investment in implementing high-impact educational practices. *Journal of the Scholarship of Teaching and Learning, 18*(3), 119–134. https://doi.org/10.14434/josotl.v18i3.22831

Winne, P. H., & Hadwin, A. F. (1998). Studying as self-regulated learning. In D. J. Hacker, J. Dunlosky, & A. C. Graesser (Eds.), *Metacognition in educational theory and practice* (pp. 277–304). Lawrence Erlbaum Associates Publishers.

World Economic Forum. (2023). *Future of Jobs Report 2023.* https://www3.weforum.org/docs/WEF_Future_of_Jobs_2023.pdf

Xia, C., & Maes, P. (2013). The design of artifacts for augmenting intellect. In *Proceedings of the 4th Augmented Human International Conference (AH '13)* (pp. 154–161). Association for Computing Machinery. https://doi.org/10.1145/2459236.2459263

Xu, X., Chen, Y., & Liu, L. (2022). Toward improving usability in human-computer interaction: Lessons from AI integration. *Computers in Human Behavior, 137*, Article 107471. https://doi.org/10.1016/j.chb.2022.107471

Yilmaz, R. (2023). The effect of generative AI on competency skills and motivation in education. *Journal of Educational Computing Research, 61*(4), 812–835. https://doi.org/10.1177/07356331231116490

Zhang, X., Wang, Y., Watson, C. E., An, N., & Zeng, Q. (2024). Impact of ChatGPT on student learning: Perceptions of faculty and students from diverse disciplines. *Journal of Applied Learning & Teaching, 7*(1). https://doi.org/10.37074/jalt.2024.7.1.15

Zhao, M., Simmons, R. G., & Admoni, H. (2022). The role of adaptation in collective human-AI teaming. *Topics in Cognitive Science, 14*(4), 822–838. https://doi.org/10.1111/tops.12633

Zielinski, D. (2024, January 24). *Employers train employees to close the AI skills gap*. SHRM. https://www.shrm.org/topics-tools/news/training-development/employers-train-employees-close-ai-skills-gap

Zimmerman, B. J. (1994). Dimensions of academic self-regulation: A conceptual framework for education. In D. H. Schunk & B. J. Zimmerman (Eds.), *Self-regulation of learning and performance: Issues and educational applications* (pp. 3–21). Lawrence Erlbaum Associates, Inc.

Part II

Optimizing Human-Centric AI Partnerships for Advancing the Ethical & Effective Implementation, Piloting, Evaluation, & Scaling of Generative AI in Higher Education: A Customizable 4-Phase Framework

6

Phase 1: Strategic Planning and Stakeholder Engagement

Introduction: Setting the Stage for Strategic AI Integration

This chapter introduces Phase 1 of an adaptable four-phase plan for navigating Artificial Intelligence (AI) adoption in Higher Education Institutions (HEIs): *Research, Stakeholder Engagement, & Strategic Planning*. Integrating AI, particularly Generative AI (GenAI), presents opportunities alongside complex challenges, requiring effective preparation before implementation or piloting (Katsamakas et al., 2024). This initial phase focuses on achieving strategic alignment between stakeholder needs and institutional goals while establishing preliminary AI ethics and governance structures.

This chapter illustrates how human-centered problem-solving frameworks like Design Thinking and organizational change frameworks like Change Management can work together effectively. These combined approaches offer tools for understanding needs, fostering collaboration, and building readiness for change within HEIs during GenAI integration.

As outlined in Part 1, diverse stakeholder engagement and a human-centric approach to AI (guided by frameworks like HCAS and RSA) are fundamental for ethical and effective integration. These core frameworks assist users, teams, and institutions in leveraging the strengths of both AI systems and humans while mitigating weaknesses. This chapter provides a solid foundation for forming smart teams (Chap. 7) and conducting detailed needs assessments (Chap. 8). Workbook Prep sections connect these concepts to practical exercises in the companion UX and PALS workbooks, where users learn advanced strategies for ethical AI integration using GenAI as a strategic partner.

Supplementary Information The online version contains supplementary material available at https://doi.org/10.1007/978-3-032-05205-6_6.

D. Woodring, *Advancing Higher Education with Human-Centric AI Partnerships*,
https://doi.org/10.1007/978-3-032-05205-6_6

Core Frameworks Reference: The frameworks mentioned below, such as HCASR, HCAS, RSA, EHCS, EAO, and AHIPs, are detailed in the companion **Human-GenAI Collaboration & Synergy (HAICS) Guide** and/or Part 1 of this textbook. Please refer to those sections for full definitions and explanations. Phase 1 planning utilizes methodologies integrating Design Thinking (Brown, 2019) and Change Management (Kotter et al., 2021), detailed further in Part 4 of the **HAICS Guide**. Planning also applies principles from the **HAICS Guide**, especially HCASR Component 1 (Ethical & Effective Goal Setting): "(HCASR C1: Defining clear, ethical, context-aware goals for AI use and preparing resources)" and HCASR Component 4 (Human-Centric Implementation & Stakeholder Engagement): "(HCASR C4: Integrating AI responsibly through participatory design and stakeholder engagement)". The core frameworks from Part 1 (HCAS, RSA, EHCS, EAO, AHIPs) provide foundational concepts referenced in this preparatory phase.

The User Experience (UX) Workbook and the Personalized Adaptive Learning Sessions (PALS) Workbook for Chap. 6 are available via download. These workbooks assist users in ethically and effectively using AI as a strategic partner for researching and implementing tools, solutions, resources, outcomes, and deliverables customized to their unique needs and contexts.

Workbook Prep 1: Foundational Preparation

Contextual Reflection: Why is foundational planning (stakeholder engagement, governance principles) critical before detailed AI assessment or team formation? How does this set the initiative's tone?

UX Workbook Link: Begin with Chap. 6 UX Workbook "Preparatory Exercises" feeding into Section 1 activities (Reviewing Mission/Values).

PALS Workbook Link & AI Partnership Guidance: Use Chap. 6 PALS Workbook (Sec 1: Articulating Prep Phase Value). Input your AI motivation; ask your AI partner to draft a rationale for Chap. 6 prep work, referencing HCASR Component 1 (Ethical & Effective Goal Setting).

Example Prompt: "Motivation: [Enhance research]. Using HCASR Component 1, explain why Ch 6 prep (engagement, governance) is vital for achieving this responsibly."

Human Oversight Reminder: AI lists general benefits; recognizing specific institutional risks (e.g., alienating groups, misaligned tools) requires human contextual understanding (as emphasized in HCASR Component 1).

The Synergy of Design Thinking and Change Management

Integrating AI effectively requires approaches that address both human needs and organizational dynamics. Design Thinking offers methods for human-centered problem definition and solution generation, emphasizing empathy, ideation, and iterative prototyping (Brown, 2019). Change Management provides structures for navigating organizational transitions, focusing on building coalitions, creating vision, and embedding change (Kotter et al., 2021). (For detailed methodologies, see *HAICS Guide*, Part 4.)

The synergy lies in using Design Thinking's methods to understand *what* needs to change—focusing on AI solutions that genuinely meet user needs—while employing Change Management frameworks to guide *how* that change is implemented across the institution. This combined approach fosters stakeholder buy-in, supports strategic goal alignment, and manages the human aspects of technological transition (Al-Freih & Chiu, 2023; Carmody-Bubb, 2023; Kotter et al., 2021). Applying this synergy during initial preparation helps ensure AI initiatives are both relevant and effectively integrated (Fig. 6.1).

Workbook Prep 2: Connecting Design Thinking and Change Management

Contextual Reflection: How can deeply understanding stakeholder needs (Design Thinking Empathy/Inspiration) fuel the genuine "Urgency" needed (Change Management Step 1) for AI exploration?

UX Workbook Link: See Chap. 6 UX Workbook, Activity 1.3 (Initial Opportunity & Risk Brainstorm) for identifying needs that fuel urgency.

PALS Workbook Link & AI Partnership Guidance: Use Chap. 6 PALS Workbook prompts (Sec 1: Linking Empathy to Urgency). Input a stakeholder pain point; ask your AI partner to help frame it as an urgency statement using Design Thinking/Change Management synergy.

Example Prompt: "Pain point: '[Faculty time on routine grading limits personalized feedback]'. Using Design Thinking/Change Management principles, reframe into 1–2 sentences highlighting urgency for AI solutions while upholding pedagogy."

Human Oversight Reminder: AI aids framing; human empathy and strategic communication are crucial for crafting messages that truly resonate and reflect authentic urgency (as emphasized in HCASR Component 4: Human-Centric Implementation & Stakeholder Engagement).

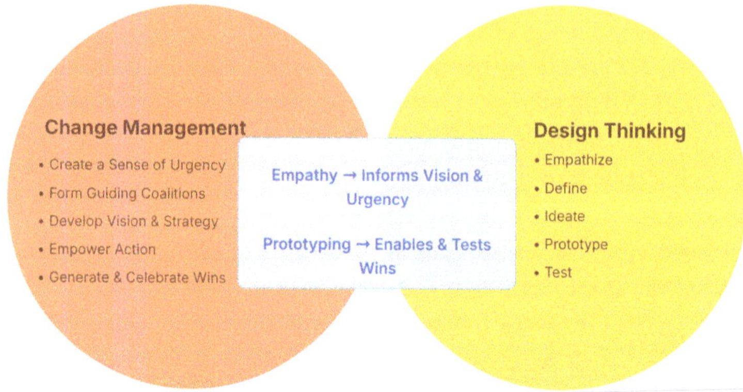

Fig. 6.1 Conceptual diagram: merging change management and design thinking for AI adoption

Aligning AI Initiatives with Institutional Mission and Values

Before evaluating specific AI tools, institutions must ensure potential AI initiatives align with their core mission, values, and strategic goals (Al-Freih & Chiu, 2023; NASPA, 2024). AI adoption should serve institutional purpose, not become an end in itself. This involves reviewing foundational documents (e.g., mission statements, strategic plans) to identify how AI could support objectives like student success, research excellence, or access (DerSimonian & Montagnino, 2025).

Leadership must articulate this alignment clearly to foster buy-in and guide ethical choices (NASPA, 2024; Salmon & Psarras, 2024). Consistency between AI exploration and institutional values (e.g., equity, integrity, academic freedom) is essential (Robert & McCormack, 2024; Wynants et al., n.d.). Documenting this alignment early provides justification, clarifies resource allocation, and communicates the strategic rationale for AI integration (DerSimonian & Montagnino, 2025; NASPA, 2024).

Workbook Prep 3: Values Alignment Check

Contextual Reflection: Review your institution's mission/strategic plan. How might specific AI applications (e.g., AI tutoring, admin automation) support or undermine 2–3 core institutional values/goals?
UX Workbook Link: See Chap. 6 UX Workbook, Activity 1.1 (Review Mission, Vision, Values, Goals) for structured analysis.

PALS Workbook Link & AI Partnership Guidance: Use Chap. 6 PALS Workbook (Sec 1: Value Alignment Prompts). Input an AI initiative (e.g., AI proctoring) and a value (e.g., student privacy); ask AI to analyze conflicts and suggest mitigation (referencing relevant frameworks like EHCS/EAO from the *HAICS Guide*, or external sources like UNESCO).

Example Prompt: "Analyze conflicts between AI proctoring and 'student privacy'. List 3 potential mitigation strategies based on UNESCO principles, considering ethical frameworks like EHCS/EAO."

Human Oversight Reminder: AI identifies potential conflicts; institutional leadership determines acceptable alignment with core values based on context and ethical deliberation (as emphasized in HCASR Component 1).

Understanding AI/GenAI Opportunities and Challenges in HE

Effective preparation requires assessing the potential benefits and difficulties associated with AI/GenAI in higher education. Opportunities include enhancing teaching and learning (e.g., personalized feedback), augmenting research (e.g., data analysis), improving student support, and increasing administrative efficiency (NASPA, 2024; Deloitte Center for Government Insights, 2024).

Simultaneously, institutions must address significant challenges proactively. Key concerns include:

- Threats to academic integrity (Bittle, 2024)
- Risk of embedding or amplifying algorithmic bias (Ramos Salazar et al., 2024)
- Data privacy and security issues (e.g., FERPA compliance) (University of Rochester, n.d.)
- Potential negative impacts on critical thinking skills development (Francis et al., 2025)
- Need for substantial technology investment, infrastructure, and comprehensive training (NASPA, 2024)

Addressing these requires aligning AI initiatives with values like equity and integrity and applying human-centric frameworks (HCAS, RSA) to guide implementation (DerSimonian & Montagnino 2025; Salmon & Psarras,

2024). Acknowledging both opportunities and challenges enables informed planning and risk management.

Resources: Crash Course in AI Ethics & Literacy: Enhance your **AI Ethics and Literacy Competencies** with Personalized Adaptive Learning Sessions (PALS). For example, you can learn more about the potential cognitive impacts of GenAI on teaching, learning, assessment, research, and/or administration. Available in Part 3 of the *Human-AI Collaboration & Synergy Guide*.

Workbook Prep 4: Balancing Potential and Risk

Contextual Reflection: For your context, which potential GenAI benefits (e.g., personalized learning, efficiency) seem most promising? Which challenges (e.g., academic integrity, bias, privacy) pose the greatest risk?

UX Workbook Link: See Chap. 6 UX Workbook, Activity 1.3 (Initial Opportunity & Risk Brainstorm) for exploring context-specific applications and challenges.

PALS Workbook Link & AI Partnership Guidance: Use Chap. 6 PALS Workbook (Sec 1: Opportunity/Challenge Analysis Prompts). Input brainstormed applications; ask AI to identify ethical risks/barriers based on research (referencing relevant frameworks like HCASR Component 3, EAO/EHCS from the *HAICS Guide*).

Example Prompt: "For 'AI writing feedback tool for first-year students', identify 3 potential ethical challenges (integrity/bias affecting multilingual learners), citing sources if possible (referencing HCASR Component 3)."

Human Oversight Reminder: Evaluate AI-identified risks using established frameworks (e.g., NIST AI RMF). Assess likelihood/impact in your context and prioritize using human judgment and stakeholder input (as emphasized in HCASR Component 4).

Planning for Diverse Stakeholder Engagement

Engaging diverse stakeholders early and transparently is essential for successful AI integration (DerSimonian & Montagnino, 2025; NASPA, 2024). This approach builds trust, reduces resistance, fosters ownership, and incorporates varied perspectives (Deloitte Center for Government Insights, 2024; Maali et al., 2020; Salmon & Psarras, 2024).

Initial planning involves identifying key stakeholder groups (e.g., faculty, staff, students, administrators) and recognizing their differing AI literacy levels, needs, and concerns. A preliminary engagement plan outlines potential methods for gathering input during Phase 1 (e.g., surveys, focus groups, advisory committees) to ensure diverse voices are heard (Robert & McCormack, 2024; Wynants et al. n.d.). A basic communication strategy outlines plans for transparent updates about goals and progress (Deloitte Center for Government Insights, 2024; Maali et al., 2020). This preparatory work ensures engagement is intentional (Table 6.1 and Fig. 6.2).

Workbook Prep 5: Planning Engagement

Contextual Reflection: How are stakeholders typically engaged in major initiatives at your institution? Which methods work best? What primary AI concerns do you anticipate from key groups (students, faculty, staff, admin)?

UX Workbook Link: See Chap. 6 UX Workbook, Activity 2.2 (Draft Preliminary Stakeholder Engagement Plan Outline) for template/guidance.

PALS Workbook Link & AI Partnership Guidance: Use Chap. 6 PALS Workbook (Sec 2: Stakeholder Engagement Prompts). Input groups/concerns; ask AI for tailored communication/engagement strategies referencing change management principles (Kotter et al., 2021) & HCASR Component 4 (Human-Centric Implementation & Stakeholder Engagement).

Example Prompt: "Suggest 2–3 engagement strategies for faculty concerns about AI's impact on integrity/workload, based on Kotter steps ('Communicate Vision', 'Empower Action') & HCASR Component 4."

Human Oversight Reminder: AI suggests generic strategies. Authentic engagement needs empathy and responsiveness tailored to institutional culture. Use RSA (Regulation, Synergy, & Augmentation) co-regulation principles when engaging teams.

Table 6.1 Preliminary stakeholder engagement plan outline (template)

Stakeholder group	Potential engagement methods/channels	Preliminary frequency/timing	Key considerations/goals
Undergraduate students	Surveys, focus groups, student Gov. reps, pilot feedback	E.g., Early Phase 1, annual	Understanding needs, concerns (privacy, integrity)
Graduate students	Surveys, dept. meetings, grad student org. reps., pilots	E.g., Phase 1, ongoing	Research uses, workload impact, specialized needs
Faculty (full-time)	Dept. meetings, faculty senate reps., workshops, pilots	E.g., Phase 1, semesterly	Pedagogy, integrity, workload, research support
Faculty (part-time/adjunct)	Surveys, targeted workshops, online forums	E.g., Phase 1, ad hoc	Access, training needs, workload balance
Staff (student support)	Team meetings, focus groups, training sessions	E.g., Phase 1, ongoing	Workflow impact, efficiency gains, training needs
Staff (IT/technical)	Project meetings, technical workshops, infra reviews	E.g., throughout	Feasibility, security, infrastructure, support

(continued)

Table 6.1 Continued

Stakeholder group	Potential engagement methods/channels	Preliminary frequency/ timing	Key considerations/ goals
Staff (library/ research)	Focus groups, workshops, pilot involvement	E.g., Phases 1 & 2	Research tools, data access, ethical use in research
Administrators/ leadership	Steering committee, regular briefings, strat. planning	E.g., ongoing	Alignment, budget, policy, risk management
Institutional Review Board (IRB)/Ethics Committee	Consultations, policy review	E.g., Phase 1 and ongoing	Ethical oversight, compliance
External partners (if any)	Targeted meetings, advisory input	E.g., as needed	Collaboration goals, data sharing agreements

Note. This table provides a template for outlining a preliminary stakeholder engagement plan. Specific methods, frequencies, and goals should be adapted based on institutional context, AI initiative scope, and resources during Phase 1 planning, as guided by the companion UX Workbook (Activity 2.2)

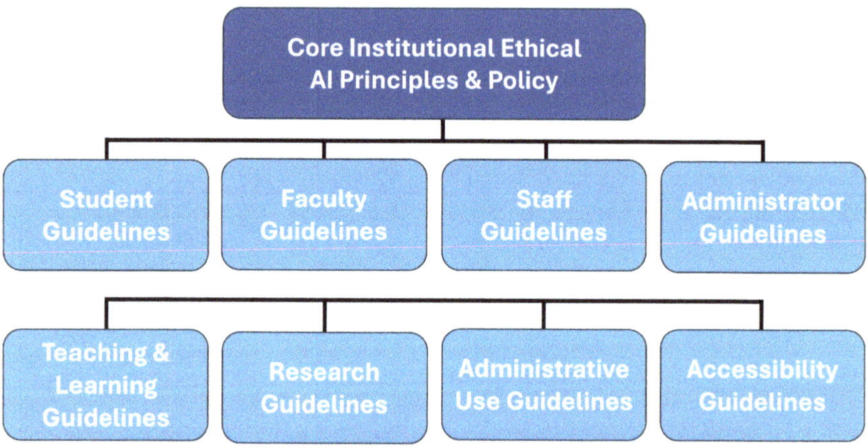

Fig. 6.2 Example structure for institutional AI guidelines

Establishing the Foundation for Ethical AI Governance

Defining foundational principles for ethical AI governance during initial planning guides the initiative responsibly (Al-Omari et al., 2024; Wynants et al., n.d.). This involves establishing core ethical principles (e.g., fairness, accountability, transparency, privacy, security, human well-being, human oversight) aligned with institutional values and informed by external guidelines (e.g., UNESCO, 2022; OECD, 2023; NIST, 2023). Stakeholder engagement helps ensure relevance (DerSimonian & Montagnino, 2025).

Planning must identify the scope for policy review, as existing policies (e.g., academic integrity, data privacy) need updates for AI (Robert & McCormack, 2024; Temper et al., 2024; University of Rochester, n.d.). Anticipating governance challenges like resource constraints is also important (Luo, 2024). Furthermore, planning should include adopting a structured risk management approach, such as the NIST AI RMF (NIST, 2023; SREB, 2024). This cyclical process involves four key functions, as illustrated in Fig. 6.3. This includes planning bias mitigation (Kim et al., 2025) and ensuring transparency (Al-Omari et al., 2024; Ramos Salazar et al., 2024). This foundational work supports ethical planning in subsequent phases.

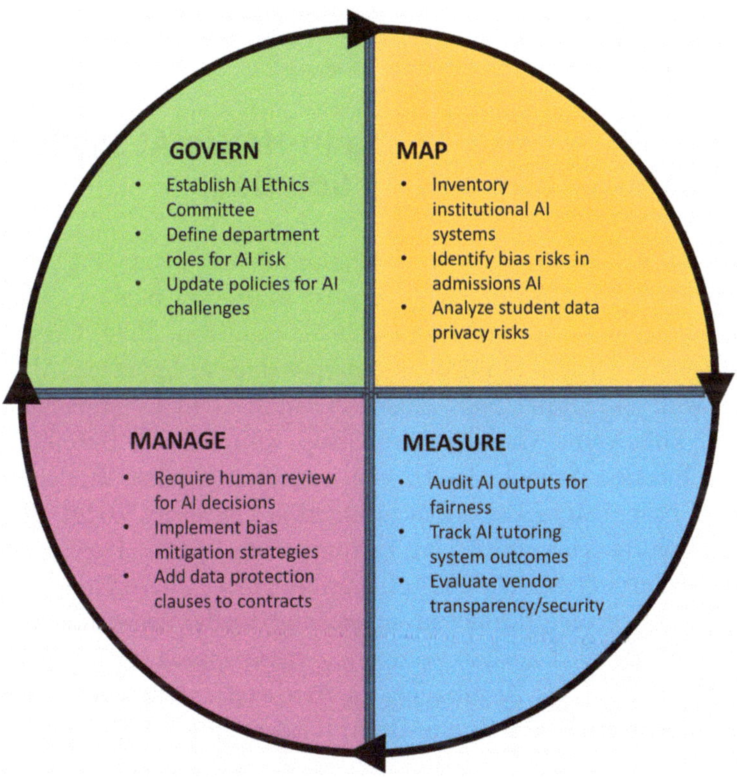

Fig. 6.3 NIST AI risk management framework cycle

Workbook Prep 6: Selecting Core Ethical Principles

Contextual Reflection: Review common AI ethical principles (fairness, accountability, transparency, privacy, etc.). Which 2–3 are most critical to emphasize initially at your institution? Why?

UX Workbook Link: See Chap. 6 UX Workbook, Activity 1.2 (Identify Guiding Ethical Principles) & Activity 2.3 (Outline Initial Governance Scope).

PALS Workbook Link & AI Partnership Guidance: Use Chap. 6 PALS Workbook (Sec 1: Defining Ethical Principles). Select a principle; ask AI to help draft a brief operational definition for AI in HE, referencing HCASR Component 1 or external principles (e.g., OECD).

Example Prompt: "Draft a 1–2 sentence operational definition for 'Accountability' regarding AI in university administrative decisions, citing HCASR Component 1 or OECD principles."

Human Oversight Reminder: AI provides definitions. Articulating principles reflecting institutional values and creating meaningful guidelines requires human ethical deliberation and consensus (as emphasized in HCASR Component 1 & 4).

Strategic Decision: Building In-House AI Systems vs. Integrating Third-Party Solutions

An early consideration is the strategic choice between developing proprietary AI systems ("build"), integrating third-party solutions ("buy"), or adopting a hybrid approach (Brighteye Ventures, 2024; Cameron, 2024; Grajek et al., 2024). Recent examples of third-party partnerships include OpenAI's collaboration with Wharton (Fore, 2024), ChatGPT EDU implementations (Rispens, 2024), and Google's partnership with Rutgers Business School (Rutgers Business School, 2024), while institutions like Vanderbilt have developed proprietary AI solutions (Vanderbilt University, 2024). This decision impacts strategic control, cost, customization, technical needs, and governance (Culotta & Mattei, 2023; Inclusion Cloud, 2024; Nguyen, 2024). Each approach presents distinct advantages and disadvantages, summarized in Table 6.2. A detailed analysis comparing these approaches is provided in **Appendix A: Strategic Approaches to Artificial Intelligence Systems in Higher Education: Build, Buy, or Both?** Readers should consult this resource when evaluating different AI acquisition strategies (Table 6.3).

Table 6.2 Stakeholder engagement strategies by HEI stakeholder group

HEI stakeholder Groups	Surveys	Focus groups	Workshops	Pilot testing	Newsletters	Town halls	Feedback forms
Undergraduates	✓		✓			✓	✓
Graduates	✓					✓	✓
Faculty (full-time)	✓					✓	✓
Faculty (part-time)	✓		✓			✓	
Staff (student support)	✓		✓				✓
Staff (IT/technical)	✓		✓			✓	✓
Staff (library/research)	✓					✓	✓
Admin/leadership	✓					✓	
Ethics committee/IRB	✓						✓

Note. The table displays which engagement methods are employed for each higher-education stakeholder group. A ✓ indicates that the listed strategy—surveys, focus groups, workshops, pilot testing, newsletters, town halls, or feedback forms—is used with that group. Undergraduates and graduates, for example, participate in surveys, town halls, and feedback forms, whereas part-time faculty and student-support staff also attend workshops

Table 6.3 Comparison of considerations for proprietary versus third-party AI solutions in higher education

Consideration	Proprietary ("Build") approach	Third-party ("Buy") approach	Hybrid approach
Strategic control	High: full control over features, roadmap, updates.	Low: dependent on vendor roadmap, features, updates.	Moderate: control over integration, potential customization.
Data privacy/ security	High: data typically stays in-house (FERPA easier).	Moderate/low: requires vetting; data may leave; vendor reliance.	Varies: depends on mix; vetting still needed.
Customization	High: tailored to specific institutional needs, workflows.	Low: generally standardized; limited options.	Moderate: can integrate tools; focus on integration.
Initial cost	High: significant investment in infrastructure, personnel.	Low/moderate: subscription-based; lower upfront cost.	Varies: lower than full build, higher than pure buy.
Ongoing cost	High: maintenance, updates, personnel, infrastructure upkeep.	Moderate/high: subscription fees, integration costs.	Moderate/high: mix of internal upkeep and vendor fees.
Time to deployment	Slow: requires significant development time.	Fast: relatively quick implementation.	Moderate: faster than build, slower than buy.
Expertise required	High: needs specialized AI/ML, data science, IT staff.	Low: primarily vendor management, integration skills.	Moderate: requires integration expertise and vendor management.
Scalability	Moderate: depends on internal infrastructure/ resources.	High: vendor typically handles scaling.	Moderate/high: depends on vendor and internal capacity.
Vendor dependency	Low: independent of external vendors for core functionality.	High: reliant on vendor viability, support, pricing.	Moderate: dependent on multiple vendors; less single reliance.
Access equity	Potentially high: can design for broad access.	Potential risk: tiered pricing may limit access (McIntosh, 2024).	Varies: depends on licensing/ integration choices.
Innovation speed	Moderate/slow: pace set by internal capacity.	Potentially fast: leverages vendor R&D.	Moderate: can leverage vendor innovation; integration time.

Note. This table summarizes key factors influencing the strategic decision between "Build," "Buy," or hybrid approaches in higher education. The optimal choice depends on institutional context, resources, risk tolerance, strategic goals, and specific use cases. Refer to Appendix A for a detailed discussion (e.g., Attridge, 2024; Coffey, 2024a, 2024b; Grajek et al., 2024; O'Connell, 2024; Paykamian, 2023)

Workbook Prep 7: Weighing Build vs. Buy vs. Both

Contextual Reflection: Based on initial readiness thoughts (tech expertise, infrastructure, risk tolerance), what's your preliminary leaning on building vs. buying AI or combining both for priority uses? What factors are most critical?

UX Workbook Link: See Chap. 6 UX Workbook, Activity 3.4 (Compare Build vs. Buy vs. Both) for a comparative analysis template.

PALS Workbook Link & AI Partnership Guidance: Use Chap. 6 PALS Workbook (Sec 3: Build vs. Buy vs. Both Analysis Prompts). Input priorities; ask AI for vendor vetting questions (data use, security, bias) or proprietary resource needs (personnel, infrastructure), referencing HCASR Component 1.

Example Prompt: "Generate 5 key FERPA-based privacy/security questions for third-party AI vendors offering student progress monitoring tools (referencing HCASR Component 1)." OR "Outline typical personnel/infrastructure needed for an HEI to develop a proprietary GenAI writing tool (referencing HCASR Component 1)."

Human Oversight Reminder: AI aids comparison. The final strategic decision requires careful deliberation by leadership (legal, IT, finance, academic), considering sustainability, risk, goals, and ethics (as emphasized in HCASR Component 1 & 5). Verify vendor claims rigorously.

Conclusion: Laying the Foundation for Responsible AI Integration

The Phase 1 preparatory activities outlined establish the necessary foundation for ethical and effective AI adoption. Merging Change Management and Design Thinking provides a strategic, human-centric basis. Key tasks include aligning AI with mission and values, assessing opportunities and challenges, planning stakeholder engagement, considering build-versus-buy options, and defining foundational ethical governance and risk management approaches.

There is no single correct solution for AI integration (Carmody-Bubb, 2023). Each HEI operates within a unique sociocultural context, requiring strategic planning centered on engaging and empowering users across the institution (Kotter et al., 2021). When guided by the core frameworks in this book (e.g., HCASR, RSA, detailed in the *Human-AI Collaboration & Synergy Guide*), AI can serve as a strategic partner during this phase, assisting in developing customized solutions tailored to specific needs, goals, and contexts. This preparatory work, guided by the companion workbooks, produces the initial User Outcomes and Deliverables (UODs) (e.g., Initial AI Readiness Assessment Report (V1 scope), Preliminary Stakeholder Engagement Plan (V1), Initial AI Ethics & Governance Framework (V1), and Preliminary Institutional AI Integration Plan (V1)) required for forming Smart Teams

(Chap. 7) and conducting detailed needs assessments (Chap. 8), grounding subsequent efforts in institutional context and ethical commitments.

Looking Ahead & Engaging the Workbook

Having established the foundational context and preparatory considerations in Chap. 6, the next step is building the human infrastructure for planning. Chapter 7 details forming diverse "Smart Teams" and conducting systematic stakeholder mapping.

The Chap. 6 Companion UX Workbook provides structured activities to execute the preparatory tasks discussed here. Engaging with the workbook—starting with "Getting Started: Preparatory Exercises"—enables collaborative definition of principles, initial opportunity/risk brainstorming, outlining engagement plans, and documenting alignment. By actively using the workbook and leveraging AI as a strategic partner under human oversight (guided by frameworks like HCASR), your institution can enhance the process of producing the key preparatory UODs required to effectively launch the team formation and stakeholder mapping activities detailed in Chap. 7.

Long-Term Scaling, Sustainability & AI Governance: Laying Ethical TLARA Foundations

As Chap. 6 concludes, the foundational work in preparing for diverse stakeholder engagement and initial AI governance is pivotal for long-term AI success. Effective AI integration hinges on strong leadership that champions adoption by fostering trust and ensuring transparent communication, aligning AI initiatives with the unique institutional mission and values (DerSimonian & Montagnino, 2025), and developing comprehensive strategies including clear guidelines and robust training (Ellucian, 2024). Both the Chap. 6 UX and PALS Workbooks are designed to help you ethically partner with AI, enhancing your planning and engagement.

Higher Education Institutions (HEIs) are complex systems within even larger systems, each operating in diverse sociocultural contexts (Instructional Design, Engagement, and Support [IDEAS], 2025). This reality demands a holistic, multi-level approach to AI, addressing varied stakeholder needs, goals, and objectives across different levels and spaces over time. The PALS Workbook for this chapter, under the theme "Long-Term Scaling, Sustainability & AI Governance: Iteratively Developing Your Custom TLARA-Informed AI Guidelines," introduces a customizable prompt. It transforms your GenAI model into an expert advisor, enabling you to use your strategic goals and contextual data to begin building tailored frameworks and policies for ethical AI integration across the TLARA (Teaching, Learning, Assessment, Research, Administration) domains (DerSimonian & Montagnino, 2025). This iterative process is key as many institutions are still developing unified approaches and require guidance for lasting strategies (IDEAS, 2025).

References

Al-Freih, M., & Chiu, T. K. F. (2023). Strategic leadership for responsible artificial intelligence adoption in higher education. *Sustainability, 15*(16), 12599. https://doi.org/10.3390/su151612599

Al-Omari, O., Alyousef, A., Fati, S., Shannaq, F., & Omari, A. (2024). Governance and ethical frameworks for AI integration in higher education: Enhancing personalized learning and legal compliance. *Journal of Ecohumanism, 4*(1), 139–155. https://doi.org/10.33182/joe.v4i1.5781

Attridge, M. (2024, April 24). Campus-GPT: How 2 University of California campuses are designing their own specialized AI tools. *BestColleges*. https://www.bestcolleges.com/news/uc-san-diego-irvine-design-specialized-ai-tools/

Bittle, K. (2024). Generative AI and academic integrity in higher education: A systematic review and research agenda. *Information, 15*(4), 206. https://doi.org/10.3390/info15040206

Brighteye Ventures. (2024, October 10). AI's impact on the future of higher education. *Brighteye Ventures Blog*. https://brighteyevc.com/post/ai-s-impact-on-the-future-of-higher-education

Brown, T. (2019). *Change by design: How design thinking transforms organizations and inspires innovation* (Rev. ed.). Harper Business.

Cameron, J. (2024, January 7). Integrating artificial intelligence into higher education governance. Association of Governing Boards of Universities and Colleges (AGB). https://agb.org/blog-post/integrating-artificial-intelligence-into-higher-education-governance/

Carmody-Bubb, M. (2023). *Cognition and decision making in complex adaptive systems: The human factor in organizational performance*. Springer.

Coffey, L. (2024a, January 19). Arizona State joins ChatGPT in first higher ed partnership. *Inside Higher Ed*. https://www.insidehighered.com/news/quick-takes/2024/01/19/arizona-state-joins-chatgpt-first-higher-ed-partnership

Coffey, L. (2024b, March 21). Universities build their own ChatGPT-like tools. *Inside Higher Ed*. https://www.insidehighered.com/news/tech-innovation/artificial-intelligence/2024/03/21/universities-build-their-own-chatgpt-ai

Culotta, A., & Mattei, N. (2023, November 20). Use open source for safer generative AI experiments. *MIT Sloan Management Review*. https://sloanreview.mit.edu/article/use-open-source-for-safer-generative-ai-experiments/

Deloitte Center for Government Insights. (2024, September 25). How higher education can realize the potential of Generative AI. *Deloitte Insights*. https://www2.deloitte.com/us/en/insights/industry/public-sector/generative-ai-higher-education.html

DerSimonian, R., & Montagnino, C. (2025, March 26). Crafting thoughtful AI policy in higher education: A guide for institutional leaders. *Campus Technology*. Retrieved May 7, 2025, from https://campustechnology.com/articles/2025/03/26/

crafting-thoughtful-ai-policy-in-higher-education-a-guide-for-institutional-leaders.aspx

Ellucian. (2024, December 16). AI in higher education: Implementation strategies for institutions and tech partners. *Higher Ed Dive*. Retrieved May 7, 2025, from https://www.highereddive.com/spons/ai-in-higher-education-implementation-strategies-for-institutions-and-tech/735434/

Fore, P. (2024, November 21). OpenAI partners with Wharton for a new course focused on leveraging ChatGPT for teachers. *Fortune*. https://fortune.com/education/articles/openai-chatgpt-university-of-pennsylvania-wharton-course/

Francis, N. J., Jones, S., & Smith, D. P. (2025). Generative AI in higher education: Balancing innovation and integrity. *British Journal of Biomedical Science, 81*, 14048. https://doi.org/10.3389/bjbs.2024.14048

Grajek, S., Pelletier, K., & Freeman, A. (2024, March 11). AI procurement in higher education: Benefits and risks of emerging tools. *EDUCAUSE Review*. https://er.educause.edu/articles/2024/3/ai-procurement-in-higher-education-benefits-and-risks-of-emerging-tools

Inclusion Cloud. (2024, February 28). What you need to know before choosing open-source or proprietary AI models. *Inclusion Cloud Blog*. https://inclusioncloud.com/insights/blog/open-source-llm-vs-proprietary-models/

Instructional Design, Engagement, and Support (IDEAS). (2025, January 27). AI's impact on higher education through the eyes of faculty: 4 key trends. University of Massachusetts Amherst. Retrieved May 7, 2025, from https://www.umass.edu/ideas/news/ais-impact-higher-education-through-eyes-faculty-4-key-trends

Katsamakas, E., Pavlov, O. V., & Saklad, R. (2024). Artificial intelligence and the transformation of higher education institutions [Preprint]. arXiv. https://doi.org/10.48550/arXiv.2402.08143

Kim, J. S., Thu, K. Y., Ismayilzada, J., Park, J., Kim, E., Ahmad, H., An, N. M., Thorne, J., & Oh, A. (2025). When Tom eats kimchi: Evaluating cultural bias of multimodal large language models in cultural mixture contexts. *arXiv preprint* arXiv:2503.16826. https://arxiv.org/abs/2503.16826

Kotter, J. P., Akhtar, V., & Gupta, G. (2021). *Change: How organizations achieve hard-to-imagine results in uncertain and volatile times*. Wiley.

Luo, J. (Jess). (2024). A critical review of 2 GenAI policies in higher education assessment: A call to reconsider the "originality" of students' work. *Assessment & Evaluation in Higher Education, 49*(5), 651–664. https://doi.org/10.1080/02602938.2024.2309963

Maali, O., Lines, B., Smithwick, J., Hurtado, K., & Sullivan, K. (2020). Change management practices for adopting new technologies in the design and construction industry. *Journal of Information Technology in Construction (ITcon), 25*, 344–358. https://doi.org/10.36680/j.itcon.2020.019

McIntosh, A. (2024, October 23). *EDUCAUSE 2024: Institutions embark on cross-campus generative AI initiatives*. EdTech: Focus on Higher Education. https://edtechmagazine.com/higher/article/2024/10/educause-2024-institutions-embark-cross-campus-generative-ai-initiatives

NASPA - Student Affairs Administrators in Higher Education. (2024, December). *The transformative potential of AI.* https://www.naspa.org/files/dmfile/NASPA-Report%2D%2D-The-Transformative-Potential-of-AI%2D%2DDecember-2024.pdf

National Institute of Standards and Technology. (2023, January). *Artificial intelligence risk management framework (AI RMF 1.0)* (NIST AI 100-1). U.S. Department of Commerce. https://doi.org/10.6028/NIST.AI.100-1

Nguyen, H. V. (2024, January 9). Open source vs. proprietary AI: A comprehensive guide to making the right choice. *SmartDev.* https://smartdev.com/open-source-vs-proprietary-ai/

O'Connell, A. J. (2024, February 7). How (and why) the University of Michigan built its own closed generative AI tools. *EDUCAUSE Review.* https://er.educause.edu/articles/2024/2/how-and-why-the-university-of-michigan-built-its-own-closed-generative-ai-tools

Organisation for Economic Co-operation and Development. (2023). *OECD AI Principles overview.* OECD.AI Policy Observatory. https://oecd.ai/en/ai-principles

Paykamian, B. (2023, November 16). Should universities build their own custom AI tools in-house? *Government Technology.* https://www.govtech.com/education/higher-ed/should-universities-build-their-own-custom-ai-tools-in-house

Ramos Salazar, L., Peeples, S. F., & Brooks, M. E. (2024). Generative AI ethical considerations and discriminatory biases on diverse students within the classroom. In S. Elmoudden & J. Wrench (Eds.), *The role of generative AI in the communication classroom* (pp. 191–213). IGI Global. https://doi.org/10.4018/979-8-3693-0831-8.ch010

Rispens, S. (2024, June 4). OpenAI builds ChatGPT system for colleges and universities. *EdScoop.* https://edscoop.com/openai-chatgpt-edu-colleges-universities-artificial-intelligence/

Robert, J., & McCormack, M. (2024, May 23). 2024 EDUCAUSE Action Plan: AI policies and guidelines. *EDUCAUSE.* https://www.educause.edu/research/2024/2024-educause-action-plan-ai-policies-and-guidelines

Rutgers Business School. (2024, November 21). Rutgers Business School partners with Google to enhance teaching and classroom learning with Generative AI. *Rutgers Business School News.* https://www.business.rutgers.edu/news/rutgers-business-school-partners-google-enhance-teaching-and-classroom-learning-generative-ai

Salmon, I., & Psarras, A. (2024). Facilitating change management of higher education institutions by integrating artificial intelligence tools in their balanced scorecards. *Quest Journal of Research in Business and Management, 12*(10), 20–26. https://questjournals.org/jrbm/papers/vol12-issue10/12102026.pdf

Southern Regional Education Board. (2024, February 18). *AI Rec 7 risk management.* Southern Regional Education Board. https://www.sreb.org/ai-rec-7-risk-management

Temper, M., Tjoa, S., & David, L. (2024). Higher Education Act for AI (HEAT-AI): A framework to regulate the usage of AI in higher education institutions. *Frontiers in Education, 9*, 1505370. https://doi.org/10.3389/feduc.2024.1505370

UNESCO. (2022). *Recommendation on the ethics of artificial intelligence.* UNESCO Publishing. https://unesdoc.unesco.org/ark:/48223/pf0000381137

University of Rochester, Office of the Provost. (n.d.). *Generative AI use in education.* University of Rochester. Retrieved May 1, 2025, from https://www.rochester.edu/provost/gen-ai-education/

Vanderbilt University. (2024, November 14). Amplify, Vanderbilt's new custom generative AI software, is now open to all faculty, staff, and students. *Vanderbilt University News.* https://news.vanderbilt.edu/2024/11/14/amplify-vanderbilts-new-custom-generative-ai-software-is-now-open-to-all-faculty-staff-and-students/

Wynants, S., Childers, G., De La Torre Roman, Y., Budar-Turner, D., & Vasquez, P. (n.d.). Ethical principles AI framework for higher education. CSU AI Commons, California State University, Fullerton. Retrieved May 1, 2025, from https://genai.calstate.edu/communities/faculty/ethical-and-responsible-use-ai/ethical-principles-ai-framework-higher-education/

7

Forming Smart Teams and Stakeholder Mapping (Phase 1)

Introduction: Building the Collaborative Foundation

After the initial planning detailed in Chap. 6, the crucial next step in integrating AI involves organizing the necessary human resources. This chapter focuses on creating dedicated "Smart Teams" with diverse membership and systematically identifying all individuals and groups who need involvement or will be affected by AI initiatives. Establishing this human infrastructure correctly from the outset builds a solid foundation for collaboration and helps mitigate potential issues such as confusion, resistance, or misaligned project directions later in the process.

Forming effective Smart Teams aligns directly with change management principles emphasizing the need for a strong "Guiding Coalition," composed of diverse individuals—often initially those enthusiastic about leading innovation—to navigate complex organizational transitions (Kotter et al., 2021). Simultaneously, thorough stakeholder mapping ensures AI initiatives are informed by varied needs, perspectives, and concerns across the institution, reflecting a human-centric approach fundamental to frameworks like HCASR (DerSimonian & Montagnino, 2025; NASPA, 2024). These activities create the collaborative intelligence required to plan and implement GenAI responsibly and effectively (Jaskyte et al., 2024).

Supplementary Information The online version contains supplementary material available at https://doi.org/10.1007/978-3-032-05205-6_7.

D. Woodring, *Advancing Higher Education with Human-Centric AI Partnerships*,
https://doi.org/10.1007/978-3-032-05205-6_7

Workbook Prep 1: The Value of Foundational Teamwork & Mapping

Contextual Reflection: Why is establishing a collaborative foundation—diverse teams and deep stakeholder understanding—essential before detailed strategic planning (Chap. 9)? How do principles from change management (Kotter et al., 2021) and design thinking (Brown, 2019) underscore this necessity?

UX Workbook Link: Review the Chap. 7 UX Workbook Introduction and "Getting Started" exercises. Activities in Section 1 ("Planning") help operationalize this foundational work. (See Activities 1.1 and 1.2.)

PALS Workbook Link & AI Partnership Guidance: Engage with Chap. 7 PALS Workbook prompts (Section 1). Input your initial thoughts on the value of this chapter's work. Ask your GenAI partner (using HCASR Component 1; see HAICS Guide) to help articulate a concise rationale for why thorough team formation and stakeholder mapping are critical prerequisites for ethical and effective AI strategy development in your specific institutional context.

Example Prompt: "Context: [Our HEI faces stakeholder skepticism about AI]. Based on HCASR Component 1 principles, help draft a 2–3 sentence rationale explaining why establishing diverse Smart Teams and conducting thorough stakeholder mapping (Chap. 7) is vital before developing our AI strategy (Chap. 9) to build trust and ensure relevance."

Human Oversight Reminder: While your GenAI partner can structure arguments, understanding the specific political climate, cultural nuances, and strategic importance of trust-building requires human insight and leadership judgment (HCASR Component 4).

Core Frameworks Reference: The frameworks mentioned below, such as HCASR, HCAS, RSA, EHCS, EAO, and AHIPs, are detailed in the companion *Human-GenAI Collaboration & Synergy (HAICS) Guide* and/or Part 1 of this textbook. Please refer to those sections for full definitions and explanations.

This chapter operationalizes principles from the synergy between design thinking and change management, particularly regarding forming guiding coalitions (Smart Teams) and mobilizing diverse groups (Kotter et al., 2021; Brown, 2019). It applies Human-Centric AI Synergy & Regulation (HCASR) Component 4 (Human-Centric Implementation & Stakeholder Engagement) to the processes of team formation and engagement planning. The underlying principles of HCAS, RSA, EHCS, and EAO inform the emphasis on diversity, ethical considerations, human-centric skill alignment, and outcome focus within team structures and stakeholder interactions.

The User Experience (UX) Workbook and the Personalized Adaptive Learning Sessions (PALS) Workbook for Chap. 7 provide practical activities and AI partnership guidance. These resources assist users in ethically and effectively using AI as a strategic partner for researching and implementing tools, solutions, resources, and user outcomes and deliverables (UODs) customized to their unique needs and contexts.

Forming Diverse "Smart Teams"

Rationale and Strategic Importance

Successfully guiding GenAI integration requires more than traditional committees; it necessitates dedicated, cross-functional "Smart Teams" (Harvard Business Publishing Education, 2025; Robert & McCormack, 2024). The rationale for these teams stems from both strategic need and established change management principles. Strategically, AI integration touches nearly every aspect of an HEI. Smart Teams provide focused leadership, facilitate collaborative problem-solving across silos, ensure initiatives align with institutional goals and ethical principles (HCASR Component 1), and manage specific facets of adoption (e.g., policy development, pilot oversight).

From a change management perspective (Kotter et al., 2021), Smart Teams function as the essential "Guiding Coalition." Bringing together diverse stakeholders—from academic affairs, IT, student services, administration, libraries, ethics bodies, and crucially, student representatives—fosters a holistic view vital for navigating AI's complexities (Jaskyte et al., 2024). This diverse coalition ensures initiatives incorporate both "top-down" strategic direction and "bottom-up" stakeholder needs, mitigating the risk of disconnected management strategies (Kotter et al., 2021). This inclusive approach builds shared ownership, surfaces potential challenges early, enhances agility alongside stability (Kotter's "Dual System"), and ultimately increases the likelihood of successful, sustainable integration (DerSimonian & Montagnino, 2025; NASPA, 2024).

Team Composition and Structure

While the optimal composition and structure should adapt to institutional context, certain principles are key. Diversity is paramount—not just functional diversity (different units) but also cognitive diversity (varied thinking styles, backgrounds, experiences), which research links to innovation (Jaskyte et al., 2024). Teams need a blend of technical expertise, pedagogical understanding, ethical awareness, operational knowledge, project management skills, and direct user experience (Harvard Business Publishing Education, 2025).

A common structure involves a central AI Council providing high-level strategic oversight, supported by specific working groups or Smart Teams focused on areas like academic integrity, research support, or administrative efficiency (Robert & McCormack, 2024). Flexibility is important, allowing structures to evolve as the institution's AI strategy matures. (See Section 5.B for more on AI Councils.)

Formal Leaders and Informal "AI Ambassadors"

Effective teams leverage both formal leadership and informal influence. Formal leaders (e.g., chairs, project managers) provide structure, ensure accountability, and connect the team to institutional governance (Harvard Business Publishing Education, 2025). Equally valuable are "AI Ambassadors"—individuals from various units who lack formal authority but possess interest, early adoption experience, peer credibility, and enthusiasm for AI (Monolith, n.d.).

These ambassadors serve as vital bridges, translating team work back to their home units, gathering grassroots feedback, championing local adoption, and identifying practical challenges or resistance early. Integrating formal leaders and informal AI Ambassadors creates a dynamic similar to Kotter's (2021) "volunteer army" supporting the guiding coalition, significantly enhancing communication, buy-in, and the team's effectiveness in driving change. Identifying and empowering these ambassadors is a key aspect of team formation.

Workbook Prep 2: Designing Smart Team Composition

Contextual Reflection: Does your current or planned AI strategy team adequately represent the "diverse many" (Kotter et al., 2021) needed for generating innovative ideas (Jaskyte et al., 2024) and building broad buy-in? Are diverse perspectives (cognitive, functional, demographic) and roles (including AI Ambassadors) sufficiently considered?

UX Workbook Link: See Chap. 7 UX Workbook, Activity 1.1 ("Plan Smart Team Composition & Structure") for templates (UX Workbook Table 7.1) and guidance on planning diverse and effective teams.

PALS Workbook Link & AI Partnership Guidance: Engage with Chap. 7 PALS Workbook prompts (Section 2). Describe your current/proposed team composition; ask your GenAI partner to identify potential underrepresented perspectives or missing expertise critical for your team's specific AI focus (referencing collaborative intelligence principles from Jaskyte et al. (2024) or HCASR Component 4).

Example Prompt*:* "Team Composition: [List members/roles/units]. Focus: [AI in student support]. Based on Jaskyte et al. (2024) on team innovation or HCASR Component 4, what are 2–3 potentially missing perspectives (e.g., specific student groups, frontline staff roles, accessibility experts) crucial for this focus?"

Human Oversight Reminder: AI suggestions are general. Assessing actual team composition impact, navigating internal politics, and ensuring genuine inclusivity require nuanced human judgment sensitive to your specific institutional context (HCASR Component 4).

Defining Roles, Responsibilities, and Team Charters

Developing the Smart Team Charter

For Smart Teams to function effectively, clarity regarding purpose, scope, goals, individual roles, and operating procedures is essential. Ambiguity can lead to confusion, duplicated effort, missed tasks, and reduced accountability (Asana, n.d.; Miro, n.d.). A Smart Team Charter serves as the foundational operating document, co-created by the team to establish this clarity and mutual understanding, guiding its work (Asana, n.d.; Miro, n.d.).

Collaboratively developing the charter is a critical team-building process that fosters cohesion and shared commitment (Harvard Business Publishing Education, 2025). The charter typically defines key elements outlining the team's operational framework (Robert & McCormack, 2024). The UX Workbook (Activity 2.1) provides guidance and templates (see UX Workbook Table 7.3 for steps). Table 7.1 presents a template outlining common sections for a comprehensive charter tailored to the team's specific needs (Fig. 7.1).

Table 7.1 Smart Team charter template

Section	Description/content prompt
1. Mission/purpose	"[State the primary reason for the team's existence. What problem does it solve or opportunity does it address regarding AI within the institution?]"
2. Scope	"[Define the boundaries of the team's work. What specific areas, processes, or AI applications are included? What is explicitly excluded?]"
3. Goals/objectives	"[List the specific, measurable outcomes the team aims to achieve. SMARTIE format (Specific, Measurable, Achievable, Relevant, Time-bound, Inclusive, Equitable) is recommended.]"
4. Team members and roles	"[List the names and primary roles of core team members. Specify the Chair/Lead. Note any designated AI Ambassadors or liaisons included.]"
5. Responsibilities	"[Outline the key duties and responsibilities of the team as a whole and, if applicable, specific roles within the team.]"
6. Decision-making process	"[Describe how the team will make decisions (e.g., consensus, majority vote, lead decides after consultation). Specify any escalation paths.]"
7. Operating norms	"[Define the team's ground rules and expectations for collaboration, meeting attendance, participation, conflict resolution, and respectful communication.]"

(continued)

Table 7.1 (continued)

Section	Description/content prompt
8. Communication plan	"[Describe how the team will communicate internally (e.g., meeting frequency, platform used) and externally (e.g., reporting, stakeholder updates).]"
9. Resources	"[Identify the necessary resources for the team to succeed, including budget, tools, technology, personnel support, and physical space, if applicable.]"
10. Reporting structure	"[Specify who the team reports to (e.g., AI Steering Committee, specific Dean or VP) and the expected frequency and format of reporting.]"
11. Initial deliverables/ tasks	"[List the key initial outputs, work products, or tasks the team is responsible for delivering, particularly in the near term (e.g., first 3–6 months).]"
12. Timeline/ milestones	"[Outline key dates, deadlines, phases, and checkpoints for the team's work and deliverables.]"
13. Success metrics	"[Define how the team's success and effectiveness will be measured. What does success look like for this team?]"
14. Ethical AI pledge	"[Include a statement affirming the team's commitment to upholding institutional ethical AI principles and responsible practices in all its work and recommendations.]"

Note. This template provides a structure for defining a Smart Team's purpose and operations. Sections can be adapted, combined, or expanded based on the specific needs, complexity, and context of the team and the institution

SMARTIE Goals and Ethical Pledges

Team charters should feature specific, measurable goals. The SMARTIE framework (Specific, Measurable, Achievable/Ambitious, Relevant, Time-bound, Inclusive, Equitable) ensures goals are well-defined and explicitly address inclusivity and equity (Learning Forward, 2021; Washington OSPI, 2025). This deliberate focus helps embed human-centric values into the team's objectives from the start.

Incorporating an "Ethical AI Pledge" within the charter is also highly recommended. This formally commits the team to upholding specific ethical principles—such as those derived from institutional frameworks (like the Chap. 6 V0 framework), HCASR Component 1, or external guides (e.g., Wynants et al., 2025)—in all its work and recommendations. This public commitment reinforces accountability and aligns team actions with institutional values.

Workbook Prep 3: Developing the Smart Team Charter

Contextual Reflection: How can a collaboratively developed Team Charter, defining mission, scope, roles, SMARTIE goals, operating norms, and ethical commitments, enhance your Smart Team's effectiveness, accountability, and alignment with institutional values and frameworks (e.g., HCASR, ethical principles from Chap. 6)?

UX Workbook Link: See Chap. 7 UX Workbook, Activity 2.1 ("Develop the Smart Team Charter (Co-creation)") for a detailed template (referenced in Table 7.1) and process guide (UX Workbook Table 7.3).

PALS Workbook Link & AI Partnership Guidance: Engage with Chap. 7 PALS Workbook prompts (Section 3). Input draft team goals; ask your GenAI partner to help evaluate them against SMARTIE criteria, suggesting refinements for clarity, measurability, and strengthening the Inclusive/Equitable dimensions (referencing HCASR Component 1). You might also ask it to research examples of team charter components or ethical pledge statements from other HEI AI initiatives.

Example Prompt*:* "Draft Goal: 'Explore AI tools for faculty'. Evaluate against SMARTIE criteria. Suggest 2 refined versions incorporating specific measures and inclusive/equitable considerations (HCASR Component 1)."

Human Oversight Reminder: While AI can refine language or provide examples, the charter's substance—mission, values, ethical commitments, processes—must emerge from genuine human collaboration and consensus (HCASR Component 4).

Section 3: Mapping Stakeholders for AI Initiatives

Understanding and engaging the diverse individuals and groups affected by or influencing AI initiatives requires systematic stakeholder mapping and analysis. This process develops a nuanced understanding of stakeholder perspectives, interests, and influence, which is fundamental for designing effective engagement strategies and ensuring alignment with the institutional context (Project Management Institute, n.d.; Borealis Software, 2024).

Identifying Stakeholders

The first step involves comprehensive brainstorming to identify all relevant internal and external stakeholders for the specific AI initiative (e.g., students, faculty by discipline/rank, staff by unit, leadership levels, community partners, vendors). Utilizing institutional knowledge, organizational charts, committee lists, and insights from knowledgeable individuals helps ensure completeness (Project Management Institute, n.d.; Harvard Business Publishing Education, 2025). The UX Workbook (Activity 1.2, leading into 3.1) guides this identification planning and execution process (UX Workbook Table 7.2).

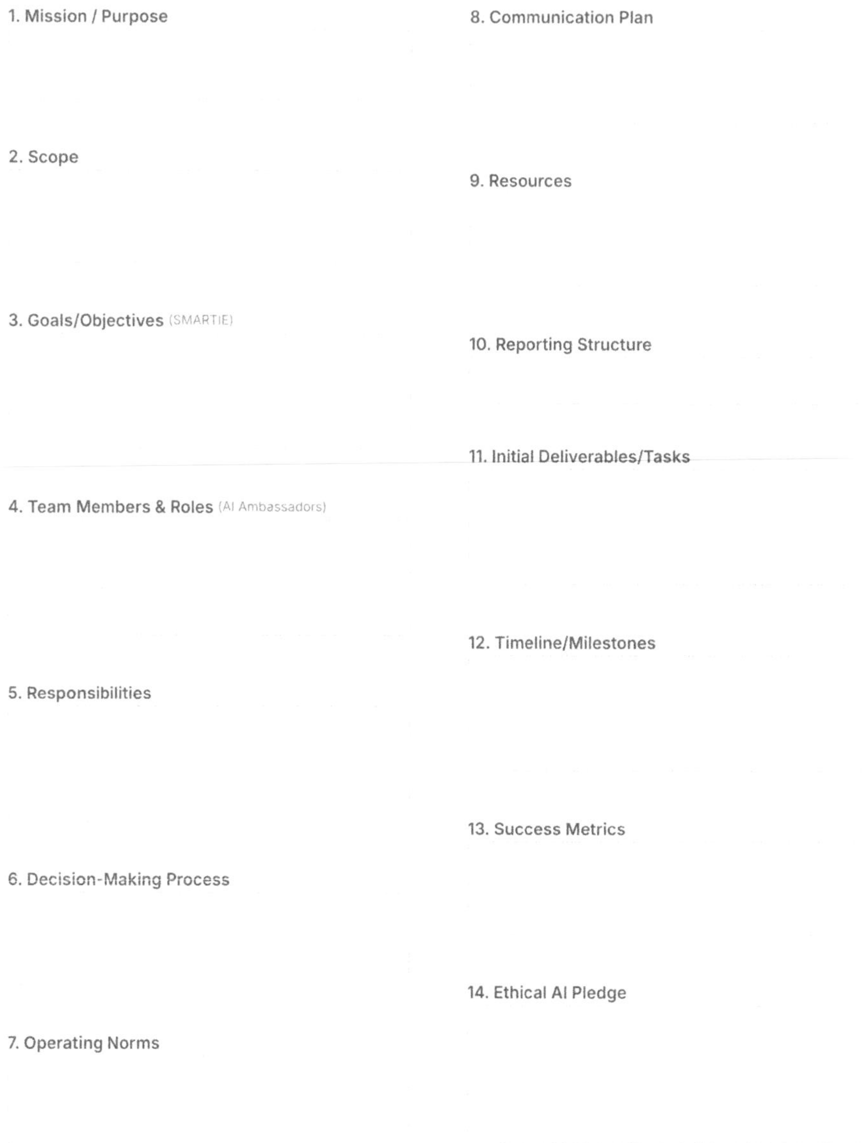

Fig. 7.1 Smart Team charter template—customizable

Analyzing Stakeholders (Interest, Influence, Stance)

Once identified, stakeholders need analysis to prioritize engagement efforts. Key dimensions include their level of **Interest** in the initiative and their **Influence** or power to affect it (Project Management Institute, n.d.; Borealis Software, 2024). Tools like the Power/Interest Grid ([Fig. 7.2]) help

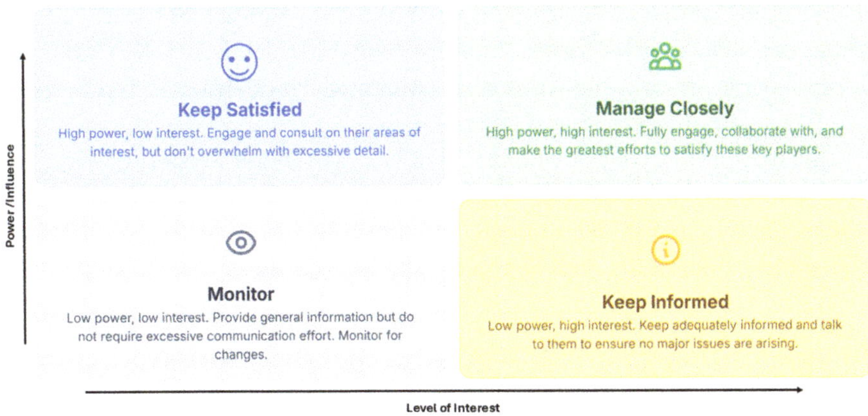

Fig. 7.2 Stakeholder power/interest grid—analysis tool

Stakeholder Group	Interest (H/M/L)	Influence (H/M/L)	Stance (+/-/?)	Key Interests & Concerns	Proposed Engagement
Senior Administrators	H	H	+	Policy alignment, risk management, institutional goals	**Manage Closely:** Direct involvement in AI Council, regular briefings
Faculty (Tenured)	H	H	?	Academic integrity, workload impact, pedagogical support	**Collaborate:** Involve in Smart Teams, targeted workshops, policy co-creation
Undergraduate Students	H	M	?	Fairness in assessment, data privacy, skill development	**Consult & Involve:** Representation on teams, focus groups, surveys
IT Staff	H	H	+	Security, data integration, vendor management, infrastructure readiness	**Collaborate:** Deep involvement in technical planning and vetting processes
Alumni	L	L	?	Institutional reputation, value of degree	**Monitor:** Keep informed via general newsletters, optional updates

Fig. 7.3 Stakeholder map example—adaptable format

categorize stakeholders to guide engagement strategy (e.g., high power/high interest stakeholders require close management).

Further analysis should explore their likely **Stance** (supportive, opposed, neutral) and specific **Concerns** or **Needs** (Borealis Software, 2024). Understanding these factors allows for more tailored and effective communication and engagement. The UX Workbook (Activity 3.2) provides tools and templates (UX Workbook Table 7.6) for this deeper analysis, potentially incorporating empathy maps and persona refinement (Activity 3.3, UX Workbook Figs. 7.2 and 7.3).

Creating the Stakeholder Map

The results of identification and analysis are consolidated into a Stakeholder Map. This serves as a reference tool for the Smart Team, documenting key stakeholders, their attributes (interest, influence, concerns), and potential relationships (Project Management Institute, n.d.; Borealis Software, 2024). Maps can take various forms, such as detailed tables (see UX Workbook Table 7.6 for an example structure), annotated Power/Interest Grids ([Fig. 7.2]), or relationship diagrams ([Fig. 7.3]).

This map is a "living document" (see UX Workbook Section 6, Fig. 7.4) that should be updated as the project progresses and the stakeholder landscape evolves. The UX Workbook (Activities 3.1–3.3) provides detailed templates and guidance for these mapping and analysis steps.

AI as a Strategic Partner in Contextual Adaptation

Every HEI operates within a unique sociocultural context, meaning no single approach to AI integration fits all institutions (Carmody-Bubb, 2023). As institutions globally adopt GenAI, contextual data—gathered ethically through stakeholder analysis and engagement—becomes invaluable. It can

Workbook Prep 4: Understanding Stakeholder Perspectives

Contextual Reflection: How can visually mapping stakeholders (e.g., using a Power/Interest Grid ([Fig. 7.2]) or relationship map ([Fig. 7.3])) help your Smart Team prioritize engagement efforts and anticipate potential dynamics related to your specific AI initiative? What other stakeholder analysis frameworks might be relevant in your context?

UX Workbook Link: See Chap. 7 UX Workbook, Section 3 activities ("Stakeholder Identification & Mapping," "Stakeholder Analysis," "Applying Empathy & Personas") for detailed guidance and templates (UX Workbook Tables 7.2, 7.6; Figs. 7.1, 7.2, and 7.3).

PALS Workbook Link & AI Partnership Guidance: Engage with Chap. 7 PALS Workbook prompts (Section 4). Input your identified stakeholder list and initial interest/influence assessment; ask your GenAI partner to help categorize them (e.g., into Power/Interest quadrants) or research alternative stakeholder analysis models (e.g., Salience Model). Use AI ethically (see HAICS Guide) to brainstorm probing questions for deeper analysis of key stakeholder concerns.

Example Prompt*:* "Stakeholders assessed: [List with H/M/L interest/influence]. Categorize into Power/Interest Grid quadrants and briefly explain engagement focus for each. Also, suggest one alternative stakeholder analysis framework we might explore."

Human Oversight Reminder: AI categorization relies solely on your input. Validating stakeholder interest/influence, understanding nuanced relationships, and interpreting analysis results require human insight and contextual knowledge. Frameworks are tools; choose and adapt what works best (HCASR Component 1 & 4).

guide the tailoring of tools, resources, and solutions, and help mitigate inherent biases (e.g., Western-centric perspectives) as models are deployed worldwide. Leveraging AI as a strategic partner, guided by human-centric frameworks like HCASR (see HAICS Guide), can assist in analyzing this contextual data to inform culturally relevant team formation, stakeholder engagement, and governance strategies.

Planning for Stakeholder Engagement

Effective stakeholder mapping provides the necessary foundation for developing a tailored Stakeholder Engagement Plan. Proactive and strategic engagement builds trust, fosters buy-in, manages expectations, gathers valuable feedback, mitigates resistance, and ensures alignment between AI initiatives and stakeholder needs (DerSimonian & Montagnino, 2025; Partnership on AI, 2024).

Tailoring Engagement Strategies

Engagement strategies must be customized based on stakeholder analysis (Interest/Influence, Stance, Needs). Different stakeholders require different levels and types of engagement, potentially ranging from simply keeping them Informed to actively Consulting, Involving, or Collaborating with them (IAP2, n.d.; Bammer, 2020). High-influence, high-interest stakeholders typically require collaboration or close involvement, while low-influence, low-interest groups might only need to be informed. The engagement plan outline developed in the UX Workbook (Activity 2.2, UX Workbook Table 7.4), refined using Section 3 outputs, details specific strategies tailored to key stakeholder groups.

Communication Channels and Feedback Loops

A robust engagement plan specifies clear communication channels appropriate for different groups (e.g., town halls, targeted emails, website updates, workshops, committee representation) and establishes effective two-way feedback loops (e.g., dedicated email address, online feedback forms, regular consultations with representatives) (Borealis Software, 2024; Co-Action, Penn State University, n.d.). Crucially, demonstrating responsiveness to feedback is vital for building and maintaining trust throughout the AI integration process (DerSimonian & Montagnino, 2025). The UX workbook guides the outlining of these channels and mechanisms (Activity 2.2, UX Workbook Table 7.4).

Workbook Prep 5: Planning Engagement Tactics & Feedback

Contextual Reflection: How can insights from stakeholder mapping and analysis inform the selection of appropriate engagement levels (e.g., Inform, Consult, Involve, Collaborate via IAP2 spectrum) and communication methods for different groups? How can you ensure feedback loops are genuinely two-way and lead to actionable responses?

UX Workbook Link: See Chap. 7 UX Workbook, Activity 2.2 ("Draft Initial Stakeholder Engagement Plan Outline," UX Workbook Table 7.4) and Activity 2.3 ("Conduct Initial Team & Stakeholder Process Risk Identification," UX Workbook Table 7.5) for planning engagement and anticipating challenges.

PALS Workbook Link & AI Partnership Guidance: Engage with Chap. 7 PALS Workbook prompts (Section 5). Input key stakeholder categories and their analysis summaries; ask your GenAI partner to brainstorm tailored engagement tactics or communication channel ideas for specific groups, referencing engagement frameworks (e.g., IAP2) or HCASR Component 4 principles. Consider asking it to research culturally relevant communication strategies if applicable.

Example Prompt: "For 'Mid-Career Faculty' (High Influence/High Interest, Concern: Workload) and 'Undergraduate Students' (Medium Influence/High Interest, Concern: Fairness), brainstorm 2–3 tailored engagement tactics for Phase 1, referencing HCASR Component 4 or IAP2 levels."

Human Oversight Reminder: AI suggests tactics; choosing effective, culturally appropriate methods requires human understanding of stakeholders, institutional norms, and engagement goals. Authentic engagement needs empathy and responsiveness (HCASR Component 4).

Section 5: Integrating Key Foundational Concepts

The formation of Smart Teams and stakeholder mapping connect to broader organizational principles crucial for successful AI integration: balancing leadership and management, adapting governance structures, and tailoring governance approaches based on AI sourcing.

Balancing Leadership and Management

Navigating the significant changes accompanying AI adoption requires balancing strong **leadership** (setting vision, inspiring action, driving change)

Table 7.2 Leadership versus management focus in AI integration (Kotter's dual system)

Aspect	Leadership role	Management role
Primary focus	Vision, change, innovation, inspiration, "Why?"	Stability, efficiency, process, risk management, compliance, "How?"
Drives	Agility, experimentation, navigating uncertainty	Reliability, predictability, operational execution
Key concern	Aligning AI with mission, fostering adoption, motivation	Establishing clear processes, resource allocation, managing risks
Example activities	Articulating vision, encouraging new ideas, building coalitions	Setting budgets, developing policies, monitoring progress, ensuring compliance (e.g., via NIST AI RMF)
Relation to Smart Teams	Empowers and inspires teams, connects to broader vision	Provides structure, resources, and oversight for team operations

Note. Effective AI integration requires balancing both visionary leadership and pragmatic management, creating a "dual system" where innovation and operational stability coexist (Kotter et al., 2021; DerSimonian & Montagnino, 2025)

with effective **management** (planning, budgeting, organizing, problem-solving). Kotter's (2021) "dual system" concept highlights this need: a traditional hierarchy focused on stability and efficiency must coexist with a more agile network structure (like Smart Teams and AI Ambassadors) focused on innovation and adaptation. This balance ensures AI initiatives are strategically guided and operationally sound. Table 7.2 summarizes these distinctions in the context of AI integration.

Adapting Teams and Governance

AI integration is an evolving process, requiring adaptable team structures and governance models. Initial Smart Teams might focus on specific pilots or policies. As institutional experience grows and AI use becomes more widespread, these structures may evolve into a more formalized **AI Council** or standing committee (Robert & McCormack, 2024; Wu et al., 2024).

Such councils typically provide ongoing strategic oversight, coordinate efforts, review ethical considerations, ensure policy alignment, and advise senior leadership (IAPP & Credo AI, 2025). They work synergistically with operational Smart Teams, leveraging team insights while providing direction and oversight, thus helping operationalize the dual system ([Table 7.3]). This iterative approach allows governance to mature alongside the institution's AI capabilities. Figure 7.4 illustrates a potential structure, and Fig. 7.5 depicts the iterative feedback loop essential for team and governance refinement.

Table 7.3 AI council/formal governance committee overview

Aspect	Description
Primary purpose	Provide high-level, formal oversight for institutional AI initiatives; distinct from operational Smart Teams.
Key responsibilities	Monitor ethical considerations; oversee AI policy development/revision; manage institutional AI risk; ensure compliance; promote coherence across initiatives; ensure accountability.
Typical membership	Senior leaders and representatives from critical areas: Legal Counsel, IT/Security (CIO), Academic Affairs (Provost Office), Research Admin, Equity Office, Faculty Governance, Student Representation, Ethics Officer.
Relationship w/ Smart Teams	Works synergistically: leverages Smart Team insights while providing strategic direction, policy frameworks, and oversight. Helps operationalize the "dual system."

Note. The formation of such a council often occurs as AI initiatives mature, providing necessary structure for managing complex ethical, legal, and operational issues at an institutional level (Brady, 2024; Wu et al., 2024)

Fig. 7.4 AI governance structure example

Governance for Proprietary vs. Third-Party AI

Governance must also adapt based on whether the institution is primarily developing its own proprietary AI systems or integrating third-party solutions. **Proprietary development** demands strong internal governance focused on ethical design, data handling, model validation, infrastructure, and internal risks (Dahlgren, 2025). **Integrating third-party tools** requires robust governance focused on vendor vetting (security, privacy, compliance, ethics), contract negotiation, performance monitoring, managing dependencies, and

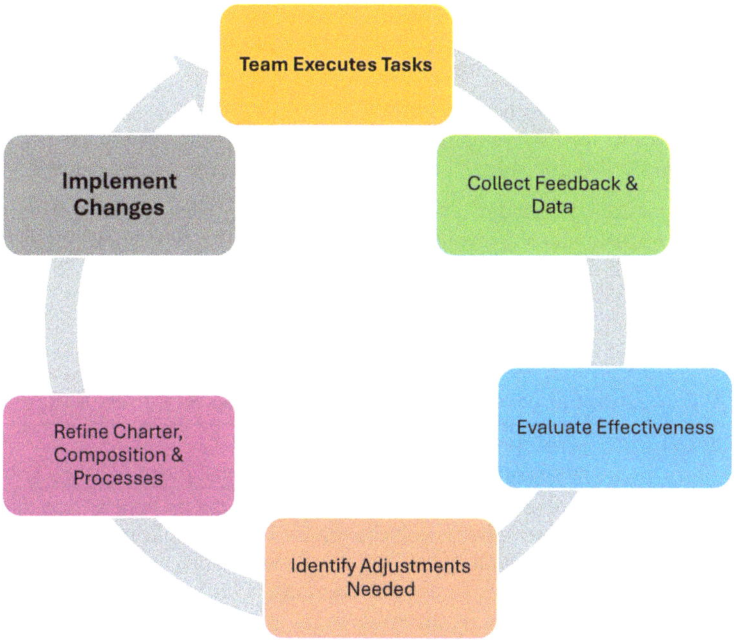

Fig. 7.5 Feedback loop for Smart Team refinement

clear data usage agreements (Grajek et al., 2025). Hybrid approaches need both internal process governance and external vendor management. Table 7.4 highlights key differences in governance emphasis. Procurement processes often require significant revision to handle the complexities of third-party AI tools effectively.

Workbook Prep 6: Planning Governance Evolution

Contextual Reflection: Does your institution have a high-level committee overseeing tech strategy, ethics, or risk? How could its mandate be adapted, or how might a new AI Council ([Table 7.3, Fig. 7.4]) be formed, for effective AI governance? What initial steps can ensure synergy between operational Smart Teams and this central oversight body (e.g., reporting lines, shared members)?

UX Workbook Link: See Chap. 7 UX Workbook, Activity 2.2, which involves identifying the AI Council (or equivalent) as a key stakeholder and planning communication. Subsequent chapters explore governance roles further.

PALS Workbook Link & AI Partnership Guidance: Use the Chap. 7 PALS Workbook prompt corresponding to UX Workbook Section 2 (Actions: Preparing for Engagement and Governance). Input potential AI Council responsibilities; ask your AI partner to draft charter elements (Scope, Membership Criteria, Reporting) based on best practices or the NIST AI RMF "Govern" function (NIST, 2023), and inquire how governance structures might adapt to different institutional cultures (e.g., shared vs. centralized).

Example Prompt: "Based on NIST AI RMF 'Govern' (NIST, 2023) & HEI needs, draft 3–4 bullet points for the 'Mandate/Responsibilities' section of an AI Council charter. How might this adapt for shared vs. centralized governance models?"

Human Oversight Reminder: Designing effective governance requires understanding institutional culture, politics, and processes. Use AI suggestions as starting points for discussion to ensure synergy, not bureaucracy (Kotter et al., 2021; HCASR C4 & 5). The goal is a functional, context-specific structure.

Table 7.4 Comparison of governance considerations: proprietary versus third-party AI

Governance focus area	Proprietary (in-house) AI development	Third-party AI integration
Data privacy and security	Focus on internal protocols, secure infrastructure, data handling	Focus on vendor due diligence, data sharing agreements, audits
IP rights	Managing internal IP creation, ownership, potential licensing	Clarifying IP ownership in vendor contracts, usage rights
Ethical review process	Internal review boards, validating models, internal guidelines	Evaluating vendor's ethical framework, transparency, alignment
Bias mitigation strategy	Internal testing, diverse datasets, algorithmic audits	Assessing vendor's methods, requesting bias reports, monitoring
Team expertise needed	Technical (AI/ML, MLOps), ethics, domain experts	Vendor evaluation, contract negotiation, integration, compliance
Vendor management/due dilig.	N/A (or minimal for underlying components)	Critical: rigorous vetting (ethics, privacy, security, bias)
Liability allocation	Primarily internal institutional responsibility	Shared responsibility; defined in contracts, potential vendor risk
Transparency requirements	Ensuring internal model/process transparency where feasible	Demanding vendor transparency on data use, algorithms (often limited)
Cost oversight	Managing development, infrastructure, personnel, maintenance costs	Managing subscription fees, integration costs, vendor price changes
Compliance (e.g., FERPA)	Ensuring internal systems and processes comply	Ensuring vendor practices and contracts meet compliance standards

Note. This table highlights differing areas of emphasis in governance depending on the AI sourcing strategy. Effective governance requires addressing these points appropriately based on the chosen approach(es). Procurement processes often need revision, especially for third-party AI (Dahlgren, 2025; Grajek et al., 2025)

User Outcomes and Deliverables (UODs) & Initial Team Actions

Completing the activities in this chapter and its companion UX Workbook yields essential UODs that structure the human elements of the AI initiative. These include process deliverables defining how teams operate and engage, and initial work product scopes outlining the first substantive tasks for the Smart Teams. Table 7.5 summarizes these key Chap. 7 outputs.

Table 7.5 Chapter 7 key deliverables summary

Deliverable type	Deliverable name	Purpose/function	Potential inputs	Link to next steps
Process	Smart Team Charter(s)	Formalizes team purpose, scope, roles, operations	Team collaboration, Chap. 6 outputs	Guides team ops, establishes accountability
Process	Stakeholder Map and Analysis	Identifies/ analyzes stakeholders, interests, influence	Team brainstorming, Chap. 6 analysis	Informs Engagement Plan, Risk Assessment
Process	Stakeholder Engagement Plan	Outlines tailored communication/ involvement strategies and feedback loops	Stakeholder map, team collaboration	Guides communication, supports Chap. 8 data gathering
Initial work product	Ethical AI Policies (V1—Scoped)	Initial draft guidelines for priority ethical concerns within team scope	Team Charter, principles (Chap. 6), stakeholder input	Refined in Chap. 8/Chap. 9, guides pilot design
Initial work product	AI Risk Mgmt. Guide (Initial Scope—V1)	Outline of scope, components, process for risk guide (e.g., using NIST AI RMF)	Team Charter, NIST AI RMF, stakeholder concerns	Developed further in Chap. 8/Chap. 9, informs governance
Initial work product	Stakeholder Use Cases Inventory (V1)	Documents user needs, tasks, goals, challenges, AI interactions, risks	Stakeholder map, team research, initial feedback	Informs Chap. 8 needs assessment, pilot requirements

Note. This table summarizes potential deliverables from Chap. 7 activities. Institutions should adapt these based on their specific needs and context

Process Deliverables

- **Smart Team Charter(s):** Formal documents defining mission, scope, SMARTIE goals, roles (incl. Ambassadors), responsibilities, operating norms, and ethical commitments (UX Workbook Activity 2.1, guided by [Table 7.1]).
- **Stakeholder Map & Analysis:** Comprehensive identification and analysis of relevant stakeholders (interest, influence, stance, concerns, needs), potentially visualized (UX Workbook Activities 3.1–3.3, guided by [Figs. 7.2, and 7.3]).
- **Stakeholder Engagement Plan Outline:** Initial plan detailing tailored strategies, communication methods, and feedback loops for key groups (UX Workbook Activity 2.2, guided by UX Workbook [Table 7.4]).

Initial Work Product Scopes (Version 1)

Newly formed Smart Teams typically begin by scoping their first major deliverables. These are explicitly "Version 1" or "Initial Scope" living documents, designed for refinement:

- **Ethical AI Policies (V1 - Scoped):** Draft concepts or initial policy statements addressing priority ethical concerns within the team's scope (UX Workbook Activity 4.1, example UX Workbook [Table 7.7]).
- **AI Risk Management Guide (Initial Scope - V1):** An outline defining the intended structure, relevant risk categories (potentially using NIST AI RMF (NIST, 2023)), and development approach for a practical risk guide tailored to the team's focus (UX Workbook Activity 4.2, example UX Workbook [Table 7.8]).
- **Stakeholder Use Cases Inventory (V1):** A consolidated inventory documenting user needs, tasks, challenges, potential AI interactions, and related ethical/risk considerations relevant to the team's scope (UX Workbook Activity 4.3, example UX Workbook [Table 7.9]).

These UODs provide structure and context-awareness for Smart Teams. The clarity of purpose and interrelation of these UODs (Charters, Maps, Engagement Plan Outline, V1 Scopes) are critical. They directly feed into the needs assessment detailed in Chap. 8, ensuring that subsequent analysis is grounded in well-defined team structures, stakeholder understanding, and clearly scoped initial objectives.

Workbook Prep 7: Reviewing Deliverables & Scoping Initial Work

Contextual Reflection: Review the key process deliverables (Charter, Map, Engagement Plan Outline) and the initial V1 work product scopes (Ethical Policies, Risk Guide, Use Cases) described above ([Table 7.5]) and detailed in the UX Workbook (Sections 4, 8). How effectively do these initial documents capture the necessary foundation for your specific AI initiative? Where might further clarification or detail be needed before proceeding to Chap. 8's needs assessment?

UX Workbook Link: See Chap. 7 UX Workbook, Section 4 ("Initial Work Product Scoping," Activities 4.1–4.3, Tables 7.7–7.9) for guidance on defining the V1 scopes, and Section 8 ("Deliverables Summary," Table 7.12) for consolidating outputs.

PALS Workbook Link & AI Partnership Guidance: Use the Chap. 7 PALS Workbook prompt corresponding to UX Workbook Section 7 (Deliverables: Finalizing Team Structures & Initial Maps). Input your draft Charter summary or key Stakeholder Map findings; ask your AI partner to suggest ways to ensure coherence across deliverables or help structure the V1 work product scopes (Policy, Risk, Use Cases) for clarity, referencing HCASR Component 4 & 5 principles.

Example Prompt: "Based on my draft Smart Team Charter mission '[Insert Mission]' and key stakeholder concerns '[List 1–2 Concerns]', help structure the V1 scope for an 'Ethical AI Policies' document, ensuring it addresses the concerns and aligns with HCASR C4."

Human Oversight Reminder: AI can assist structuring documents and checking textual alignment. However, ensuring the substance of UODs (charters, maps, scopes) truly reflects institutional context, ethical commitments, and stakeholder realities require critical human review, team consensus, and leadership validation (HCASR C4 & C5). These are foundational documents guiding subsequent work.

Long-Term Scaling, Sustainability, & AI Governance: Building Collaborative TLARA Foundations

Chapter 7 emphasizes that establishing diverse Smart Teams and conducting thorough stakeholder mapping are crucial for successful AI integration, forming the human infrastructure necessary for lasting change (Kotter et al., 2021). These steps ensure that initiatives for ethical AI are not top-down mandates but are co-created and understood across different **levels** and **spaces** within the institution. The insights gained pave the way for developing **TLARA (Teaching, Learning, Assessment, Research, Administration)** guidelines that are contextually relevant and support the institution's unique **sociocultural context** and mission (DerSimonian & Montagnino, 2025). Both the Chap. 7 UX and PALS Workbooks are designed to help you ethically partner with AI to optimize these collaborative processes.

HEIs are intricate systems, and their journey with AI requires a holistic approach that considers the interconnectedness of all parts over time (IDEAS, 2025). The PALS Workbook for this chapter, through its section "Long-Term Scaling, Sustainability & AI Governance: Iteratively Developing Your Custom TLARA-Informed AI Guidelines," offers a customizable prompt that positions your GenAI as an expert advisor. This tool facilitates using your strategic goals, stakeholder insights, and team structures to further build your TLARA frameworks, focusing on how collaborative efforts can translate into effective AI governance and promote AI literacy (Ellucian, 2024). This iterative development ensures your guidelines are robust and broadly supported.

Conclusion: Looking Ahead & Engaging the Workbook

This chapter addressed the critical tasks of forming diverse Smart Teams and systematically mapping stakeholders—essential steps in building the human foundation for successful AI integration. By establishing clear team structures and charters (Section 2) and developing a nuanced understanding of the stakeholder environment (Section 3), institutions create the collaborative capacity needed to guide AI initiatives responsibly (Section 1). Planning for tailored stakeholder engagement ensures diverse perspectives inform the process, fostering buy-in and mitigating resistance (Section 4). These activities, integrated with concepts of leadership, management, and adaptable governance (Section 5), produce key process UODs and initial work product scopes (Section 6, [Table 7.5]).

Having established this human infrastructure, the next step in Phase 1 involves conducting detailed needs assessments and readiness evaluations, as detailed in Chap. 8. The UODs produced through the Chap. 7 UX Workbook—including adopted Team Charters, Stakeholder Maps, Engagement Plan outlines, and initial V1 scopes for policies, risks, and use cases—serve as direct inputs for this subsequent assessment work. Engaging thoroughly with the Chap. 7 UX and PALS Workbooks enables Smart Teams to consolidate their foundation and prepare effectively for the comprehensive analysis required in the next stage of planning.

References

Asana. (n.d.). Team charter template: A roadmap for team success (with examples). Asana Resource Center. Retrieved May 5, 2025, from https://asana.com/resources/team-charter-template

Bammer, G. (2020, January 7). Stakeholder engagement in research: The research-modified IAP2 spectrum. Integration and Implementation Insights. https://i2insights.org/2020/01/07/research-modified-iap2-spectrum/

Borealis Software. (2024, December 10). Stakeholder mapping: When, why, and how to map stakeholders. *Borealis Blog*. https://www.boreal-is.com/blog/stakeholder-mapping-identify-stakeholders/

Brady, C. (2024). *The transformative potential of artificial intelligence: Recommendations for student affairs leaders*. NASPA – Student Affairs Administrators in Higher Education.

Brown, T. (2019). *Change by design: How design thinking transforms organizations and inspires innovation* (Rev. and updated ed.). Harper Business.

Carmody-Bubb, M. (2023). *Cognition and decision making in complex adaptive systems: The human factor in organizational performance*. Springer.

Co-Action, Penn State University. (n.d.). *Framework for stakeholder inclusion in the technology planning process*. Retrieved May 5, 2025, from https://coaction.psu.edu/inclusion-framework-phases/

Dahlgren, B. (2025, February 27). How university leaders can ethically and responsibly implement AI. *Campus Technology*. https://campustechnology.com/articles/2025/02/27/how-university-leaders-can-ethically-and-responsibly-implement-ai.aspx

DerSimonian, R., & Montagnino, C. (2025, March 26). Crafting thoughtful AI policy in higher education: A guide for institutional leaders. *Campus Technology*. https://campustechnology.com/articles/2025/03/26/crafting-thoughtful-ai-policy-in-higher-education-a-guide-for-institutional-leaders.aspx

Ellucian. (2024, December 16). AI in higher education: Implementation strategies for institutions and tech partners. *Higher Ed Dive*. https://www.highereddive.com/spons/ai-in-higher-education-implementation-strategies-for-institutions-and-tech/735434/

Grajek, S., Pelletier, K., & Freeman, A. (2025, March 11). AI procurement in higher education: Benefits and risks of emerging tools. *EDUCAUSE Review*. https://er.educause.edu/articles/2025/3/ai-procurement-in-higher-education-benefits-and-risks-of-emerging-tools

Harvard Business Publishing Education. (2025, April 3). Developing your institution's AI policy. *Inspiring Minds*. https://hbsp.harvard.edu/inspiring-minds/guidelines-effective-ai-policy

IAPP & Credo AI. (2025, April). *AI Governance Profession Report 2025*. International Association of Privacy Professionals. https://iapp.org/resources/article/ai-governance-profession-report/

Instructional Design, Engagement, and Support (IDEAS). (2025, January 27). AI's impact on higher education through the eyes of faculty: 4 key trends. University of Massachusetts Amherst. https://www.umass.edu/ideas/news/ais-impact-higher-education-through-eyes-faculty-4-key-trends

International Association for Public Participation (IAP2). (n.d.). *IAP2 spectrum of public participation*. Retrieved May 5, 2025, from https://cdn.ymaws.com/www.iap2.org/resource/resmgr/pillars/spectrum_8.5x11_print.pdf

Jaskyte, K., Hunter, A., & Mell, A. C. (2024). Predictors of interdisciplinary team innovation in higher education institutions. *Innovative Higher Education, 49*(1), 113–132. https://doi.org/10.1007/s10755-023-09676-3

Kotter, J. P., Akhtar, V., & Gupta, G. (2021). *Change: How organizations achieve hard-to-imagine results in uncertain and volatile times*. Wiley.

Learning Forward. (2021, December). Set a SMARTIE goal to target inclusion and equity. *The Learning Professional, 42*(6), 62. https://learningforward.org/journal/leading-for-equity/set-a-smartie-goal-to-target-inclusion-and-equity/

Miro. (n.d.). Team charter: The ultimate guide — With examples! *Miro Blog*. Retrieved May 5, 2025, from https://miro.com/organizational-chart/what-is-a-team-charter/

Monolith. (n.d.). How to become the AI champion for your organisation? *Monolith Blog*. Retrieved May 5, 2025, from https://www.monolithai.com/blog/ai-engineering-champion

NASPA - Student Affairs Administrators in Higher Education. (2024, December). *The transformative potential of AI*. https://www.naspa.org/files/dmfile/NASPA-Report%2D%2D-The-Transformative-Potential-of-AI%2D%2DDecember-2024.pdf

National Institute of Standards and Technology. (2023). *Artificial Intelligence Risk Management Framework (AI RMF 1.0)* (NIST AI 100-1). U.S. Department of Commerce. https://doi.org/10.6028/NIST.AI.100-1

Partnership on AI. (2024, July 25). AI needs inclusive stakeholder engagement now more than ever. *Partnership on AI Blog*. https://partnershiponai.org/ai-needs-inclusive-stakeholder-engagement-now-more-than-ever/

Project Management Institute. (n.d.). A guide to the project management body of knowledge (PMBOK® Guide) – Seventh edition and the standard for project management.

Robert, J., & McCormack, M. (2024, May 23). *2024 EDUCAUSE action plan: AI policies and guidelines*. EDUCAUSE. https://www.educause.edu/research/2024/2024-educause-action-plan-ai-policies-and-guidelines

Washington Office of Superintendent of Public Instruction (OSPI). (2025, March). *2025–2026 Step-by-step school improvement planning and implementation guide.*

https://ospi.k12.wa.us/sites/default/files/2024-03/step-step-sip-planning-and-implementation-guide.pdf

Wu, D., Dempsey, P., Lin, C.-Y., Sharif, A., & Miller, L. (2024). Charting the course for AI governance in higher education: Insights from a system-wide task force. *Journal of Postsecondary Education and Disability, 37*(1), 7–24.

Wynants, S., Childers, G., De La Torre Roman, Y., Budar-Turner, D., & Vasquez, P. (2025, February). *ETHICAL principles AI framework for higher education.* CSU AI Commons, California State University, Fullerton Faculty Development Center. https://genai.calstate.edu/communities/faculty/ethical-and-responsible-use-ai/ethical-principles-ai-framework-higher-education

8

Assessing Institutional Readiness and Capacity (Phase 1)

Introduction: Understanding the Starting Point

Following the formation of Smart Teams and initial stakeholder mapping (Chap. 7), this chapter details the next critical step in Phase 1: conducting a thorough Needs Assessment and AI Readiness Evaluation. This assessment phase is essential, providing the data-driven foundation for developing the institution's AI vision, governance framework, and strategic roadmap (Chap. 9). Proceeding without understanding stakeholder needs and institutional capabilities risks misaligned or ineffective strategies (U.S. Department of Education, Office of Educational Technology [OET], 2024; Wynants et al., 2025). This chapter outlines systematic processes for gathering this vital information.

The assessment involves two interconnected inquiries. The **Needs Assessment** focuses on understanding stakeholder challenges, goals, and contexts regarding potential AI applications, employing human-centered methods aligned with design thinking's empathize and define stages (Brown, 2019; U.S. Department of Education, OET, 2024). The **AI Readiness Evaluation** assesses institutional capacity across key dimensions like data, technology, culture, skills, and governance, potentially using structured tools like the EDUCAUSE AI readiness assessment to understand the current state (EDUCAUSE, 2024; U.S. Department of Education, OET, 2024; Kotter et al., 2021). Together, these assessments inform the strategic planning detailed in Chap. 9.

Supplementary Information The online version contains supplementary material available at https://doi.org/10.1007/978-3-032-05205-6_8.

D. Woodring, *Advancing Higher Education with Human-Centric AI Partnerships*,
https://doi.org/10.1007/978-3-032-05205-6_8

Workbook Prep 1: The Importance of Assessment

Contextual Reflection: Why are both needs assessment (understanding stakeholder needs) and readiness evaluation (understanding institutional capacity) essential before crafting an AI strategy (Chap. 9)? Consider the risks of misalignment or failure highlighted by change management principles (Kotter et al., 2021) if the current state isn't understood, or the risks of ineffective solutions if user needs aren't centered (Brown, 2019).

UX Workbook Link: See Chap. 8 UX Workbook "Getting Started" exercises and Section 1 ("Planning Needs Assessment & Readiness Evaluation") activities, which focus on defining the scope and selecting frameworks for your assessment.

PALS Workbook Link & AI Partnership Guidance: Engage with Chap. 8 PALS Workbook prompts (Section 1). Based on your institution's preliminary AI priorities, ask your GenAI partner (using HCASR Component 1; see HAICS Guide) to help brainstorm 2–3 specific, measurable assessment objectives for Chap. 8 that will yield data relevant to those priorities.

Example Prompt: "Priorities: [Improve student support equity], [Enhance research efficiency]. Help draft 3 specific Chap. 8 assessment objectives (using SMARTIE principles where applicable) to gather data relevant to these priorities, referencing HCASR Component 1."

Human Oversight Reminder: AI can assist in drafting objectives. However, defining strategic, feasible assessment objectives that yield actionable, contextually relevant data requires human judgment, institutional knowledge, and alignment with ethical considerations (HCASR Component 1 & 4).

Core Frameworks: The frameworks mentioned below, such as HCASR, HCAS, RSA, EHCS, EAO, and AHIPs, are detailed in the companion *Human-GenAI Collaboration & Synergy (HAICS) Guide* and/or Part 1 of this textbook. Please refer to those sections for full definitions and explanations.

This chapter applies principles from the synergy between design thinking (Brown, 2019) ("Empathize," "Define" stages) and change management (Kotter et al., 2021) (Understanding the current state). These frameworks guide the assessment of needs (HCASR C1), evaluation of readiness, identification of competency gaps (EHCS, EAO), and setting initial ethical goals (HCASR C1) to inform strategic planning.

The User Experience (UX) Workbook and the Personalized Adaptive Learning Sessions (PALS) Workbook for Chap. 8 provide practical activities and AI partnership guidance. These resources assist users in ethically and effectively using AI as a strategic partner for researching and implementing tools, solutions, resources, and User Outcomes and Deliverables (UODs) customized to their unique needs and contexts.

Understanding Stakeholder Needs, Goals, and Context

A thorough needs assessment requires understanding the goals, workflows, challenges, and contexts of those potentially affected by AI initiatives. This human-centered approach ensures that subsequent AI solutions address actual problems and opportunities rather than assumed ones.

Identifying Needs & Goals Through Engagement

Gathering input from diverse stakeholders—including students, faculty, staff, and administrators—is essential for establishing a comprehensive understanding of needs related to AI (U.S. Department of Education, OET, 2024; Wynants et al., 2025). Utilizing multiple data collection methods provides a richer picture. Common methods, detailed further in Table 8.1 and the UX

Table 8.1 Methods for stakeholder needs assessment in AI initiatives

Method	Description/purpose	Examples of information gathered
Surveys	Broadly distribute to gauge baseline perceptions and needs across large groups.	Perceived needs, current challenges, attitudes toward AI, digital literacy levels, desired functionalities, initial concerns.
Focus groups and interviews	Conduct targeted discussions for in-depth exploration with specific stakeholder segments.	Detailed workflows, comfort levels with tech, specific ethical concerns (privacy, bias), desired support, nuanced perspectives.
Workflow analysis	Observe or collaboratively map existing processes where AI might be implemented.	Current process steps, pain points, bottlenecks, time sinks, opportunities for AI augmentation or optimization.
PALS/feedback log analysis	Analyze existing feedback logs or collect real-time experiences and challenges related to current tools or potential AI uses.	Real-time challenges, user-generated ideas, friction points in current systems, context-specific needs and workarounds.
User experience (UX) research	Apply methods like persona refinement or user journey mapping to empathize with stakeholder experiences.	Detailed user profiles (personas), step-by-step user experiences (journey maps), unmet needs, points of friction, emotional responses.

Note. Utilizing a combination of these methods provides a more comprehensive understanding of diverse stakeholder needs, goals, challenges, and contexts, ensuring AI initiatives are grounded in reality (Source: U.S. Department of Education, OET, 2024; Wynants et al., 2025)

Workbook Prep 2: Selecting Needs Assessment Methods

Contextual Reflection: For understanding the needs of a specific group (e.g., part-time faculty, online students) regarding a potential AI initiative, which 1–2 needs assessment methods described in Table 8.1 seem most appropriate and feasible in your context? Why? Are there other methods worth exploring?

UX Workbook Link: See Chap. 8 UX Workbook, Section 1 (Activity 1.3 "Plan Data Gathering Methods") and Section 2 ("Actions: Collecting Data and Initial Analysis") for guidance on selecting and executing these methods.

PALS Workbook Link & AI Partnership Guidance: Engage with Chap. 8 PALS Workbook prompts (Section 2). Specify a stakeholder group and potential AI application area (e.g., staff using AI for scheduling). Ask your GenAI partner (using HCASR C2 & C3) to help brainstorm open-ended survey or focus group questions focused on understanding current needs, goals, and challenges, drawing on empathy principles central to design thinking.

Example Prompt: "For [Academic Advisors] regarding [AI tools for student progress monitoring], brainstorm 4 open-ended focus group questions focused on understanding their current workflow challenges, information needs, and potential concerns (apply design thinking empathy principles)."

Human Oversight Reminder: AI generates generic questions. Crafting effective, unbiased, culturally sensitive, and relevant assessment questions requires human expertise in research methodology, survey design, and deep understanding of the specific stakeholder group and institutional context (HCASR Component 4). Always ensure ethical review (e.g., IRB) if required.

Workbook (Section 2), include surveys, focus groups, interviews, workflow analysis, analysis of existing feedback (e.g., PALS logs, helpdesk tickets), and UX research techniques like persona refinement or user journey mapping (Gibbons, 2018; Oakes, 2024).

The primary aim is to gather concrete evidence on current needs, challenges (e.g., access issues, training requirements, ethical concerns), and perceived opportunities for AI integration. This process should be guided by HCASR Component 4 principles, ensuring engagement is inclusive, respectful, and genuinely seeks to understand diverse perspectives. The Smart Teams or other groups formed in Chap. 7 typically facilitate this stakeholder engagement, using the insights gathered to inform the assessment. Figure 8.1 suggests a potential dashboard format for summarizing key findings from this needs assessment phase.

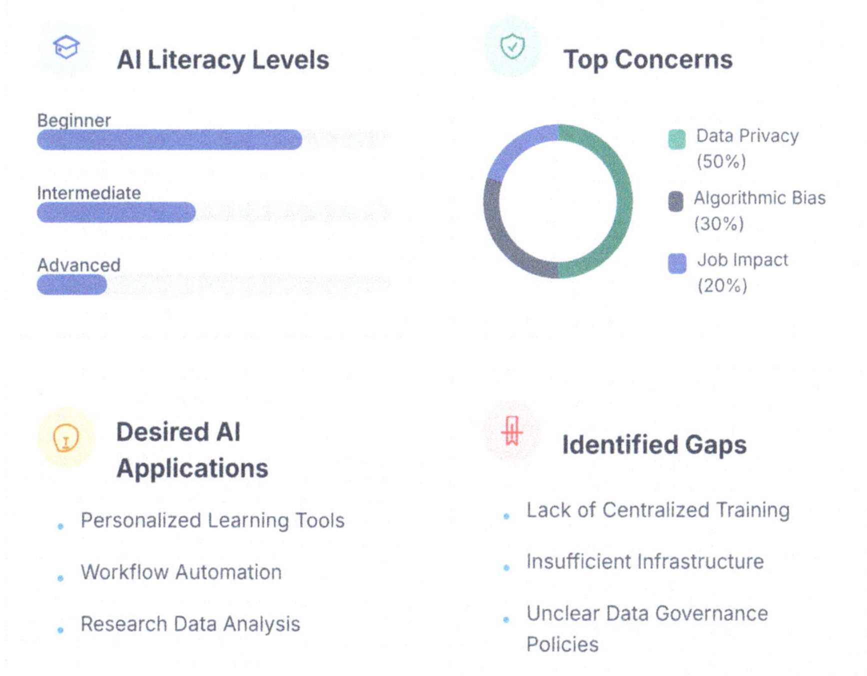

Fig. 8.1 Stakeholder needs assessment summary dashboard

Defining Initial Goals Based on Assessed Needs

The insights gathered from the needs assessment directly inform the initial definition of goals for potential AI initiatives. Vague aspirations must be translated into specific, actionable objectives. Using the SMARTIE framework—Specific, Measurable, Achievable, Relevant, Time-bound, Inclusive, Equitable—helps ensure these initial goals are well-defined, aligned with stakeholder needs and institutional context, and explicitly consider fairness and belonging ([Table 8.2]).

Table 8.3 provides a template to guide the drafting process, ensuring all SMARTIE components are addressed. Focusing on the "Inclusive" and "Equitable" dimensions from the outset helps proactively design AI initiatives that benefit all community members and mitigate potential disparities (Oregon Health Authority, 2025; The Core Collaborative, 2024). These initial goals serve as crucial reference points during the subsequent readiness evaluation and strategic planning phases.

Table 8.2 Defining SMARTIE goals for AI initiatives

Criterion	Name	Description/guiding question
S	Specific	What exactly do we want to achieve? Is the goal clear and focused?
M	Measurable	How will we know when the goal is achieved? What metrics will be used?
A	Achievable	Is the goal realistic given current resources, constraints, and timelines?
R	Relevant	Does the goal align with identified stakeholder needs and broader institutional mission/priorities? Why is this important?
T	Time-bound	What is the target date or timeframe for achieving this goal?
I	Inclusive	How does this goal intentionally promote belonging for diverse community members? Whose perspectives were considered in setting this goal?
E	Equitable	How does this goal fairly benefit all community members, particularly those from underrepresented or marginalized groups? How does it address or avoid exacerbating potential disparities or systemic inequities?

Note. Applying the SMARTIE framework helps ensure that goals for AI integration are well-defined, trackable, aligned with institutional context, and proactively designed to advance inclusion and equity (Source: Oregon Health Authority, 2025; The Core Collaborative, 2024)

Table 8.3 SMARTIE goal planning template

SMARTIE component	Guiding question	Draft goal component
Specific	What exactly do we want to accomplish? Who is involved?	"[Draft goal component here]"
Measurable	How will we know if the goal is achieved? What metrics will indicate success?	"[Draft goal component here]"
Achievable	Is the goal realistic given available resources, time, and constraints?	"[Draft goal component here]"
Relevant	How does the goal align with the institution's mission and strategic priorities?	"[Draft goal component here]"
Time-bound	What is the deadline? What are key milestones along the way?	"[Draft goal component here]"
Inclusive	How will this goal promote belonging? Whose perspectives are included?	"[Draft goal component here]"
Equitable	How will this goal address potential disparities? Who benefits, and are benefits distributed fairly?	"[Draft goal component here]"

Workbook Prep 3: Drafting an Initial SMARTIE Goal

Contextual Reflection: Based on an identified stakeholder need from your initial assessment work, attempt to draft one high-level SMARTIE goal for a potential AI initiative. Critically evaluate how well your draft addresses the "Inclusive" and "Equitable" dimensions described in Table 8.2.

UX Workbook Link: See Chap. 8 UX Workbook, Activity 1.4 ("Draft Initial SMARTIE Objectives") for practice in formulating these initial goals using templates like Table 8.3 and Fig. 8.2.

PALS Workbook Link & AI Partnership Guidance: Engage with Chap. 8 PALS Workbook prompts (Section 1/2). Input your draft SMARTIE goal; ask your GenAI partner (using HCASR Component 1) to evaluate it against the SMARTIE criteria and suggest ways to strengthen clarity, measurability, or the articulation of inclusivity/equity.

Example Prompt: "Evaluate this draft SMARTIE goal: '[Increase AI tool usage by students]'. Suggest revisions to make it more Specific, Measurable, Inclusive, and Equitable, referencing HCASR Component 1 principles."

Human Oversight Reminder: AI checks structure and keywords. Ensuring goals are genuinely Achievable within your context, Relevant to strategic priorities, and truly Inclusive/Equitable in practice requires human judgment, stakeholder consultation, and ethical deliberation (HCASR Component 1 & 4).

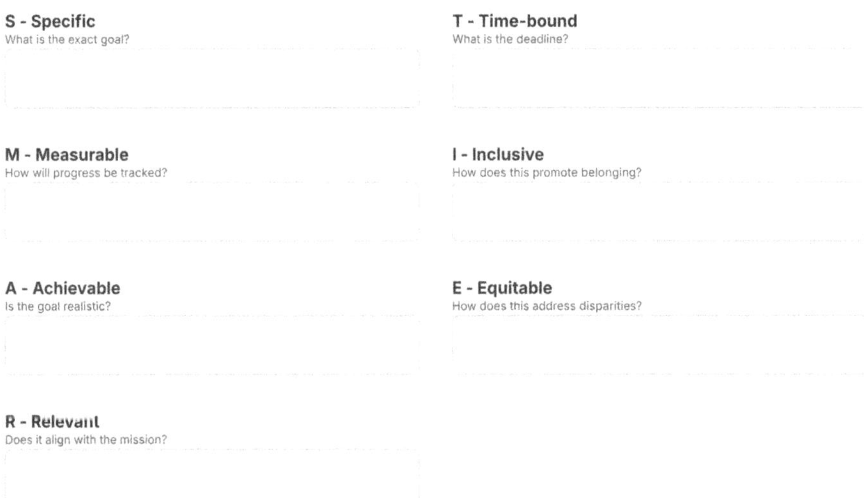

Fig. 8.2 SMARTIE goal planning template

Assessing Institutional Readiness for AI

Alongside understanding needs, evaluating the institution's current capacity to effectively and responsibly implement AI is crucial. This involves assessing readiness across several key dimensions.

Applying UX Strategies for Deepening Readiness Insights

Quantitative readiness metrics (e.g., infrastructure audits, policy counts) provide part of the picture, but qualitative insights gained through UX strategies are vital for understanding the human dimensions of readiness. Methods like empathy mapping and persona development help uncover stakeholder attitudes, assumptions, anxieties, and practical challenges related to AI adoption (Gibbons, 2018; Oakes, 2024). Empathy maps ([Table 8.4]) explore what stakeholders Say, Think, Do, and Feel regarding AI, revealing underlying motivations or barriers. Personas ([Table 8.5], [Fig. 8.3]) create representative profiles based on research, grounding readiness assessment in the lived

Table 8.4 Comparison of UX strategies for deepening AI readiness insights

Feature	Empathy mapping	Persona development
Method	Collaborative visualization exploring user experience through four quadrants.	Creating representative user profiles based on research data.
Purpose/focus	To build shared, empathetic understanding of specific user types regarding AI; reveal underlying attitudes.	To capture and communicate the distinct goals, motivations, challenges, and needs of key stakeholder segments regarding AI.
Key components/ inputs	Explores what users Say (quotes), Think (beliefs), Do (actions), and Feel (emotions). Uses data from interviews, surveys, observations.	Includes: Name, Role, Background, AI Goals, Frustrations, AI Literacy, Tech Access, AI Needs, Ethical Concerns. Grounded in qualitative/quantitative research.
Contribution to readiness	Reveals potential conflicts, unstated beliefs, anxieties, and motivations influencing AI adoption. Helps identify underlying readiness factors beyond stated opinions.	Provides concrete representations of target users, informing design of user-centric AI tools, tailored training, relevant support, and practical policies.

Note. These UX strategies provide deep qualitative insights into the lived experiences and contexts of stakeholders, complementing broader needs assessment methods and informing user-centric AI implementation (Source: Gibbons, 2018; Oakes, 2024)

Table 8.5 Sample persona template

Persona attribute	Description/placeholder
Photo/icon	Insert a representative photo or icon here.
Name	"[Draft persona name]"
Role	"Position or stakeholder group: e.g., Undergraduate Student, Faculty, Administrator."
Background	"Brief description: academic background, experience, demographics if relevant."
Goals related to AI	What do they hope AI tools will help them achieve?
Frustrations/pain points	Challenges they experience with current tech/AI solutions.
AI literacy level	"Low, Medium, High—connect to Competency Mapping."
Technology access	"Do they have reliable access to necessary devices, internet, platforms?"
Key needs for AI tools/ support	"Specific features, supports, or training they would benefit from."
Ethical concerns	"Privacy, bias, academic integrity worries, accessibility needs, etc."

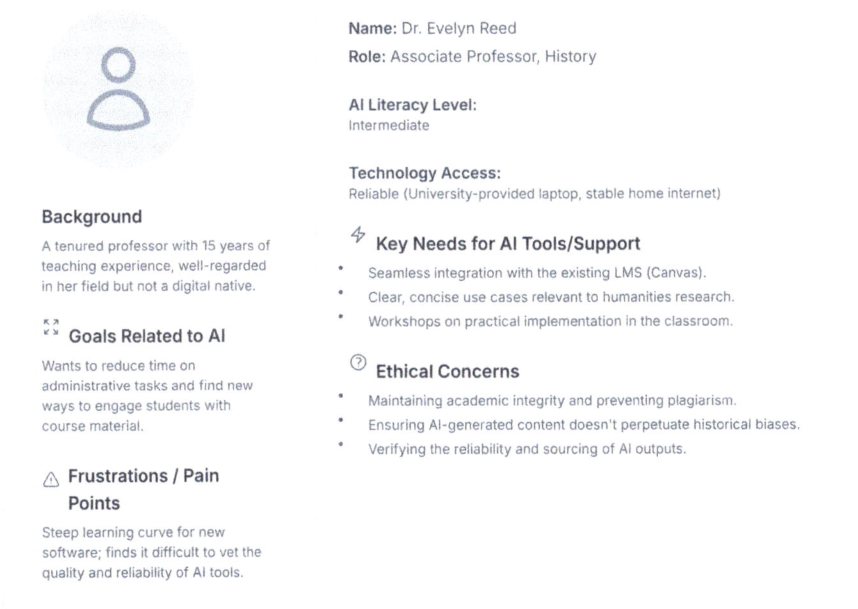

Fig. 8.3 Sample persona template for AI readiness

experiences of key user groups (U.S. Department of Education, OET, 2024). Integrating these UX insights ensures that readiness evaluations consider usability, accessibility, trust, and the need for user-centered support alongside technical and structural factors.

Workbook Prep 4: Applying User Experience Strategies

Contextual Reflection: Select a key stakeholder group relevant to a potential AI initiative. What assumptions does your team currently hold about this group's thoughts, feelings, or behaviors regarding AI? How could using empathy mapping ([Table 8.4]) or developing a data-grounded persona ([Table 8.5], [Fig. 8.3]) challenge or refine those assumptions and lead to more user-centered readiness insights?

UX Workbook Link: See Chap. 8 UX Workbook, Section 3 ("UX Strategies & Activities") which integrates empathy and persona insights into assessing literacy, competencies, and equity. Refer also to Chap. 7 UX Workbook, Section 3 for initial persona/empathy work.

PALS Workbook Link & AI Partnership Guidance: Engage with Chap. 8 PALS Workbook prompts (Section 3). Input key synthesized data points (from needs assessment or initial analysis) about a specific stakeholder group; ask your GenAI partner (using HCASR Component 3 & 4) to help structure this information into specific sections of a persona template.

Example Prompt: "Data Summary: [Approx. 40% of adjunct faculty report limited access to institutional software licenses at home; express concern about time needed to learn new tech]. Help draft the 'Technology Access' and 'Potential AI-Related Concerns' sections for an 'Adjunct Faculty' persona."

Human Oversight Reminder: Base personas on real research data, not stereotypes. Use AI primarily to structure information derived from human analysis. Ensure personas reflect diversity and foster genuine empathy, not oversimplification (HCASR Component 4; Brown, 2019).

Evaluating Data Infrastructure and Governance

Data is the fuel for many AI systems. Assessing data readiness involves evaluating data availability, quality, the infrastructure for managing it (storage, processing, integration), and the governance frameworks surrounding its use (policies for ownership, access, security, privacy/FERPA compliance, ethics) (U.S. Department of Education, OET, 2024). Table 8.6 outlines key dimensions for this critical assessment, which helps identify necessary investments, infrastructure upgrades, or policy developments needed before implementation.

Table 8.6 Key dimensions of data readiness assessment for AI

Dimension	Key aspects to evaluate
Data availability and quality	Is relevant data accessible? Is it accurate, complete, consistent, and usable for the intended AI application?
Data infrastructure	Adequacy of storage capacity, data processing power, integration capabilities (e.g., APIs, data warehouses) to support AI tools.
Data governance	Presence and effectiveness of clear policies/practices for: data ownership, access control, security, privacy (incl. FERPA compliance), data lifecycle management, ethical data use.

Note. Evaluating these dimensions identifies necessary investments, infrastructure upgrades, or policy developments required before implementing AI responsibly (Source: U.S. Department of Education, OET, 2024)

Workbook Prep 5: Assessing Data Readiness

Contextual Reflection: For a potential AI initiative (e.g., AI-powered analytics for student retention), what are the primary data sources required? Based on your current understanding, rate your institution's readiness (High/Medium/Low) regarding the dimensions in Table 8.6 (availability, quality, infrastructure, governance) for that specific data. What is the most significant data-related challenge?

UX Workbook Link: See Chap. 8 UX Workbook, Section 4 activities ("Checkpoints: Validating Assessment Process and Findings"), particularly Checkpoint 4.3 which focuses on Data & Governance Readiness Validation.

PALS Workbook Link & AI Partnership Guidance: Engage with Chap. 8 PALS Workbook prompts (Section 4). Specify the potential AI initiative and data required; ask your GenAI partner (using HCASR Component 1 & 3) to help brainstorm 3–5 key questions focused specifically on assessing the *governance* readiness (policies, access controls, privacy safeguards, ethical use guidelines) for using that type of data with AI in your HEI context.

Example Prompt: "Initiative: [AI analysis of student LMS activity for early alerts]. Data: [Detailed student activity logs]. Generate 4 key assessment questions focused only on data governance readiness (relevant policies, access protocols, student privacy, ethical oversight) for this use case, referencing HCASR Component 1."

Human Oversight Reminder: AI can suggest generic questions. Assessing policy nuances, infrastructure capabilities, compliance (e.g., FERPA), and ethical implications requires specific human expertise (data governance officers, IT security, legal counsel) and thorough internal investigation (HCASR Component 4).

Assessing Technological Capabilities

Beyond data, the broader technological environment must be assessed. This includes evaluating existing system compatibility (e.g., can a new AI tool integrate with the current LMS or SIS?), network adequacy (can bandwidth handle increased data flow?), IT staff capacity and skills (are staff prepared to support new AI tools?), and the overall cybersecurity posture relative to new AI-specific risks (U.S. Department of Education, OET, 2024). Table 8.7 details key dimensions for assessing technological capability, informing decisions about feasibility and necessary infrastructure planning.

Assessing Workforce Skills and AI Literacy

Effective and responsible AI use depends heavily on the skills and understanding of the people involved. Assessing workforce readiness requires evaluating baseline AI literacy across all stakeholder groups (students, faculty, staff, administrators) and mapping the specific competencies required for different roles against current capabilities (U.S. Department of Education, OET, 2024; UNESCO, 2025; Coursera, 2024). Competencies may range from basic digital literacy and understanding AI concepts to advanced skills in data analysis, prompt engineering, ethical evaluation of AI outputs, or developing AI tools (Jay, 2025). Table 8.8 outlines key components of this assessment. Identifying skill gaps through methods like competency mapping ([Fig. 8.4]) is crucial for informing targeted training and professional development plans (EHCS, EAO principles).

Table 8.7 Key dimensions of technological capability assessment for AI

Dimension	Key aspects to evaluate
System compatibility	Can existing enterprise systems (e.g., LMS, SIS, ERP) integrate effectively with potential AI tools or platforms?
Network infrastructure	Is the current network bandwidth and stability sufficient for potentially data-intensive AI applications?
IT staff capacity	Does the institution have sufficient IT staff with the necessary skills for AI implementation, integration, and support?
Cybersecurity posture	Are current cybersecurity measures robust enough to address new AI-specific risks and vulnerabilities?

Note. Understanding these technological capabilities helps scope feasible AI initiatives and identifies needs for infrastructure upgrades, staff training, or external support (Source: U.S. Department of Education, OET, 2024)

Table 8.8 Components of workforce skills and AI literacy assessment

Assessment area	Focus	Example methods/tools
Baseline AI literacy	General understanding across students, faculty, staff regarding AI concepts, capabilities, limitations, ethical implications.	Surveys, self-assessments, introductory workshops
Role-specific AI competencies	Identifying and assessing specific technical, analytical, and ethical AI skills needed for different roles.	Competency mapping, skills inventories, training needs analysis
Skill gaps	Identifying discrepancies between current workforce capabilities and the competencies required to achieve strategic AI goals.	Gap analysis (comparing current vs. required levels)

Note. This assessment identifies crucial skill gaps and directly informs the design of targeted training and professional development programs essential for responsible AI adoption (Source: Coursera, 2024; Jay, 2025; UNESCO, 2025; U.S. Department of Education, OET, 2024)

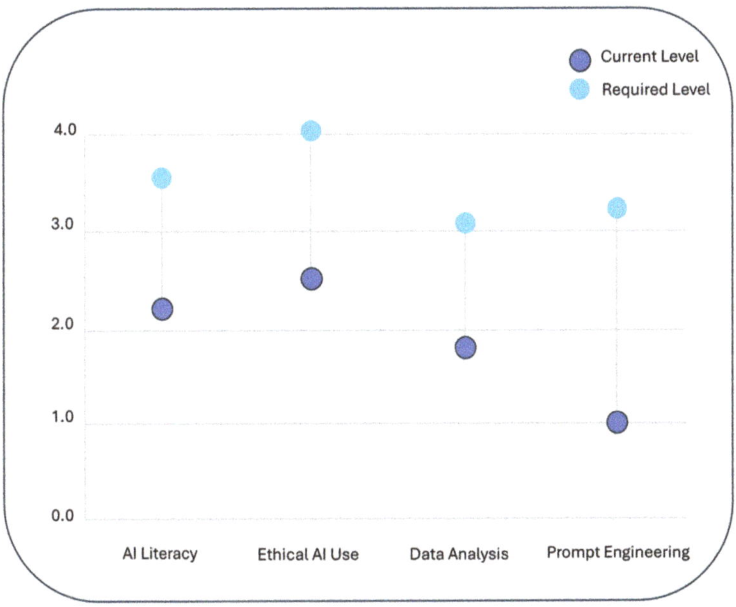

Fig. 8.4 Competency gap analysis chart

Workbook Prep 6: Planning Competency Mapping

Contextual Reflection: Consider a specific role (e.g., librarian, faculty advisor). Beyond basic AI literacy, what is one key AI-related competency they might need for planned AI initiatives? How might you begin assessing the current level of that competency?

UX Workbook Link: See Chap. 8 UX Workbook, Section 3 activities, specifically Activity 3.3 ("Map Required AI Competencies & Analyze Gaps").

PALS Workbook Link & AI Partnership Guidance: Engage with Chap. 8 PALS Workbook prompts (Section 3). Define a needed competency for a specific role; ask your GenAI partner (using HCASR Component 5) to help draft 2–3 specific, observable behavioral descriptors for that competency, useful for assessment rubrics or training objectives. Consider researching existing competency frameworks (like UNESCO's) for comparison.

Example Prompt: "Competency: 'Ethical application of AI advising tools' for Academic Advisors. Draft 3 specific behavioral descriptors (e.g., 'Explains AI tool limitations and data usage to students clearly', 'Reviews AI recommendations for potential bias before sharing'). Reference HCASR Component 5."

Human Oversight Reminder: AI drafts generic descriptors. Defining relevant, context-specific, measurable competencies requires collaborative input from domain experts, HR, and role incumbents. Adapt frameworks, don't just adopt them (HCASR Component 4 & 5).

Evaluating Organizational Culture and Change Readiness

Technology implementation is also a cultural process. Assessing cultural readiness involves understanding factors like leadership support and vision for AI, prevailing stakeholder attitudes (enthusiasm, skepticism, fear), the existing culture of collaboration across departments (vital for Smart Team success), institutional tolerance for risk and experimentation, and the effectiveness of internal communication channels (Kotter et al., 2021). Table 8.9 lists key cultural factors influencing AI adoption. Understanding these factors helps anticipate barriers and informs the change management strategies needed for successful implementation.

Workbook Prep 7: Gauging Cultural Readiness

Contextual Reflection: Briefly characterize your institution's prevailing culture regarding technological change. What norms, values, or past experiences might enable AI adoption? What cultural aspects might present barriers requiring careful change management (Kotter et al., 2021)?

Table 8.9 Key factors in assessing organizational culture and change readiness for AI

Factor	Relevance to AI implementation
Leadership support	Demonstrates commitment, provides vision, and allocates resources necessary for significant change initiatives.
Stakeholder attitudes	General disposition toward technology and AI (positive or negative) influences adoption rates and potential resistance.
Collaboration culture	Extent of cross-departmental collaboration impacts the effectiveness of interdisciplinary Smart Teams.
Risk tolerance	Institutional willingness to experiment and accept potential setbacks influences the pace and scope of AI piloting.
Internal communication	Effectiveness of existing channels impacts the ability to manage change, share information, and gather feedback.

Note. Assessing these cultural factors helps anticipate potential barriers or enablers to AI adoption and informs the development of tailored change management strategies (Source: Kotter et al., 2021)

UX Workbook Link: See Chap. 8 UX Workbook, Section 1 (Activity 1.2 includes assessing culture) and Section 4 (Checkpoint 4.2 involves stakeholder buy-in).

PALS Workbook Link & AI Partnership Guidance: Engage with Chap. 8 PALS Workbook prompts (Section 2). Describe a specific cultural aspect (e.g., "Highly decentralized decision-making"); ask your GenAI partner to analyze how this might enable or hinder AI implementation, referencing change management literature (Kotter et al., 2021).

Example Prompt*:* "Culture Aspect: ['Emphasis on traditional pedagogical methods in humanities departments']. Based on Kotter's change model, analyze how this might be a potential barrier to adopting AI teaching tools and suggest one potential change management approach to address it respectfully."

Human Oversight Reminder: AI analysis of culture is superficial. Understanding embedded norms, values, politics, and history requires nuanced human observation, qualitative inquiry, and consultation across the institution (HCASR Component 4).

Evaluating Existing AI Governance and Ethical Frameworks

Finally, readiness assessment must examine the existing governance landscape. This involves reviewing current policies (academic integrity, data privacy, acceptable use, procurement) for their adequacy in addressing AI-specific issues, assessing the effectiveness of any current AI-related governance structures (like the AI Council formed in Chap. 7), gauging how well existing institutional ethical frameworks (e.g., ETHICAL principles [Wynants et al., 2025], HCASR) are understood and applied, and evaluating the capacity of current risk management practices to handle AI-specific risks (potentially using the NIST AI RMF [NIST, 2023]) (U.S. Department of Education, OET, 2024). External tools, like the EDUCAUSE assessment ([Fig. 8.5]), can provide structure (EDUCAUSE, 2024; Weil & Forrester, 2024). Table 8.10 outlines key areas for identifying governance gaps that must be addressed in strategic planning.

Workbook Prep 8: Assessing Governance Gaps

Contextual Reflection: Based on your current understanding, what is the most significant gap in your institution's existing governance structures or policies regarding AI? How might this gap hinder responsible AI adoption if not addressed in the strategic plan (Chap. 9)?

UX Workbook Link: See Chap. 8 UX Workbook, Section 4 activity ("Checkpoints"), specifically Checkpoint 4.3 which validates governance and policy readiness findings.

PALS Workbook Link & AI Partnership Guidance: Engage with Chap. 8 PALS Workbook prompts corresponding to UX Workbook Section 4. Specify a potential AI application area; ask your GenAI partner (using HCASR Component 1 & 3) to help brainstorm 3–4 specific questions to assess the adequacy of existing institutional policies in addressing the challenges posed by this specific AI use.

Example Prompt: "Generate 4 assessment questions for using AI in admissions review: Does our current Data Privacy policy adequately address third-party AI tools analyzing applicant data? Does our Non-Discrimination policy cover potential algorithmic bias in AI screening? (Reference HCASR C1)."

Human Oversight Reminder: AI generates questions based on patterns. Determining policy adequacy or identifying legal/ethical risks requires expertise from legal counsel, policy analysts, ethicists, IT security, and stakeholders, along with careful policy review (HCASR Component 4).

Fig. 8.5 Key dimensions of institutional AI readiness

Table 8.10 Key areas for assessing existing AI governance and ethical frameworks

Area	Key aspects to evaluate
Existing policies	Adequacy and applicability of current policies (e.g., academic integrity, data privacy, acceptable use, procurement) to address AI-specific issues.
Governance structures	Existence, mandate clarity, and effectiveness of current AI-related governance bodies (e.g., AI Council, Ethics Board).
Ethical framework application	Extent to which adopted institutional ethical AI frameworks or principles (e.g., ETHICAL, HCASR) are understood, communicated, and applied in practice.
Risk management practices	Capacity of current risk management processes to identify, assess, and manage AI-specific risks (potentially leveraging NIST AI RMF or similar).
External framework alignment	(Optional) Use of established external frameworks (e.g., EDUCAUSE) to structure the assessment and benchmark readiness.

Note. This assessment identifies crucial gaps in current governance and ethical oversight that need to be addressed during strategic planning to ensure responsible AI deployment (Source: EDUCAUSE, 2024; NIST, 2023; U.S. Department of Education, OET, 2024)

Integrating Findings and Informing Strategy

The final step in the assessment phase involves synthesizing the collected data into actionable reports that directly inform subsequent strategic decisions.

Synthesizing Needs Assessment and Readiness Evaluation Data

Bringing together the findings from the needs assessment (identifying the demand and desired outcomes) and the readiness evaluation (assessing the supply and current capabilities) creates a holistic view of the institution's position regarding AI (U.S. Department of Education, OET, 2024). This synthesis explicitly highlights the critical gaps between what stakeholders need or desire and what the institution is currently prepared to deliver or support. Key UODs, detailed in Table 8.11 and the UX Workbook (Section 5), typically include:

- An **AI Readiness Assessment Findings Report (V1):** Summarizing institutional strengths and weaknesses across readiness dimensions ([Fig. 8.6]).

Table 8.11 Key outputs of needs assessment and readiness evaluation synthesis

Output name	Purpose/content	Key inputs
AI Readiness Assessment Findings Report (V1)	Summarizes institutional strengths, weaknesses, and overall readiness levels across evaluated dimensions (Data, Tech, Culture, Skills, Governance); may include scores/ratings.	Data from Readiness Evaluation (Sect. 2), including Competency Gap Analysis, Governance/Policy Gap Analysis.
Needs & Gap Analysis Report (V1)	Integrates readiness findings with needs assessment data; explicitly identifies gaps between needs and readiness; often includes a Priority Gaps versus Quick Wins Matrix.	Readiness Report (V1), Needs Assessment data (Sect. 1), Use Case Inventory, Contextual Factors.
Refined Stakeholder Map (V2)	Updated map incorporating deeper understanding of stakeholder needs, AI literacy/readiness levels, specific concerns, and potential AI impact revealed during assessment.	Original Stakeholder Map (Chap. 7), Needs Assessment findings (Sect. 1), Readiness findings (Sect. 2).

Note. These synthesized reports provide a holistic view of the institution's AI landscape, identifying the crucial gaps between stakeholder needs and institutional capabilities, thereby directly informing strategic planning (Source: EDUCAUSE, 2024; Kotter et al., 2021; U.S. Department of Education, OET, 2024)

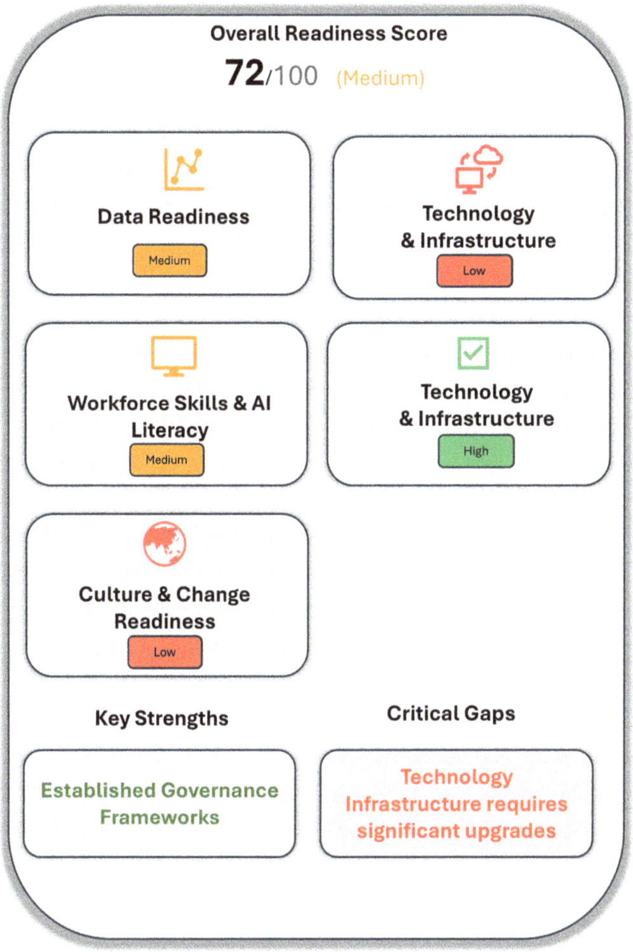

Fig. 8.6 Institutional AI readiness dashboard

- A **Needs & Gap Analysis Report (V1):** Integrating readiness and needs data to pinpoint specific gaps. This often includes a **Priority Gaps vs. Quick Wins Matrix** ([Fig. 8.7]) to help prioritize efforts.
- A **Refined Stakeholder Map (V2):** Updating the Chap. 7 map with deeper insights gained during assessment.

Workbook Prep 9: Synthesizing Assessment Data

Contextual Reflection: Consider a specific readiness gap (e.g., "Limited data access") and a related competency gap (e.g., "Staff lack data prep skills"). How does understanding the connection help brainstorm more holistic solutions (e.g., combined infrastructure investment and training)?

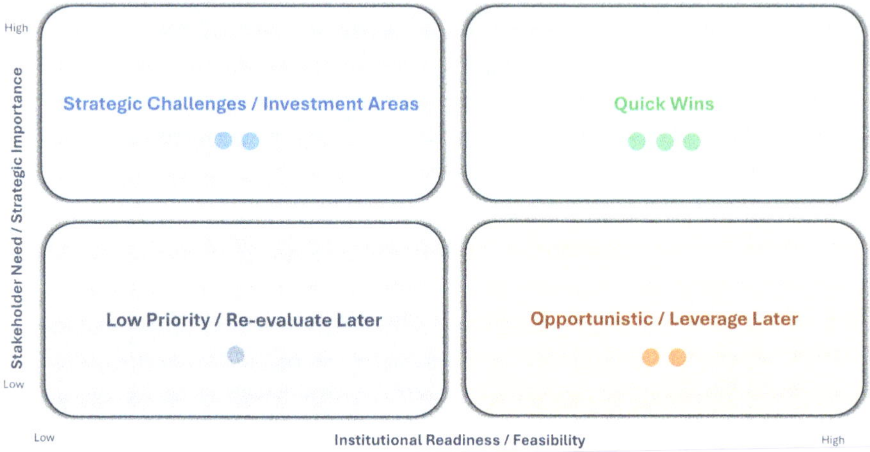

Fig. 8.7 Priority gaps versus quick wins matrix

UX Workbook Link: See Chap. 8 UX Workbook, Section 5 ("Living Documents: Consolidating Assessment Findings") and Section 7 ("Deliverables: Finalizing Assessment Documentation"), focusing on creating integrated Needs & Gap and Readiness reports.

PALS Workbook Link & AI Partnership Guidance: Engage with Chap. 8 PALS Workbook prompts (Section 5/7). Input a linked readiness gap and competency gap; ask your GenAI partner (using HCASR Component 5) to brainstorm potential strategic interventions addressing both simultaneously.

Example Prompt: "Readiness Gap: ['Lack of clear guidelines on ethical AI use in research']. Competency Gap: ['Researchers lack training in evaluating AI tools for bias']. Brainstorm 3 potential interventions addressing both gaps, referencing HCASR Component 5."

Human Oversight Reminder: AI suggests strategies. Prioritizing, designing, assessing feasibility, securing resources, and managing implementation require strategic human judgment, planning, and leadership (HCASR Component 4 & 5).

These synthesized reports provide the necessary evidence base for informed strategic planning in Chap. 9 (EDUCAUSE, 2024; Kotter et al., 2021).

Informing the AI Sourcing Strategy: Proprietary vs. Third-Party/Open-Source

The comprehensive assessment findings critically inform the strategic AI sourcing decision—whether to primarily build proprietary tools, integrate third-party solutions, leverage open-source models, or use a hybrid approach.

This decision requires weighing trade-offs based on assessed internal technical expertise, infrastructure capacity, data governance maturity, budget constraints, risk tolerance, and identified needs for customization versus standardization (Nguyen, 2025; Culotta & Mattei, 2023; Grajek et al., 2025; Inclusion Cloud, 2025; Numbers Station, n.d.). The UX Workbook for Chap. 9 guides the application of assessment data to evaluate these sourcing trade-offs, which were initially introduced in Chap. 6 (Table 6.2).

Workbook Prep 10: Connecting Readiness to Sourcing

Contextual Reflection: Based on your Chap. 8 readiness findings (e.g., internal expertise level, budget, data governance maturity, risk tolerance, customization needs), which factors weigh most heavily in considering proprietary ("build") vs. third-party/open-source ("buy/adapt") AI solutions? Does this data shift your preliminary leaning?

UX Workbook Link: The finalized Chap. 8 assessment deliverables (Needs & Gap Report V1, Readiness Findings V1 - see UX Workbook Section 7) are primary inputs for the AI sourcing strategy discussion in the Chap. 9 UX Workbook (Section 3.B).

PALS Workbook Link & AI Partnership Guidance: Engage with Chap. 8 PALS Workbook prompts (Section 7/Output links). Input key summary findings from your readiness assessment; ask your GenAI partner to analyze the pros and cons of proprietary vs. third-party/open-source solutions specifically in light of those findings, referencing Chap. 9, Section 2.C trade-offs.

Example Prompt: "Key Findings: [Strong data governance, high risk aversion (student data), budget constraints, low internal AI dev expertise]. Analyze pros/cons of proprietary vs. third-party AI advising tool specifically given these institutional findings."

Human Oversight Reminder: AI analysis weighs documented trade-offs against findings. The sourcing strategy is a complex institutional decision requiring leadership judgment on strategic priorities, long-term vision, risk appetite, and ethical alignment, using AI analysis as one input (HCASR Component 1 & 4).

User Outcomes and Deliverables (UODs): Concluding Phase 1 Assessment

The Needs Assessment and AI Readiness Evaluation phase culminates in key UODs providing the evidence base for Chap. 9's strategic planning. These finalized reports document the institution's current state and stakeholder needs. Key UODs typically include:

- Needs & Gap Analysis Report (V1)
- AI Readiness Assessment Findings Report (V1) (including Competency Gap Analysis and Equity & Accessibility Review findings)
- Priority Gaps vs. Quick Wins Matrix
- Refined Stakeholder Map (V2)
- Governance Guide Notes (Assessment Implications)

Guidance for consolidating these UODs is in the Chap. 8 UX Workbook (Section 7).

Conclusion: From Assessment to Informed Strategy

Chapter 8 outlined the essential assessment work required to understand the institutional landscape before embarking on AI strategic planning. Conducting systematic needs assessments and evaluating readiness across key dimensions create the necessary data-driven foundation. Synthesizing these findings into actionable reports ([Table 8.11]) allows institutions to identify strategic priorities, potential quick wins, necessary investment areas, and make informed AI sourcing decisions. This ensures the strategic roadmap developed in Chap. 9 is grounded in institutional reality and stakeholder needs, setting the stage for responsible and effective AI integration.

Looking Ahead & Engaging the Workbook

With a comprehensive understanding of needs and readiness established through the activities in this chapter and its companion workbooks, the institution is prepared to translate these insights into a coherent strategic direction in Chap. 9: AI Vision, Governance & Strategic Planning. The Chap. 8 UX and PALS Workbooks provide the structured activities and AI partnership guidance to execute this assessment phase effectively, producing the crucial evidence base required for the next stage.

Long-Term Scaling, Sustainability & AI Governance: Grounding TLARA Guidelines in Reality

Chapter 8 has equipped you with the methodologies to conduct a thorough needs assessment and AI readiness evaluation, which are essential for successful AI adoption (U.S. Department of Education, Office of Educational Technology [OET], 2024). This data-driven understanding of your institution's current capabilities, stakeholder requirements, and potential gaps across various **levels** and **spaces** is critical. It ensures that your strategic AI initiatives, particularly the development of custom **TLARA (Teaching, Learning, Assessment, Research, Administration)** guidelines, are grounded in reality and tailored to your specific **sociocultural context**, rather than based on assumptions, and prepares your institution to navigate the complexities of AI integration over **time** (Wynants et al., 2025). Both the Chap. 8 UX and PALS Workbooks guide you in using AI ethically to analyze this crucial data and inform your strategic planning.

Higher Education Institutions (HEIs) are dynamic systems, and a holistic, evidence-based approach is vital for implementing AI effectively (IDEAS, 2025). The PALS Workbook for Chap. 8, through its section "Long-Term Scaling, Sustainability & AI Governance: Iteratively Developing Your Custom TLARA-Informed AI Guidelines," provides a customizable prompt. This tool enables your GenAI model to act as an expert advisor, helping you synthesize your assessment findings and explore how they should directly inform the principles and policies within each TLARA domain. Engaging with the "Topics for PALS TLARA Focus"—such as "Assessing AI Readiness for Sustainable Scaling" and "Identifying AI Literacy Gaps"—will ensure your "Living Document" of TLARA guidelines becomes increasingly practical, strategically aligned, and supportive of long-term AI sustainability and effective governance (Ellucian, 2024).

References

Brown, T. (2019). *Change by design: How design thinking transforms organizations and inspires innovation* (Revised and Updated ed.). Harper Business.

Coursera. (2024, February 19). *What is competency mapping? Methods and why it matters.* Coursera for Business. https://www.coursera.org/enterprise/articles/competency-mapping

Culotta, A., & Mattei, N. (2023, November 20). Use open source for safer generative AI experiments. *MIT Sloan Management Review.* https://sloanreview.mit.edu/article/use-open-source-for-safer-generative-ai-experiments/

EDUCAUSE. (2024, April 10). *Higher education generative AI readiness assessment.* EDUCAUSE Library. https://library.educause.edu/resources/2024/4/higher-education-generative-ai-readiness-assessment

Ellucian. (2024, December 16). AI in higher education: Implementation strategies for institutions and tech partners. *Higher Ed Dive.* https://www.highereddive.com/spons/ai-in-higher-education-implementation-strategies-for-institutions-and-tech/735434/

Gibbons, S. (2018, January 14). *Empathy mapping: The first step in design thinking*. Nielsen Norman Group. https://www.nngroup.com/articles/empathy-mapping/

Grajek, S., Pelletier, K., & Freeman, A. (2025, March 11). AI procurement in higher education: Benefits and risks of emerging tools. *EDUCAUSE Review*. https://er.educause.edu/articles/2025/3/ai-procurement-in-higher-education-benefits-and-risks-of-emerging-tools

Inclusion Cloud. (2025, February 28). What you need to know before choosing open-source or proprietary AI models. *Inclusion Cloud Blog*. https://inclusioncloud.com/insights/blog/open-source-llm-vs-proprietary-models/

Instructional Design, Engagement, and Support (IDEAS). (2025, January 27). *AI's impact on higher education through the eyes of faculty: 4 key trends*. University of Massachusetts Amherst. https://www.umass.edu/ideas/news/ais-impact-higher-education-through-eyes-faculty-4-key-trends

Jay, S. (2025). What is competency mapping? Your ultimate 2025 guide. *AIHR Blog*. https://www.aihr.com/blog/competency-mapping/

Kotter, J. P., Akhtar, V., & Gupta, G. (2021). *Change: How organizations achieve hard-to-imagine results in uncertain and volatile times*. Wiley.

National Institute of Standards and Technology. (2023, January). *Artificial intelligence risk management framework (AI RMF 1.0)* (NIST AI 100-1). U.S. Department of Commerce. https://doi.org/10.6028/NIST.AI.100-1

Nguyen, H. V. (2025, January 9). *Open source vs. proprietary AI: A comprehensive guide to making the right choice*. *SmartDev*. https://smartdev.com/open-source-vs-proprietary-ai/

Numbers Station. (n.d.). Private vs. third-party foundation models: Navigating the path to an AI-driven future. *Numbers Station Blog*. Retrieved May 5, 2025, from https://www.numbersstation.ai/private-vs-third-party-foundation-models-navigating-the-path-to-an-ai-driven-future/

Oakes, S. (2024, November 8). *How to develop effective edTech personas*. Backpack Interactive. https://backpackinteractive.com/insights/develop-effective-edtech-personas/

Oregon Health Authority. (2025). *SMARTIE goals*. Retrieved May 5, 2025, from https://www.oregon.gov/oha/HPA/dsi-tc/Documents/SMARTIE-goals-handout.pdf

The Core Collaborative. (2024, August 20). *Setting SMARTIE goals*. The Core Collaborative. https://thecorecollaborative.com/setting-smartie-goals/

U.S. Department of Education, Office of Educational Technology. (2024). *Navigating Artificial Intelligence in postsecondary education: Building capacity for the road ahead*. https://portal.ct.gov/das/-/media/das/ctedtech/publications/2025/2025-used-oet-archive/os-24-002833-ai-in-postsecondary-ed.pdf

UNESCO. (2025). *What you need to know about UNESCO's new AI competency frameworks for students and teachers*. Retrieved May 5, 2025, from https://www.unesco.org/en/articles/what-you-need-know-about-unescos-new-ai-competency-frameworks-students-and-teachers

Weil, D., & Forrester, J. (2024, October 17). A road map for leveraging AI at a smaller institution. *EDUCAUSE Review*. https://er.educause.edu/articles/2024/10/a-road-map-for-leveraging-ai-at-a-smaller-institution

Wynants, S., Childers, G., De La Torre Roman, Y., Budar-Turner, D., & Vasquez, P. (2025, February). *ETHICAL principles AI framework for higher education*. CSU AI Commons, California State University, Fullerton Faculty Development Center. https://genai.calstate.edu/communities/faculty/ethical-and-responsible-use-ai/ethical-principles-ai-framework-higher-education

9

Establishing AI Vision and Governance (Phase 1)

Introduction: Charting the Course for AI Integration

Following comprehensive readiness assessments and detailed stakeholder mapping, the initial phase of AI integration focuses on translating this foundational understanding into a clear strategic direction. This involves developing a shared institutional AI vision, setting aligned strategic goals, and creating a comprehensive implementation roadmap. These elements are designed to guide Higher Education Institutions (HEIs) in defining their desired future state for human-AI synergy and outlining the path to achieve it.

Crafting a compelling vision is crucial, as it provides direction for governance and subsequent implementation. This process requires synthesizing leadership aspirations with identified stakeholder needs and clearly communicating how AI integration aligns with the institutional mission. The vision is then operationalized through SMARTIE goals (Specific, Measurable, Achievable, Relevant, Time-bound, Inclusive, Equitable) and a detailed strategic plan.

This approach leverages principles from the synergy between design thinking and change management, emphasizing collaboration, adaptability, and ethical considerations guided by frameworks such as the Human-Centric AI Synergy & Regulation (HCASR) framework.

The User Experience (UX) Workbook and the Personalized Adaptive Learning Sessions (PALS) Workbook for Chap. 9 offer practical activities and AI partnership guidance. These resources assist users in ethically and effectively using AI as a strategic partner for researching and implementing tools, solutions, resources, and User Outcomes and Deliverables (UODs) customized to their unique needs and contexts.

Supplementary Information The online version contains supplementary material available at https://doi.org/10.1007/978-3-032-05205-6_9.

D. Woodring, *Advancing Higher Education with Human-Centric AI Partnerships*,
https://doi.org/10.1007/978-3-032-05205-6_9

Workbook Prep 1: Reflecting on Strategic Importance

Contextual Reflection: Why is establishing a shared vision and defining strategic goals (as discussed in this chapter) crucial before selecting specific AI tools or platforms? Consider the risks of long-term misalignment highlighted in change management principles (Kotter et al., 2021) if strategy doesn't guide technology choices.

UX Workbook Link: See Chap. 9 UX Workbook "Getting Started" exercises and Section 1 ("Planning") activities, which focus on defining strategic goals and aligning them with competency frameworks derived from Chap. 8 assessments.

PALS Workbook Link & AI Partnership Guidance: Engage with Chap. 9 PALS Workbook prompts (Section 1). Input key findings from your Chap. 8 Needs & Readiness reports and your institutional mission statement. Ask your GenAI partner (using HCASR Component 1 principles; see HAICS Guide) to help brainstorm potential themes or core elements for an AI vision statement that connects these inputs, referencing principles of effective vision creation (e.g., Bowen, 2018; Kotter et al., 2021).

Example Prompt: "Key Findings: [Faculty research interest in AI], [Student equity concerns re: access]. Mission Element: [Commitment to innovation & inclusion]. Help brainstorm 3–5 potential AI vision themes connecting these points, guided by HCASR Component 1 and principles of effective visioning."

Human Oversight Reminder: Your GenAI partner can brainstorm thematic connections. Crafting a vision statement that truly resonates, is strategically sound, and reflects institutional values requires human leadership, collaborative deliberation, and ethical judgment (HCASR Component 4).

Core Frameworks Reference: The frameworks mentioned below, such as HCASR, HCAS, RSA, EHCS, EAO, and AHIPs, are detailed in the companion *Human-GenAI Collaboration & Synergy (HAICS) Guide* and/or Part 1 of this textbook. Please refer to those sections for full definitions and explanations.

This chapter leverages principles from the synergy between design thinking (Brown, 2019) and change management (Kotter et al., 2021), emphasizing collaboration, adaptability, and ethical considerations guided by frameworks like HCASR.

The User Experience (UX) Workbook and the Personalized Adaptive Learning Sessions (PALS) Workbook for Chap. 9 provide practical activities and AI partnership guidance. These resources assist users in ethically and effectively using AI as a strategic partner for researching and implementing tools, solutions, resources, and user outcomes and deliverables (UODs) customized to their unique needs and contexts.

Developing a Shared Vision for Change

A clear, shared vision directs and motivates AI integration efforts. Developing this vision requires collaborative input informed by prior assessment phases and effective strategic communication.

Building Strategic Vision & Alignment with Collaborative "Smart Teams"

The diverse, interdisciplinary "Smart Teams" established in Chap. 7 are fundamental to developing a shared AI vision. By bringing together representatives from faculty, research, students, administration, IT, libraries, ethics, and potentially external partners, these teams ensure varied perspectives inform the visioning process (Jaskyte et al., 2024). While cognitive diversity fuels innovation, it requires careful management to harness its full potential (Jaskyte et al., 2024). These teams, blending formal leaders and informal "AI Ambassadors" (Chap. 7), provide the collaborative intelligence essential for responsible strategy development. Aligning with change management principles (Kotter et al., 2021; see HAICS Guide), these teams embody the "diverse many" needed to drive the initiative. Ensuring team functionality and role clarity (via Team Charters developed in Chap. 7) is a prerequisite for effective visioning work.

Workbook Prep 2: Assessing Team Readiness for Visioning

Contextual Reflection: Does your AI strategy team, as formed in Chap. 7, possess the necessary diversity of perspectives (cognitive, functional, demographic) and collaborative intelligence (Jaskyte et al., 2024) to effectively develop a shared vision? Are roles defined clearly enough (via the Team Charter) to facilitate this strategic work?

UX Workbook Link: Refer back to your Chap. 7 UX Workbook deliverables (Team Charter, Stakeholder Map & Analysis). Review Chap. 9 UX Workbook Section 1 activities which rely on functional Smart Teams.

PALS Workbook Link & AI Partnership Guidance: Engage with Chap. 9 PALS Workbook prompts (Section 1). Describe your team's current composition and defined roles. Ask your GenAI partner (using HCASR C3 & C4) to identify potential gaps in perspectives or needed expertise specifically for the task of visioning and strategic goal setting, perhaps researching common roles involved in institutional strategic planning.

Example Prompt: "Team: [List members/roles]. Task: Develop institutional AI Vision & SMARTIE goals. Based on Jaskyte et al. (2024) or HCASR C4, identify 2–3 missing perspectives (e.g., long-term strategic planning expert, budget authority liaison, dedicated ethicist) potentially crucial for this task."

Human Oversight Reminder: AI suggests generic roles or gaps. Assessing the functional effectiveness and collaborative potential of your specific team requires human insight into interpersonal dynamics, institutional politics, and actual member engagement levels (HCASR Component 4).

Formulating a Shared Vision (Vision/ Opportunity Statements)

Informed by readiness assessments (Chap. 8) and stakeholder mapping (Chap. 7), Smart Teams collaboratively craft the AI Vision Statement and related Opportunity Statements. The **Vision Statement** articulates the desired long-term future state for ethical Human-AI collaboration, aligned with the HEI's mission and values (Bowen, 2018; David, 2020; Kotter et al., 2021). **Opportunity Statements** specify anticipated benefits in key institutional areas like teaching, research, or operations (David, 2020). An effective vision possesses several key characteristics, summarized in Table 9.1. It should be strategically sound, desirable to stakeholders, feasible given assessed capabilities, and easily communicable (Kotter et al., 2021). Furthermore, it should create a sense of urgency by addressing both the need to adapt ("survive") and the potential to innovate ("thrive") through AI integration (Kotter et al.,

Table 9.1 Characteristics of effective vision and opportunity statements

Characteristic	Description	Link to frameworks/ sources
Strategically sound	Aligns with institutional mission, values, and readiness assessment findings.	Kotter et al. (2021); David (2020)
Desirable	Appeals to the long-term interests of key stakeholders (students, faculty, staff, etc.).	Bowen (2018); HCASR Comp. 4
Feasible	Achievable within the institution's capabilities and resources (assessed in Chap. 8).	Kotter et al. (2021)
Communicable	Easy to communicate and explain quickly (e.g., in five minutes or less).	Bowen (2018); Kotter et al. (2021)
Urgency-creating	Addresses both the need to adapt ("survive") and the potential to innovate ("thrive").	Kotter et al. (2021)
Benefit-focused	Clearly articulates the expected positive outcomes and value creation for the institution.	David (2020)

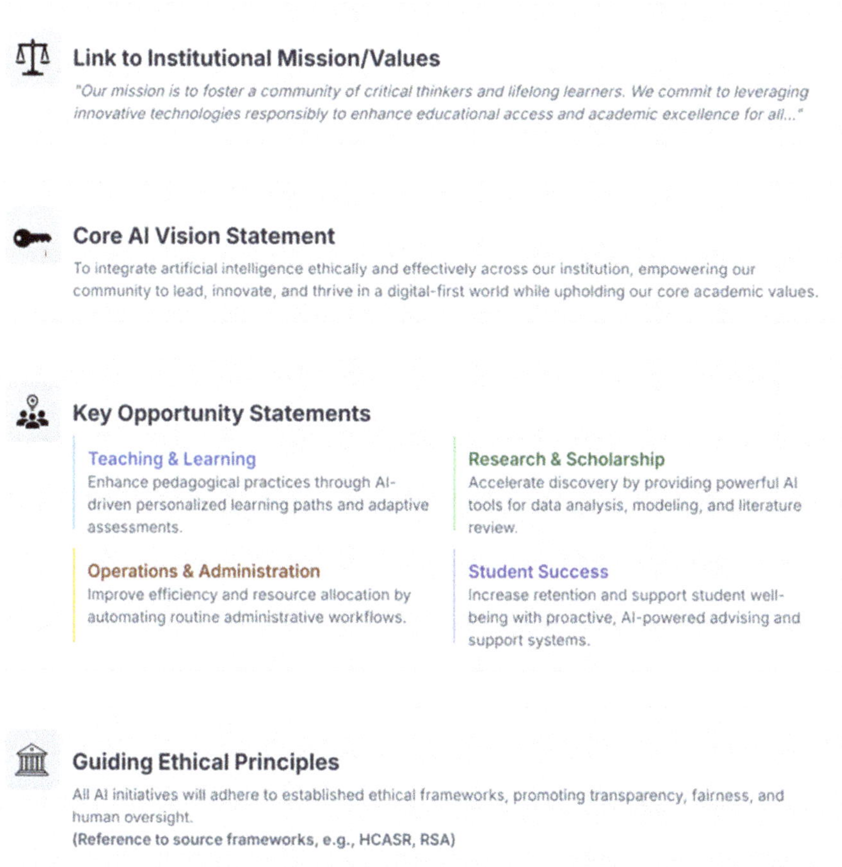

Fig. 9.1 Vision and opportunity statement template

2021). Clear communication of this vision is vital for mobilizing stakeholders and addressing potential resistance. Figure 9.1 offers a template structure.

Vision in Practice: Formulation & Adjustment

The AI vision serves as a living guide, not a static declaration. Its initial formulation requires careful collaboration, integrating assessment findings (Chap. 8) and stakeholder feedback to ensure it is well-grounded and commands broad commitment. Importantly, the vision must be revisited and potentially adjusted over time based on feedback from pilot projects (Chaps. 10, 11, and 12), ongoing monitoring data (Chap. 13+), significant technological advancements, regulatory shifts, or evolving institutional priorities. Established governance structures, such as the AI Council discussed in Chap.

7, should facilitate this periodic review process, ensuring the vision remains relevant, ethically sound, and strategically aligned. This capacity for strategic adaptation is crucial for navigating the dynamic landscape of technological disruption (Kotter et al., 2021).

Workbook Prep 3: Making the Vision Adaptable

Contextual Reflection: Considering AI's rapid evolution, what specific mechanisms (e.g., annual reviews by AI Council, triggers based on tech shifts, pilot evaluation feedback channels) could ensure your institution's AI vision remains a "living guide"? Who should facilitate these reviews?

UX Workbook Link: See Chap. 9 UX Workbook activities related to establishing governance structures (Section 5, Activity 5.1) and incorporating feedback loops (Section 8, Activity 8.3), which support vision adaptability.

PALS Workbook Link & AI Partnership Guidance: Engage with Chap. 9 PALS Workbook prompts (Section 1/5). Input your draft AI vision statement. Ask your GenAI partner (using HCASR Component 5 principles) to brainstorm 3–5 potential internal or external triggers (e.g., significant change in GenAI capabilities, new regulations, major shift in institutional strategic priorities, findings from annual evaluations) that might necessitate a formal review and potential adjustment of the AI vision.

Example Prompt: "Vision: '[Insert concise Vision Statement]'. Brainstorm 4 distinct triggers (technological, institutional, ethical/regulatory, or evaluation-based) that should prompt a formal review and potential revision of this AI vision, referencing HCASR Component 5."

Human Oversight Reminder: AI can suggest common triggers. Determining appropriate review frequency, triggers, and processes requires human strategic judgment considering institutional capacity, stability, and the practicalities of reconvening stakeholders (HCASR Component 5).

Defining Goals, Strategic Planning & Innovation with GenAI

This section focuses on translating the shared vision into concrete strategic goals and a clear roadmap, while fostering an environment supportive of innovation using GenAI. (Note: SMARTIE goals defined in Chap. 8, Table 8.2)

Table 9.2 Asana/anthropic AI maturity model stages (example)

Stage #	Stage name	Key characteristics/focus
1	AI Skepticism	Limited or no AI usage; focus on manual processes; potential resistance or lack of awareness.
2	Activation	Small-scale, isolated AI experiments; individual or small group adoption; focus on learning basic capabilities.
3	Experimentation	Broader, more intentional piloting of AI tools for specific tasks; focus on exploring value and workflows.
4	Scaling	Formalized AI initiatives; expanding successful pilots; focus on integration, standards, and measuring impact.
5	Maturity	AI deeply embedded in strategy and operations; focus on continuous optimization, innovation, and advanced synergy.

Note. Adapted from Marvilano (2024)

Building a Roadmap for AI Value Creation Using Maturity Models

A strategic roadmap operationalizes the vision, outlining the path from the current state (assessed in Chap. 8) to the desired future. AI maturity models can provide a useful structure for assessing progress and planning next steps (e.g., Asana/Anthropic [Marvilano, 2024], Jisc [2024], Brady [2024]). These models typically depict stages of AI adoption, from initial exploration to full integration. Institutions should select or adapt a model that aligns with their context and use it to map their planned progression. The roadmap should define phases, key activities, milestones, resource considerations, and expected value creation aligned with the vision (Marvilano, 2024). Table 9.2 presents an example based on the Asana/Anthropic model, illustrating typical stages. Ongoing evaluation against the roadmap is essential. Figure 9.2 depicts how such a roadmap might be visualized. The Chap. 9 UX Workbook provides activities for developing a tailored roadmap.

Workbook Prep 4: Selecting a Roadmap Framework

Contextual Reflection: Review the AI maturity model concepts presented ([Table 9.2], Jisc, Brady). Which approach (staged progression, continuous capability building, etc.) best aligns with your institution's change culture (assessed in Chap. 8)? Why? Are other roadmap frameworks used that could be adapted?

UX Workbook Link: See Chap. 9 UX Workbook activities focused on developing the Implementation Plan (Section 1, Activity 1.1) and the overall Strategy & Roadmap (Section 7), which involves selecting/adapting a model.

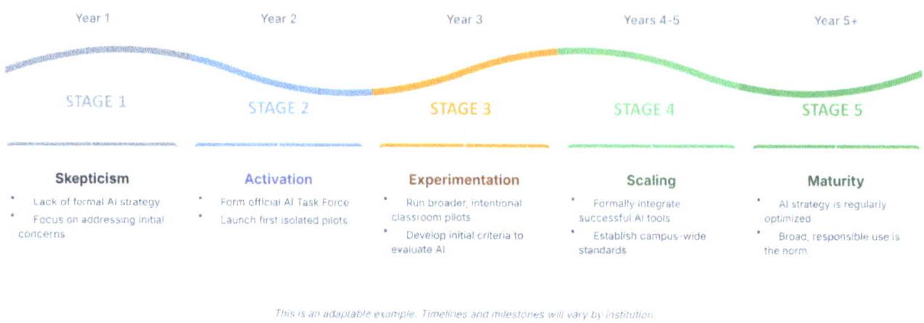

Fig. 9.2 Institutional AI maturity roadmap (adapted)

PALS Workbook Link & AI Partnership Guidance: Engage with Chap. 9 PALS Workbook prompts (Section 2). Describe your institution's typical tech adoption approach. Ask your GenAI partner (using HCASR Component 1 & 3) to compare philosophical approaches of two maturity models (e.g., Asana vs. Jisc) and suggest which might align better with your style.

Example Prompt: "Our institution adopts tech cautiously, preferring extensive piloting. Compare the Asana/Anthropic 5-stage model and the Jisc AI Maturity Toolkit approaches. Which model structure seems more aligned with a cautious, pilot-heavy style (HCASR C1)? Why?"

Human Oversight Reminder: AI compares documented features. Selecting/adapting the best framework requires human understanding of goals, resources, processes, leadership preferences, and cultural nuances (HCASR Component 1). The model informs structure, not prescription.

Key Factors for Success

Successful AI integration, as guided by the roadmap, requires strong **Leadership Commitment** (championing the vision, allocating resources), a supportive **Institutional Culture** (embracing change, ethical experimentation, continuous learning), and a clear **Strategic Focus** (aligning initiatives with mission, values, SMARTIE goals) (Deloitte, 2024; Dahlgren, 2025). Cultivating a **Mindset Geared Towards Adaptation** and learning is essential for navigating the evolving AI landscape (Kotter et al., 2021). Facilitating **Collaboration and Experimentation**, potentially through dedicated "innovation spaces" or controlled "sandboxes," encourages the hands-on prototyping, testing, and iteration needed to unlock human-AI potential responsibly (Brown, 2019). These factors should be considered when developing the roadmap and supporting plans.

Applying Insights: Revisiting the AI Sourcing Strategy

The strategic planning process necessitates revisiting the proprietary versus third-party/open-source AI sourcing decision, first considered in Chap. 6. This re-evaluation analyzes options through the lens of the finalized vision, SMARTIE goals, roadmap structure, and detailed Chap. 8 readiness assessment data (Cameron, 2025; Culotta & Mattei, 2023; Grajek et al., 2025; Nguyen, 2025). Key trade-offs—such as data control, customization needs, deployment speed, total cost of ownership, required internal expertise, risk profile, and IP considerations—must be weighed against strategic priorities and assessed institutional capabilities (Grajek et al., 2025; Inclusion Cloud, 2025; Numbers Station, n.d.). The SMARTIE framework, especially its "Achievable," "Inclusive," and "Equitable" components, provides an essential evaluative lens (The Core Collaborative, 2024; Oregon Health Authority, 2025; U.S. Department of Education, OET, 2024). Table 9.3 guides this strategic re-evaluation by prompting comparison across key factors informed by the readiness assessment and strategic goals.

Table 9.3 Institutional AI maturity roadmap (adapted)

Evaluation factor	Proprietary development (build)	Third-party/open-source integration (buy/adapt)
Alignment with vision	High potential for deep alignment if designed for purpose.	May require compromising on vision to fit tool capabilities.
Data control and security	Maximum control; data stays in-house.	Requires significant vetting; potential for data to leave institutional control.
Customization needs	Fully customizable to specific pedagogy, workflows, and user needs.	Limited to vendor options; may not fit unique contexts.
Deployment speed	Slower; requires full development cycle.	Faster; leverages existing, market-ready solutions.
Total cost of ownership	High upfront and ongoing cost for personnel and infrastructure.	Lower upfront cost, but recurring subscription fees can be high.
Internal capacity match	Requires high internal expertise (AI/ML, Dev, Ethics).	Requires strong vendor management and integration skills.
Ethical risk profile	Full responsibility for ethical design, bias mitigation, and transparency.	Relies on vendor transparency; risk of inheriting bias or opaque processes.
SMARTIE alignment (I&E)	Can be explicitly designed for inclusivity and equity goals.	Depends on vendor's commitment; may have one-size-fits-all approach.

Workbook Prep 5: Evaluating AI Sourcing Against Goals

Contextual Reflection: Select one key institutional SMARTIE goal (Activity 2.1). How does this goal, particularly its Inclusive/Equitable dimensions, influence thinking on proprietary ("build") vs. third-party/open-source ("buy/adapt") sourcing? Which approach seems better positioned to achieve that goal's equity aims, considering Chap. 8 readiness findings?

UX Workbook Link: See Chap. 9 UX Workbook Section 3.B ("Scenario-Based Planning") which directly facilitates comparing proprietary vs. third-party scenarios against strategic goals and readiness data, using frameworks like Table 9.3.

PALS Workbook Link & AI Partnership Guidance: Engage with Chap. 9 PALS Workbook prompts (Section 2/3). Input a specific SMARTIE goal and key constraints from your Chap. 8 assessment (e.g., budget limits, expertise level, data sensitivity). Ask your GenAI partner (using HCASR Component 1) to help analyze the pros and cons of proprietary vs. third-party/open-source solutions for achieving that goal within those constraints.

Example Prompt: "SMARTIE Goal: '[Ensure equitable access to AI tutoring for all first-year students...]'. Constraints: [Limited budget, high data privacy requirements]. Analyze alignment & pros/cons of Proprietary vs. Third-Party solutions for achieving this equity-focused goal within our constraints (HCASR C1)."

Human Oversight Reminder: AI analysis uses provided inputs and documented trade-offs. The final sourcing strategy is complex, balancing priorities, risk, vision, ethics, and resources, requiring human leadership (HCASR Component 4).

Strategic Planning & Roadmap Development

This final stage of planning focuses on developing the comprehensive strategic plan and implementation roadmap, ensuring alignment with broader institutional initiatives.

Components of the Strategic AI Plan

A robust institutional AI strategic plan integrates the vision, goals, sourcing decisions, and readiness findings from Phase 1. It serves as the primary blueprint guiding implementation and typically encompasses several essential components. These include: the updated Vision & Goals; a detailed Governance Framework; defined Priority Initiatives within a Phased Roadmap; supporting plans for Technology & Infrastructure, Talent & Capacity Building, and Resource Allocation; strategies for Change Management & Communication; a comprehensive Risk Management Plan; and a clear Monitoring & Evaluation (M&E) Plan. Table 9.4 provides an overview of these key components. The Chap. 9 UX Workbook (Section 7) guides the

Table 9.4 Key components of an institutional strategic AI plan

Component	Key elements/purpose
1. Updated Vision & Goals	The approved AI Vision statement and overarching institutional SMARTIE goals.
2. Governance Framework	Detailed structure, roles, responsibilities, policies, and processes for AI oversight.
3. Priority Initiatives/ Roadmap	Phased plan of specific AI projects and initiatives, aligned with goals.
4. Technology/ Infrastructure Plan	Plan for necessary hardware, software, platforms, and network upgrades.
5. Talent/Capacity Building Plan	Strategy for developing necessary AI literacy and skills across the institution.
6. Resource Allocation Plan	Details on budget, personnel, and time allocation for strategic initiatives.
7. Change Management/ Comm. Plan	Strategy for managing adoption, communicating progress, and engaging stakeholders.
8. Risk Management Plan	Comprehensive plan for identifying, assessing, and mitigating AI-related risks.
9. Monitoring & Evaluation (M&E) Plan	Framework for tracking progress against goals and measuring the impact of AI initiatives.

detailed development and consolidation of these components into the final strategic plan document.

Ensuring Strategic Alignment and Integration

The AI strategic plan cannot exist in isolation. It must be deliberately aligned with the institution's overall strategic plan, academic priorities, budget cycles, existing IT roadmap, research agenda, data governance policies, enterprise risk management frameworks, and diversity, equity, and inclusion (DEI) commitments (Deloitte, 2024; Grajek et al., 2025). Achieving and maintaining this integration requires ongoing communication and collaboration between AI governance bodies (e.g., AI Council, Smart Teams) and other institutional planning committees. Diverse representation on Smart Teams can facilitate these connections (Jaskyte et al., 2024).

Phases to Transition from Planning to Pilot

The strategic planning deliverables operationalize the vision and set the stage for Phase 2: Implementation & Piloting (starting with Chap. 10). This transition aligns with design thinking's move towards "Implementation" (Brown, 2019) and change management's focus on operationalizing strategy and generating "Short-Term Wins" through pilots (Kotter et al., 2021). Figure 9.3 depicts this transition conceptually.

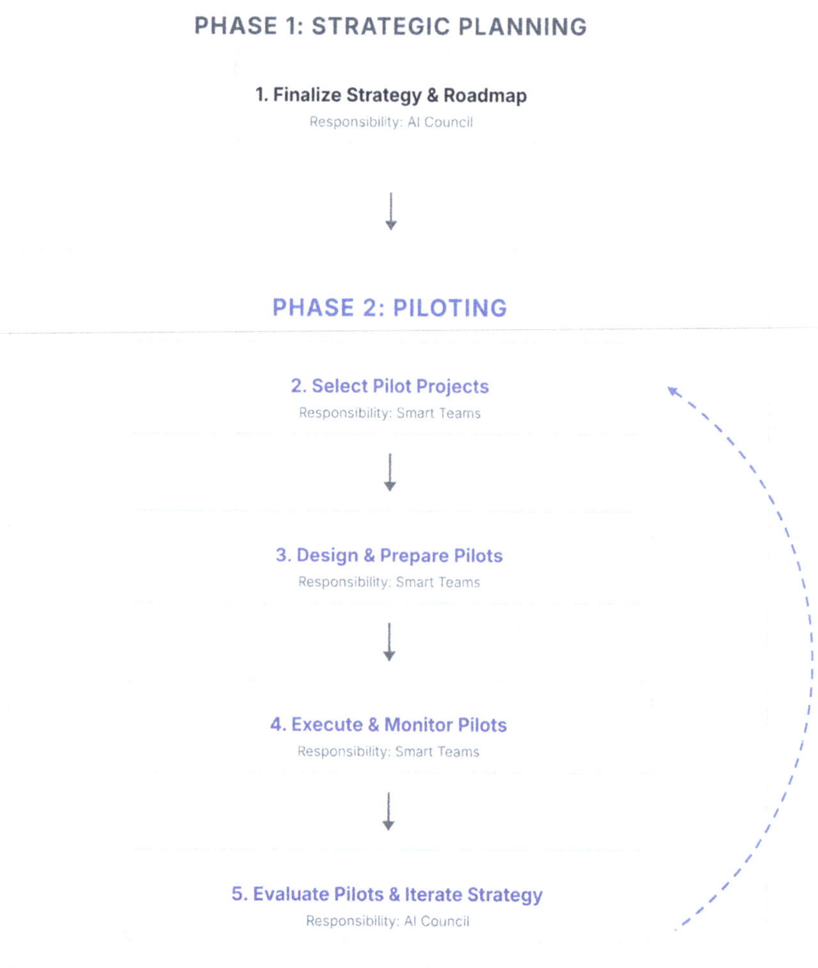

Fig. 9.3 Phased transition plan (planning to pilot)

Workbook Prep 6: Planning for Strategic Alignment

Contextual Reflection: Looking at the key components of a strategic AI plan ([Table 9.4]), which seems most challenging for your institution to develop comprehensively? Which is most critical for ensuring alignment between AI initiatives and broader institutional priorities?

UX Workbook Link: See Chap. 9 UX Workbook activities covering the Implementation Plan (Section 1), Resource Allocation (Section 2.B), Scenario Planning (Section 3.B), Governance Framework (Section 5), and the final consolidated Strategy & Roadmap (Section 7).

PALS Workbook Link & AI Partnership Guidance: Engage with Chap. 9 PALS Workbook prompts (Section 3/7). Select 1–2 strategic plan components (e.g., "Talent & Capacity Building," "Risk Management"); ask your GenAI partner (using HCASR Component 1 & 5) to help brainstorm 3–5 essential sub-sections or key questions for each component to ensure it's comprehensive and actionable.

Example Prompt: "For the 'Talent & Capacity Building Plan' component, brainstorm 4 key sub-sections/questions we must address, considering HCASR Component 5 (e.g., lifelong AI literacy, equitable training access, measuring competency development)."

Human Oversight Reminder: AI structures outlines. Defining specific content, priorities, metrics, resource commitments, and implementation details requires deep institutional knowledge, stakeholder input, collaboration, and strategic decision-making (HCASR Component 4).

User Outcomes & Deliverables (UODs): Finalizing Phase 1 of AI Integration

This strategic planning phase produces a suite of formal User Outcomes and Deliverables (UODs) that guide Phase 2 implementation and subsequent activities, synthesizing planning efforts based on iterative design and change management principles (Brown, 2019; Kotter et al., 2021). Key UODs concluding Phase 1 typically include finalized versions (V1 or V2 as appropriate) of documents covering:

- Smart Team Charter & Governance Framework (Refined)
- Stakeholder Alignment Plan & AI Readiness Report (Consolidated)
- Institutional AI Needs & Gap Analysis (V1 Final)
- AI Task Allocation Summary & Competency Mapping (Initial)

- AI Vision & Opportunity Statement (Approved)
- Institutional AI Strategy & Roadmap (Comprehensive V1), which explicitly details the collective contribution of all User Outcomes and Deliverables (UODs).
- Decision-Making Authority Framework & Pilot Feedback Process (Defined)

The Chap. 9 UX Workbook (Section 7) guides the consolidation and finalization of this UOD package.

Conclusion: Laying the Foundation for Action

Chapter 9 provided the strategic blueprint for AI integration, guiding the development of a shared vision, SMARTIE goals, a phased roadmap, AI sourcing strategies, and key User Outcomes and Deliverables (UODs). It emphasized combining the synergy between design thinking and change management, leveraging collaborative intelligence from Smart Teams. Grounding the vision in readiness data (Chap. 8) and focusing on ethical, inclusive planning (guided by frameworks like HCASR; see HAICS Guide) creates a robust foundation. The comprehensive UODs ensure clarity and alignment for transitioning to Phase 2 implementation and piloting.

Looking Ahead & Engaging the Workbook

Having established the strategic vision, goals, roadmap, and key User Outcomes and Deliverables (UODs), the institution is now prepared to operationalize these plans. The Chap. 9 Companion UX Workbook provides the practical steps and collaborative activities necessary to finalize these strategic documents and prepare for pilot implementation (Phase 2). Engaging with the workbook activities, starting with the "Getting Started: Workbook Prep Exercises," will enable your Smart Teams to translate the strategic framework discussed here into concrete actions, effectively utilizing AI as a strategic partner under human oversight to confidently enter the AI piloting phase described in Chap. 10: Designing Pilots from the Strategic Roadmap.

Conclusion: Laying the Foundation for Action

Chapter 9 provided the strategic blueprint for AI integration, guiding the development of a shared vision, SMARTIE goals, a phased roadmap, AI sourcing strategies, and key deliverables. It emphasized combining the synergy between design thinking and change management, leveraging collaborative intelligence from Smart Teams. Grounding the vision in readiness data (Chap. 8) and focusing on ethical, inclusive planning (guided by frameworks like HCASR; see HAICS Guide) creates a robust foundation. The comprehensive deliverables ensure clarity and alignment for transitioning to Phase 2 implementation and piloting.

Looking Ahead & Engaging the Workbook

Having established the strategic vision, goals, roadmap, and key deliverables, the institution is now prepared to operationalize these plans. The Chap. 9 Companion UX Workbook provides the practical steps and collaborative activities necessary to finalize these strategic documents and prepare for pilot

Workbook Prep 7: Reviewing Strategic Deliverables

Contextual Reflection: Review the list of key Phase 1 concluding deliverables above. Which deliverable do you believe is most crucial for enabling a smooth transition into Phase 2 (Implementation & Piloting)? Why?

UX Workbook Link: See v9 UX Workbook Section 7 ("Deliverables: Finalizing the Strategic Blueprint") and Section 9 ("Deliverables & Outcomes by Role") for guidance on finalizing and understanding each deliverable's purpose.

PALS Workbook Link & AI Partnership Guidance: Engage with Chap. 9 PALS Workbook prompts (Section 7). Select a key deliverable (e.g., "Institutional AI Strategy & Roadmap"); ask your GenAI partner (using HCASR Component 1) to generate a checklist of 5–7 essential elements/characteristics (drawing from components listed in Table 9.4) indicating the document is comprehensive, actionable, and ready for Phase 2.

Example Prompt: "Generate a checklist of 6 essential components/characteristics for the final 'Institutional AI Strategy & Roadmap' deliverable to ensure it is comprehensive and actionable for guiding Phase 2, referencing Chap. 9 Section 3.A and HCASR C1 principles."

Human Oversight Reminder: AI checklists are starting points. The Smart Team and leadership must determine if deliverables adequately address all necessary components based on institutional context, strategic decisions, and implementation readiness (HCASR Component 4).

implementation (Phase 2). Engaging with the workbook activities, starting with the "Getting Started: Workbook Prep Exercises," will enable your Smart Teams to translate the strategic framework discussed here into concrete actions, effectively utilizing AI as a strategic partner under human oversight to confidently enter the AI piloting phase described in Chap. 10: Designing Pilots from the Strategic Roadmap.

Long-Term Scaling, Sustainability & AI Governance: Aligning TLARA with Institutional Strategy

Completing Chap. 9 signifies a major milestone: the articulation of your institution's AI vision, the design of its initial governance structures, and the development of a comprehensive strategic plan. These high-level directives are essential for ensuring that AI integration is purposeful, ethically managed, and aligned with your institutional mission over time (Bowen, 2018). The challenge now is to translate this overarching strategy into actionable, domain-specific guidelines within your custom **TLARA (Teaching, Learning, Assessment, Research, Administration)** framework, ensuring coherence across all operational spaces and levels (Kotter et al., 2021). Both the Chap. 9 UX and PALS Workbooks are structured to assist you in partnering with AI to make these critical strategic connections.

HEIs, as complex adaptive systems, require that strategic plans for innovations like AI are not static documents but living blueprints that inform practice within diverse sociocultural contexts (IDEAS, 2025). The PALS Workbook for Chap. 9, in its section "Long-Term Scaling, Sustainability & AI Governance: Iteratively Developing Your Custom TLARA-Informed AI Guidelines," offers a customizable prompt to facilitate this. It positions your GenAI as a strategic advisor to help you explore how your AI vision, governance model, and strategic priorities should concretely shape the policies and practices within each TLARA domain. Utilizing the "Topics for PALS TLARA Focus"—such as "Crafting a Long-Term AI Vision for Institutional Sustainability"—will ensure your "Living Document" of TLARA guidelines directly supports and enacts your institution's strategic objectives for ethical and effective AI integration (DerSimonian & Montagnino, 2025; Ellucian, 2024).

References

Bowen, S. A. (2018). Mission and Vision. In R. L. Heath & W. Johansen (Eds.), *The international encyclopedia of strategic communication* (pp. 1–6). John Wiley & Sons. https://doi.org/10.1002/9781119010722.iesc0111

Brady, C. (2024, December 12). A strategic framework for action for leading AI integration in higher education. *Dr. Claire Brady Blog*. https://www.drclairebrady.com/post/a-strategic-framework-for-action-for-leading-ai-integration-in-higher-education

Brown, T. (2019). *Change by design: How design thinking transforms organizations and inspires innovation* (Revised and Updated ed.). Harper Business.

Cameron, J. (2025, January 7). Integrating Artificial Intelligence into higher education governance. *Association of Governing Boards of Universities and Colleges (AGB)*. https://agb.org/blog-post/integrating-artificial-intelligence-into-higher-education-governance/

Culotta, A., & Mattei, N. (2023, November 20). Use open source for safer generative AI experiments. *MIT Sloan Management Review*. https://sloanreview.mit.edu/article/use-open-source-for-safer-generative-ai-experiments/

Dahlgren, B. (2025, February 27). How university leaders can ethically and responsibly implement AI. *Campus Technology*. https://campustechnology.com/articles/2025/02/27/how-university-leaders-can-ethically-and-responsibly-implement-ai.aspx

David, F. R. (2020). Analysis of vision and mission statements characteristics and their association with organizational performance: A guide to writing effective vision and mission statements. *Applied Studies in Agribusiness and Commerce – APSTRACT, 14*(1–2), 87–95. https://doi.org/10.19041/APSTRACT/2020/1-2/11

Deloitte Center for Government Insights. (2024, September 25). How higher education can realize the potential of Generative AI. *Deloitte Insights*. https://www2.deloitte.com/us/en/insights/industry/public-sector/generative-ai-higher-education.html

DerSimonian, R., & Montagnino, C. (2025, March 26). Crafting thoughtful AI policy in higher education: A guide for institutional leaders. *Campus Technology*. https://campustechnology.com/articles/2025/03/26/crafting-thoughtful-ai-policy-in-higher-education-a-guide-for-institutional-leaders.aspx

Ellucian. (2024, December 16). AI in higher education: Implementation strategies for institutions and tech partners. *Higher Ed Dive*. https://www.highereddive.com/spons/ai-in-higher-education-implementation-strategies-for-institutions-and-tech/735434/

Grajek, S., Pelletier, K., & Freeman, A. (2025, March 11). AI procurement in higher education: Benefits and risks of emerging tools. *EDUCAUSE Review*. https://er.educause.edu/articles/2025/3/ai-procurement-in-higher-education-benefits-and-risks-of-emerging-tools

Inclusion Cloud. (2025, February 28). What you need to know before choosing open-source or proprietary AI models. *Inclusion Cloud Blog*. https://inclusioncloud.com/insights/blog/open-source-llm-vs-proprietary-models/

Instructional Design, Engagement, and Support (IDEAS). (2025v *AI's impact on higher education through the eyes of faculty: 4 key trends*. University of Massachusetts Amherst. https://www.umass.edu/ideas/news/ais-impact-higher-education-through-eyes-faculty-4-key-trends

Jaskyte, K., Hunter, A., & Mell, A. C. (2024). Predictors of interdisciplinary team innovation in higher education institutions. *Innovative Higher Education, 49*(1), 113–132. https://doi.org/10.1007/s10755-023-09676-3

Jisc. (2024). *AI maturity toolkit for tertiary education*. Retrieved May 5, 2025, from https://www.jisc.ac.uk/ai-maturity-toolkit-for-tertiary-education

Kotter, J. P., Akhtar, V., & Gupta, G. (2021). *Change: How organizations achieve hard-to-imagine results in uncertain and volatile times*. Wiley.

Marvilano. (2024, June 21). The five-stage journey to AI maturity: Asana's roadmap for organizations. *Marvilano Blog*. https://www.marvilano.com/post/the-five-stage-journey-to-ai-maturity-asana-s-roadmap-for-organizations

Nguyen, H. V. (2025, January 9). Open source vs. proprietary AI: A comprehensive guide to making the right choice. *SmartDev*. https://smartdev.com/open-source-vs-proprietary-ai/

Numbers Station. (n.d.). Private vs. third-party foundation models: Navigating the path to an AI-driven future. *Numbers Station Blog*. Retrieved May 5, 2025, from https://www.numbersstation.ai/private-vs-third-party-foundation-models-navigating-the-path-to-an-ai-driven-future/

Oakes, S. (2024, November 8). How to develop effective edTech personas. *Backpack Interactive*. https://backpackinteractive.com/insights/develop-effective-edtech-personas/

Oregon Health Authority. (2025). *SMARTIE goals*. Retrieved May 5, 2025, from https://www.oregon.gov/oha/HPA/dsi-tc/Documents/SMARTIE-goals-handout.pdf

The Core Collaborative. (2024, August 20). *Setting SMARTIE goals*. The Core Collaborative. https://thecorecollaborative.com/setting-smartie-goals/

U.S. Department of Education, Office of Educational Technology. (2024). *Navigating Artificial Intelligence in postsecondary education: Building capacity for the road ahead*. https://portal.ct.gov/das/-/media/das/ctedtech/publications/2025/2025-used-oet-archive/os-24-002833-ai-in-postsecondary-ed.pdf (Note: The original URL was a google search link; a more direct link is preferable if found. The content from this source is cited with (U.S. Department of Education, OET, 2024)).

Wu, D., Dempsey, P., Lin, C.-Y., Sharif, A., & Miller, L. (2024). Charting the course for AI governance in higher education: Insights from a system-wide task force. *Journal of Postsecondary Education and Disability, 37*(1), 7–24. https://files.eric.ed.gov/fulltext/EJ1410943.pdf

10

Designing Context-Specific Pilot Programs (Phase 1)

Introduction: From AI Strategy to Pilot Design

Artificial Intelligence (AI), particularly Generative AI (GenAI), continues to reshape Higher Education Institutions (HEIs), presenting both significant opportunities and complex challenges (U.S. Department of Education, Office of Educational Technology [OET], 2024). Successfully navigating this transformation requires more than technological adoption; it demands careful strategic planning, ethical consideration, and a commitment to building institutional capacity (Askew et al., 2024; U.S. Department of Education, OET, 2024).

Following the comprehensive strategic planning, vision setting, and governance framework development detailed in Phase 1 (Chaps. 6, 7, 8, and 9), this chapter transitions HEIs into the initial stage of Phase 2: Implementation and Piloting. The focus here is on the critical design and preparation for initial, controlled AI pilot projects. This pilot design phase acts as an essential bridge, translating the strategic blueprint established in Chap. 9 into operational reality, which will be explored in Chap. 11. Effective pilot design is crucial for converting broad strategies into specific, testable hypotheses. It involves defining how success will be measured, planning for ethical execution, proactively anticipating barriers, and ensuring pilots are structured to yield valuable learning before any wider scaling is considered (Askew et al., 2024; Brown, 2019). This process translates strategic intent into detailed operational plans, guided by the established roadmap, integrated ethical foresight, and a commitment to structured learning and principles from the synergy between design thinking and change management (Kotter et al., 2021).

Supplementary Information The online version contains supplementary material available at https://doi.org/10.1007/978-3-032-05205-6_10.

D. Woodring, *Advancing Higher Education with Human-Centric AI Partnerships*,
https://doi.org/10.1007/978-3-032-05205-6_10

This chapter integrates the synergy between design thinking (Brown, 2019) and change management (Kotter et al., 2021), now consolidated within the HAICS Guide Part 4. These frameworks guide the design of purposeful pilots, focusing on defining clear, ethical, context-aware goals for AI use and preparing resources (HCASR C1). They also steer planning for ethical implementation and stakeholder engagement, emphasizing human-centric approaches (HCASR C4). Furthermore, they support establishing feedback loops aimed at continuous improvement and advanced human–AI synergy (HCASR C5), and anticipating human-AI collaboration dynamics (HCAS, RSA) during the pilot design phase.

The User Experience (UX) Workbook and the Personalized Adaptive Learning Sessions (PALS) Workbook for Chap. 10 provide practical activities and AI partnership guidance. These resources assist users in ethically and effectively using AI as a strategic partner for researching and implementing tools, solutions, resources, and User Outcomes and Deliverables (UODs) customized to their unique needs and contexts.

Workbook Prep 1: The Rationale for Pilot Design

Reflection: Before launching an AI tool identified in the strategic plan, why is dedicating time to meticulous pilot design—defining specific metrics, planning for ethical governance, and anticipating risks—more beneficial for institutional learning and responsible innovation than simply deploying it broadly? Consider how preparatory work aligns with Design Thinking's prototyping/testing emphasis and Change Management's focus on readiness and managing transitions.

UX Workbook Link: Begin with Chap. 10 UX Workbook "Preparatory Exercises" and Section 1 ("Planning the Pilot") activities, guiding the definition of pilot scope and strategic alignment based on Phase 1 outputs.

PALS Workbook Link & AI Partnership Guidance: Use the Chap. 10 PALS Workbook prompt for UX Section 1. Input relevant strategic goals (Chap. 9) and readiness findings (Chap. 8); ask your GenAI partner to articulate a rationale for this design phase, emphasizing risk mitigation, goal alignment, and learning, referencing HCASR Component 4 (Ethical Co-Creation).

Example Prompt: "Goals: [G1, G2]. Risk/Gap: [Area]. Draft 2–3 sentences explaining why a dedicated design phase for an AI pilot in [Area] is critical before implementation (HCASR C4)."

Human Oversight Reminder: While AI can articulate rationales, the strategic commitment to thorough, ethical pilot design requires human leadership valuing evidence-based practice and foresight (HCASR Component 4). Ensure the rationale fits your institutional context.

Core Frameworks Reference: The frameworks mentioned below, such as HCASR, HCAS, RSA, EHCS, EAO, and AHIPs, are detailed in the companion *Human-GenAI Collaboration & Synergy (HAICS) Guide* and/or Part 1 of this textbook. Please refer to those sections for full definitions and explanations.

This chapter integrates the synergy between design thinking (Brown, 2019) and change management (Kotter et al., 2021), now consolidated within the HAICS Guide Part 4. These frameworks guide the design of purposeful pilots (HCASR Component 1), planning for ethical implementation and stakeholder engagement (HCASR Component 4), establishing feedback loops (HCASR Component 5), and anticipating human-AI collaboration dynamics (HCAS, RSA) during the pilot design phase.

The User Experience (UX) Workbook and the Personalized Adaptive Learning Sessions (PALS) Workbook for Chap. 10 provide practical activities and AI partnership guidance. These resources assist users in ethically and effectively using AI as a strategic partner for researching and implementing tools, solutions, resources, and user outcomes and deliverables (UODs) customized to their unique needs and contexts.

Strategic Roadmap & AI Pilot Design

This section focuses on the critical design elements required before launching AI pilot programs, ensuring they are strategically aligned, purposeful, measurable, and ethically sound. The goal is to translate the comprehensive strategic plan developed in Phase 1 into concrete, actionable pilot blueprints.

Designing Tailored AI Pilot Programs

The first step involves translating the institutional AI strategic roadmap (Chap. 9) into specific, tailored pilot program designs. Pilot selection should strategically address high-priority needs or opportunities identified during Phase 1 assessments (Chap. 8) or target areas offering potential for significant "short-term wins" to build momentum (Kotter et al., 2021). Pilots must be customized for the specific context and stakeholder group(s) involved rather than adopting generic implementations (Askew et al., 2024).

Table 10.1 AI pilot strategy document template

Element	Description/Purpose
1. Problem/Opportunity	Clearly defines the specific institutional challenge or opportunity the pilot aims to address.
2. AI Tool(s)	Specifies the particular AI tool, platform, or model to be used in the pilot.
3. Target Users/ Participants	Identifies the specific stakeholder groups (e.g., students in Course X, staff in Dept Y) who will participate.
4. Scope and Duration	Defines the boundaries of the pilot, including its timeline, what is included, and what is excluded.
5. Pilot Objectives	Outlines 2–3 specific, measurable objectives aligned with institutional SMARTIE goals.
6. Success Metrics	Details the Key Performance Indicators (KPIs) and qualitative measures used to evaluate success.
7. Key Activities & Timeline	Provides a high-level plan of major activities (e.g., training, deployment, feedback sessions) and milestones.
8. Ethical Considerations & Risk Assessment	Describes how the pilot will adhere to ethical guidelines and manage identified risks.
9. Data Collection Plan	Outlines how data for success metrics and user feedback will be collected, managed, and analyzed.
10. Responsible Team	Names the Smart Team or individuals responsible for overseeing the pilot.

The detailed planning for each pilot should be captured in an **AI Pilot Strategy Document**. This document serves as the core blueprint, outlining key areas such as the specific problem/opportunity, the AI tool(s) involved, target users/participants, scope and duration, specific pilot objectives aligned with institutional SMARTIE goals, success metrics, key activities and timelines, how ethical considerations and risks will be addressed, the data collection plan, and the responsible team. Table 10.1 summarizes these essential elements. Developing this comprehensive document ensures pilots are purposeful and contribute directly to the institution's strategic learning objectives (Fig. 10.1).

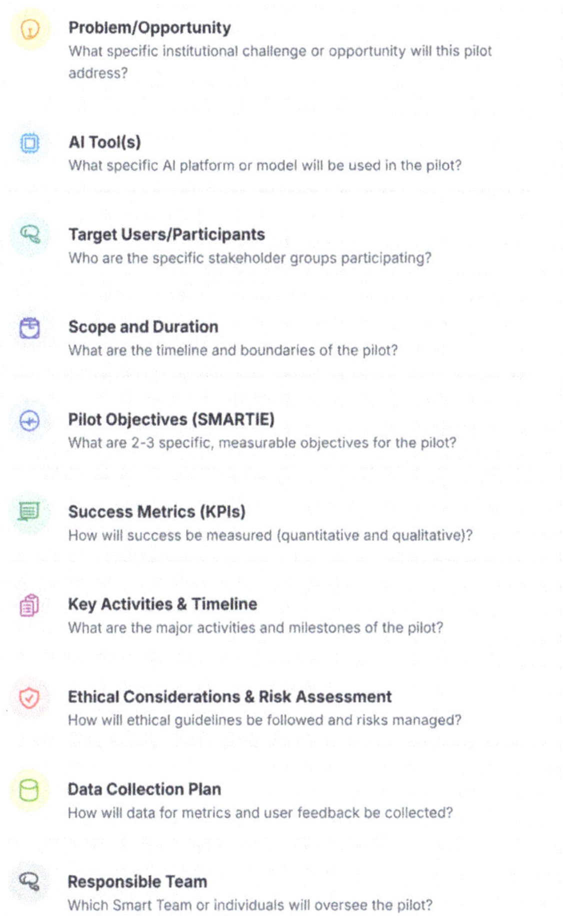

Fig. 10.1 AI pilot design template

Workbook Prep 2: Selecting a Pilot Project

Reflection: Consider your strategic AI roadmap (Chap. 9) and readiness findings (Chap. 8), particularly the Priority Gaps vs. Quick Wins matrix. Based on this data, what specific high-priority area or potential "quick win" seems most suitable for an initial AI pilot? Justify why it meets criteria like addressing a significant need, potential impact, and feasibility given readiness.

UX Workbook Link: Use insights when working through Chap. 10 UX Workbook, Section 1 activities, specifically summarizing pilot alignment with strategy and defining pilot scope/objectives.

PALS Workbook Link & AI Partnership Guidance: Use the Chap. 10 PALS Workbook prompt for UX Section 1. Input the potential pilot area and its connection to institutional goals; ask your GenAI partner to help brainstorm elements for the AI Pilot Strategy Document (e.g., target users, potential tools, activities, timelines), referencing HCASR Component 1 (Ethical & Effective Goal Setting) and Component 4 (Human-Centric Implementation).

Example Prompt: "Pilot: ['AI tool for library research support'] for goal [Goal]. Brainstorm 3–4 pilot timeline activities & list 2–3 participating stakeholder groups (HCASR C1 & C4)."

Human Oversight Reminder: AI can generate ideas. Defining realistic scope, selecting tools, outlining feasible activities, and identifying participants requires human expertise, contextual knowledge, and collaborative input (HCASR Component 4).

Establishing Benchmarks, Success Metrics, and AI Impact Assessment Scope

Defining how pilot success will be measured is a critical component of the design phase. This requires establishing clear, measurable **Success Metrics** aligned with pilot objectives and institutional SMARTIE goals. Key Performance Indicators (KPIs) should track objective outcomes related to efficiency, performance, adoption, financial impact, or specific learning improvements (Cloud Security Alliance, 2025). Examples include user adoption rates, task completion time savings, error rate reductions, system uptime, cost savings, or improvements in specific learning outcomes (Cloud Security Alliance, 2025; Mah & Pietsch, 2025). Crucially, **Baseline Institutional Readiness Metrics**—capturing the pre-pilot state of relevant quantitative KPIs and qualitative data (e.g., user satisfaction, AI literacy levels from Chap. 8)—must be documented before the pilot begins for comparison.

Table 10.2 Categories of pilot evaluation metrics & assessment areas

Category	Type/Focus	Examples
Pilot Metrics	**Quantitative KPIs**	User adoption rates, task completion time, error rate reduction, system uptime, cost savings, specific learning outcome improvements.
	Qualitative Data	User experience feedback, perceived value, satisfaction scores, contextual insights, ethical concerns, suggestions for improvement.
AI Impact Assessment Scope	**Broader Evaluation Areas**	Pedagogical effectiveness, ethical compliance (fairness, bias), accessibility, impact on roles/skills, alignment with institutional values.

Pilot metrics should encompass both quantitative and qualitative dimensions (Table 10.2). Quantitative KPIs provide objective measures. **Qualitative Data** planning focuses on capturing user experience, perceived value, satisfaction, context, and potential ethical concerns, often using methods like surveys, feedback logs (e.g., PALS logs), or interviews (Digital Promise, 2023). Comparing pre- and post-implementation data provides a clearer picture of impact (Cloud Security Alliance, 2025).

Beyond specific metrics, the design phase must also define the scope of the **AI Impact Assessment** planned for later evaluation (e.g., Chap. 13). This involves determining how the pilot will gather data relevant to broader areas such as pedagogical effectiveness, ethical compliance (fairness, transparency, privacy, bias mitigation), accessibility, impact on roles/skills, and alignment with institutional values (Askew et al., 2024; U.S. Department of Education, OET, 2024; Wynants et al., 2025). Defining metrics and assessment scopes upfront ensures the pilot generates meaningful evaluation evidence.

Workbook Prep 3: Defining Meaningful Metrics

Reflection: For the selected pilot project, what single quantitative metric (e.g., time saved) AND single qualitative indicator (e.g., user feedback theme, satisfaction score) would be most essential for evaluating success? How do these align with pilot objectives and institutional SMARTIE goals?

UX Workbook Link: Use reflections when completing Chap. 10 UX Workbook, Section 1 (defining initial success metrics) and Section 3 (outlining Usability Testing Plan and establishing baseline metrics).

PALS Workbook Link & AI Partnership Guidance: Use the Chap. 10 PALS Workbook prompt for UX Section 2. Input pilot objectives and target users; ask your AI partner to brainstorm potential quantitative KPIs and qualitative methods, referencing HCASR Component 1 (Measurability) and Component 5 (Monitoring & Evaluation).

Example Prompt: "Objective: '[Reduce faculty grading time by 20%]'. Users: [Faculty Dept X]. Suggest 2 KPIs & 2 qualitative methods for measuring success (HCASR C1 & C5)."

Human Oversight Reminder: AI suggests standard metrics. Selecting meaningful, measurable, relevant metrics that capture intended impacts (including equity) requires critical human evaluation and alignment with the planned AI Impact Assessment scope (HCASR Component 1 & 5).

Integrating Ethical Governance & Risk Management into Pilot Design

AI pilot projects must operate within the ethical governance structures established in Phase 1 (Chap. 9) and incorporate proactive risk management. Applying customized institutional frameworks (developed in Part 1/Phase 1) or established external guidelines during design enhances this integration. Aligning internal principles (e.g., ETHICAL Principles [Wynants et al., 2025]) with recognized external frameworks like the NIST AI Risk Management Framework (AI RMF), UNESCO recommendations, and OECD AI Principles provides a robust foundation (Tables 10.3 and 10.5).

The NIST AI RMF offers a practical structure (Govern, Map, Measure, Manage) for identifying, assessing, and managing risks (NIST, 2023). UNESCO and OECD provide high-level ethical guidance (human rights, fairness, transparency, accountability) (OECD, 2025; UNESCO, 2022). Considering risk-based approaches (e.g., EU AI Act [Nguyen, 2025]) and national guidance (e.g., Canada's ISED framework [Montreal AI Ethics Institute, 2025]) offers valuable benchmarks. Institutions should adapt relevant principles guided by HCASR Component 1.

Therefore, the AI Pilot Strategy Document must detail planned operational protocols (Table 10.4) for key areas:

Table 10.3 External ethical & risk frameworks informing AI pilot design

Framework	Key Contribution to Pilot Design
NIST AI RMF	Provides a structured process (Govern, Map, Measure, Manage) for identifying, assessing, and managing pilot-specific risks.
UNESCO & OECD Principles	Offer high-level ethical guidance on human rights, fairness, transparency, and accountability to shape pilot goals and oversight.
EU AI Act Considerations	Provides benchmarks for risk classification (e.g., high-risk applications) that can inform the level of scrutiny and governance required for a pilot.

Table 10.4 Required ethical & governance protocols in AI pilot design

Protocol Area	Purpose/Focus in the Pilot Plan
Data Governance	Ensuring compliance with regulations (e.g., FERPA), secure data handling, and clear data usage policies for the pilot.
Bias Monitoring & Mitigation	Planning for how potential biases in the AI tool's outputs or operation will be assessed and addressed during the pilot.
Transparency & Explainability	Defining how participants will be informed about the AI's use, its capabilities, its limitations, and how it uses their data.
Recourse & Reporting	Establishing clear channels for users to report issues, raise concerns, or appeal AI-driven recommendations or decisions.
Human Oversight	Specifying the required levels and methods of human involvement in reviewing AI outputs or making final decisions.

Table 10.5 Alignment of internal (ETHICAL example) and external ethical frameworks for pilot design

Internal Principle (ETHICAL)	Corresponding External Framework Principle(s) (UNESCO/OECD/NIST)	Application Focus in Pilot Design
Exploration	Innovation, Research	Designing pilots to test novel applications and gather data on effectiveness.
Transparency	Transparency & Explainability	Ensuring pilot participants understand how the AI works and how their data is used.
Human-Centered	Human Oversight & Control, Human Rights	Prioritizing user well-being and ensuring human agency in the pilot process.
Integrity	Accountability & Responsibility	Maintaining academic and operational integrity; establishing clear lines of responsibility.
Continuous Learning	Robustness, Security & Safety	Using pilot feedback to iteratively improve the AI tool and support processes.
Accessibility	Fairness & Non-discrimination	Designing pilots to be inclusive and accessible to all participants, including those with disabilities.
Legal Compliance	Rule of Law & Accountability	Ensuring the pilot adheres to all relevant laws and institutional policies (e.g., FERPA, privacy).

- **Data Governance:** Compliance with regulations (FERPA/GDPR), data handling.
- **Bias Monitoring & Mitigation:** Assessing and addressing potential bias.
- **Transparency & Explainability:** Communicating AI use and limitations.
- **Recourse & Reporting:** Channels for issues, concerns, appeals.
- **Human Oversight:** Required levels of human involvement.

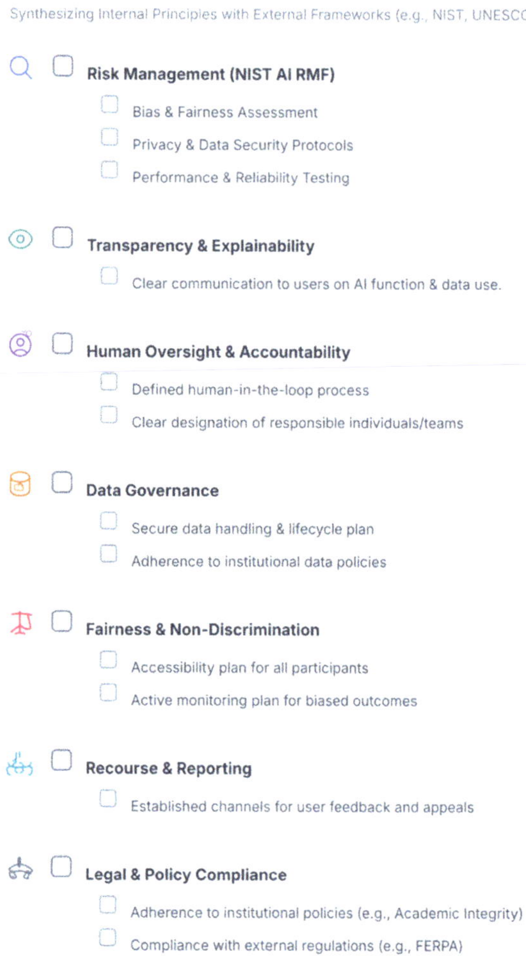

Fig. 10.2 Integrated governance checklist for AI pilots

The designated AI Ethics Board or Governance Council (Phase 1) must review and approve these protocols and the initial Risk Register before pilot launch (FAMU Enterprise Risk Management, n.d.; Montreal AI Ethics Institute, 2025). Figure 10.2 conceptualizes an integrated checklist.

User Outcomes & Deliverables (UODs) for Pilot Design

The pilot design phase culminates in three essential planning User Outcomes and Deliverables (UODs) serving as the blueprint for implementation (Chap. 11) and evaluation (Chaps. 12 and 13). The **AI Pilot Strategy Document** is the comprehensive plan detailing objectives, scope, metrics, and ethical protocols. The **Institutional Readiness Metrics (Baseline)** document provides the crucial pre-pilot data against which impact will be measured. The **Risk Register (Initial)** identifies potential pilot risks and mitigation strategies. These three UODs are interrelated: the strategy document outlines the "what" and "how," the baseline metrics provide the "starting point" for measurement defined in the strategy, and the risk register addresses potential "what ifs" that could affect the strategy's execution. (See UX Workbook Activity 10.1 and 10.2 for development details.)

- AI Pilot Strategy Document
- Institutional Readiness Metrics (Baseline)
- Risk Register (Initial)

Preparing for Pilot Implementation: Governance, Barriers, and Engagement

This section addresses practical preparations before launching pilots, focusing on anticipating barriers, confirming governance readiness, and planning stakeholder engagement during the pilot.

Anticipating and Planning for Barriers During Implementation

Launching pilots often encounters barriers. Proactive planning can mitigate common technical hurdles (integration issues), human factors (resistance, skill gaps, time constraints), or procedural challenges (policy ambiguity, infrastructure limitations) (Mah & Pietsch, 2025; Mueller & Chiou, 2025). Understanding faculty/staff perspectives, potential resistance, and varying AI literacy is crucial (Mah & Pietsch, 2025). Continuous stakeholder engagement, planned during design, is vital for identifying barriers early (Askew et al., 2024).

Establishing robust feedback loops (e.g., PALS logs, check-ins, communication channels) allows teams to adapt based on user experience (Digital Promise, 2023). Diverse "Smart Teams" (Kotter et al., 2021) anticipate how feedback will be analyzed and problems troubleshooted collaboratively (Jaskyte et al., 2024). Effective change management involves clear communication, fostering innovation culture, and adapting strategies based on feedback (Askew et al., 2024).

Workbook Prep 4: Proactive Barrier Management Planning

Reflection: For the selected pilot, what is one significant potential barrier (technical, human/cultural, procedural)? What concrete, proactive step could the design team plan now to mitigate it (e.g., targeted training, communication plan, technical workaround, stakeholder workshop)?

UX Workbook Link: Use insights when working through Chap. 10 UX Workbook, Section 2 activities (Barrier Analysis, Pilot Change Management Plan outline).

PALS Workbook Link & AI Partnership Guidance: Use the Chap. 10 PALS Workbook prompt for UX Section 2. Input anticipated barriers; ask AI partner to brainstorm mitigation strategies emphasizing stakeholder engagement/feedback, referencing HCASR Component 4.

Example Prompt: "Barrier: ['Faculty resistance due to perceived time constraints for learning new AI tool']. Brainstorm 2 mitigation strategies involving faculty engagement/feedback (HCASR C4)."

Human Oversight Reminder: AI suggests generic tactics. Developing effective, culturally appropriate, resource-feasible plans requires deep understanding of context, concerns, resources, and collaborative problem-solving (HCASR Component 4).

Preparing for Ethical Compliance & AI Governance in Practice

Ethical principles and governance frameworks require clear implementation and monitoring plans before pilot launch. Preparation involves detailing how ethical practices will be operationalized (Chap. 11). Key activities include:

- Detailing execution/monitoring of risk mitigation strategies (Initial Risk Register) (Montreal AI Ethics Institute, 2025)
- Scheduling audits/compliance checks (bias, data protection) (FAMU Enterprise Risk Management, n.d.)
- Confirming feedback channels for ethical concerns are ready
- Clarifying procedures for required human oversight (Al-Omari et al., 2025; Montreal AI Ethics Institute, 2025)
- Establishing reporting protocols for performance/incidents to governance bodies
- Planning how pilot experiences inform future framework updates (HCASR Component 5) (FAMU Enterprise Risk Management, n.d.)

Diverse teams should participate (Jaskyte et al., 2024). Considering relevant case studies (Wu et al., 2024), risk classifications (Temper et al., 2025), and human-AI collaboration ethics research (Al-Omari et al., 2025; Ferk Savec & Jedrinović, 2025; Owoc et al., 2021) informs robust planning.

Workbook Prep 5: Planning for Ethical Operations

Reflection: Considering institutional principles (Ch 6/9) and the specific pilot, what one crucial action must the team plan now to operationalize an ethical principle like transparency or fairness from day one (Chap. 11)? (e.g., How inform participants? How monitor bias initially?)

UX Workbook Link: Use reflection when engaging with Chap. 10 UX Workbook, Section 3 (Developing Initial Pilot Risk Register, Planning Ethical AI Testing/Reviews) and Section 4 (Planning Compliance Monitoring).

PALS Workbook Link & AI Partnership Guidance: Use the Chap. 10 PALS Workbook prompt for UX Section 3. Input a specific ethical risk; ask AI partner to suggest 2–3 concrete monitoring actions or operational protocols for implementation (Chap. 11), referencing HCASR Component 4 & 5.

Example Prompt*:* "Risk: ['Potential for AI grading feedback to perpetuate bias against non-native English speakers']. Suggest 2 practical monitoring actions/checks for implementation (HCASR C4 & C5)."

Human Oversight Reminder: AI suggests standard techniques. Designing effective ethical oversight requires human judgment, awareness of tool limitations/biases, context consideration, and team commitment (HCASR Component 4). Framework updates require human deliberation.

Conclusion

Phase 2 commences not with immediate implementation, but with meticulous pilot design and preparation. This chapter emphasized translating the strategic roadmap into tailored, measurable, and ethically grounded pilot blueprints. Key activities included defining pilot objectives, establishing benchmarks, scoping impact assessment, integrating governance and risk management (drawing from institutional and external frameworks like NIST (2023), UNESCO (2022), and OECD (2025)), anticipating barriers, and establishing feedback loops. The core UODs—the AI Pilot Strategy Document, Baseline Readiness Metrics documentation, and the Initial Risk Register—provide the essential foundation for launching effective and responsible pilots in Chap. 11. This careful preparation ensures subsequent implementation is purposeful and yields the data needed for evaluation and decision-making in Phase 3.

Long-Term Scaling, Sustainability & AI Governance: Piloting TLARA-Informed Innovations

Chapter 10 has guided you through the crucial process of translating your strategic AI roadmap into tangible pilot designs. Well-designed pilots are essential learning **spaces** that allow institutions to test AI applications within specific **TLARA (Teaching, Learning, Assessment, Research, Administration)** domains before committing to full-scale implementation (Ellucian, 2024). This phase is critical for understanding how AI tools function in your unique sociocultural context, identifying potential challenges, and gathering data to inform decisions about long-term sustainability and ethical scaling over time. Both the Chap. 10 UX and PALS Workbooks provide frameworks for ethically partnering with AI to design these vital learning experiences and anticipate implementation barriers.

HEIs, as complex systems, benefit immensely from such iterative, evidence-gathering stages (IDEAS, 2025). The PALS Workbook for Chap. 10, through its section "Long-Term Scaling, Sustainability & AI Governance: Iteratively Developing Your Custom TLARA-Informed AI Guidelines," prompts you to consider how pilot design itself can be a step in refining your TLARA framework. By engaging with the "Topics for PALS TLARA Focus"—such as "Designing Pilots for Scalability and Sustainability Insights" and "Integrating Ethical Governance into Pilot Design"—your GenAI partner will assist you in developing pilot-specific ethical considerations. This ensures your "Living Document" of TLARA guidelines evolves with practical insights, preparing for effective governance and the potential for wider, responsible AI adoption (DerSimonian & Montagnino, 2025).

Looking Ahead & Engaging the Workbook

Chapter 10 provided the critical bridge from strategic planning (Phase 1) to action by detailing the essential steps for designing effective AI pilot programs. The Chap. 10 Companion UX Workbook offers the practical activities needed to execute this design phase. Engaging with the workbook will enable Smart Teams to collaboratively create the AI Pilot Strategy Document, define metrics, develop the initial Risk Register, and plan for stakeholder engagement. This produces the key planning UODs required to confidently proceed to the active implementation stage detailed in Chap. 11: Implement & Pilot.

References

Al-Omari, O., Alyousef, A., Fati, S., Shannaq, F., & Omari, A. (2025). Governance and ethical frameworks for AI integration in higher education: Enhancing personalized learning and legal compliance. *Journal of Ecohumanism, 4*(2), 80–86. https://doi.org/10.62754/joe.v4i2.5781

Askew, T., Mathew, R., Fishman, T., Kunkel, D., & Caron, B. (2024, September 25). How higher education can realize the potential of Generative AI. *Deloitte Insights*. https://www2.deloitte.com/us/en/insights/industry/public-sector/generative-ai-higher-education.html

Brown, T. (2019). *Change by design: How design thinking creates new alternatives for business and society* (Revised and Updated ed.). Harper Business.

Cloud Security Alliance (CSA). (2025, March 28). A guide on how AI pilot programs are shaping enterprise adoption. *Cloud Security Alliance Blog*. https://cloudsecurityalliance.org/articles/a-guide-on-how-ai-pilot-programs-are-shaping-enterprise-adoption

DerSimonian, R., & Montagnino, C. (2025, March 26). Crafting thoughtful AI policy in higher education: A guide for institutional leaders. *Campus Technology*. https://campustechnology.com/articles/2025/03/26/crafting-thoughtful-ai-policy-in-higher-education-a-guide-for-institutional-leaders.aspx

Digital Promise. (2023, December 7). Principles for building effective feedback loops. https://digitalpromise.org/2023/12/07/principles-for-building-effective-feedback-loops/

Ellucian. (2024, December 16). AI in higher education: Implementation strategies for institutions and tech partners. *Higher Ed Dive*. https://www.highereddive.com/spons/ai-in-higher-education-implementation-strategies-for-institutions-and-tech/735434/

FAMU Enterprise Risk Management. (n.d.). *Risk Alert - Generative AI*. Florida A&M University. [Institutional Document - Verify Publication Date if possible] Retrieved May 5, 2025, from https://www.google.com/search?q=https://www.

famu.edu/administration/chief-operating-officer/enterprise_risk_management/_includes/docs/Risk%2520Alert%2520-%2520Generative%2520AI.pdf

Ferk Savec, V., & Jedrinović, S. (2025). The role of AI implementation in higher education in achieving the sustainable development goals: A case study from Slovenia. *Sustainability, 17*(1), 183. [Verify Publication Status/Details]. https://doi.org/10.3390/su17010183

Instructional Design, Engagement, and Support (IDEAS). (2025, January 27). *AI's impact on higher education through the eyes of faculty: 4 key trends*. University of Massachusetts Amherst. https://www.umass.edu/ideas/news/ais-impact-higher-education-through-eyes-faculty-4-key-trends

Jaskyte, K., Hunter, A., & Mell, A. C. (2024). Predictors of interdisciplinary team innovation in higher education institutions. *Innovative Higher Education, 49*(1), 113–132. https://doi.org/10.1007/s10755-023-09676-3

Kotter, J. P., Akhtar, V., & Gupta, G. (2021). *Change: How organizations achieve hard-to-imagine results in uncertain and volatile times*. Wiley.

Mah, D., & Pietsch, M. (2025). Perspectives of academic staff on artificial intelligence in higher education: Exploring areas of relevance. *Frontiers in Education, 10*. [Verify Publication Status/Details] https://www.google.com/search?q=https://doi.org/10.3389/feduc.2025.1484904

Montreal AI Ethics Institute. (2025, March 16). *ISED launches AI Risk Management Guide based on voluntary code*. Retrieved May 5, 2025, from https://montrealethics.ai/ised-launches-ai-risk-management-guide-based-on-voluntary-code/

Mueller, B., & Chiou, E. K. (2025). A work system team perspective for AI in higher education. *Communication Education, 74*(1), 84–103. https://doi.org/10.1080/03634523.2024.2449059

National Institute of Standards and Technology (NIST). (2023, January). *Artificial Intelligence Risk Management Framework (AI RMF 1.0)* (NIST AI 100-1). U.S. Department of Commerce. https://doi.org/10.6028/NIST.AI.100-1

Nguyen, N. (2025, February 12). *From regulation to innovation: What the EU AI Act means for EdTech*. FeedbackFruits. [Verify Date - Likely Error] Retrieved May 5, 2025, from https://feedbackfruits.com/blog/from-regulation-to-innovation-what-the-eu-ai-act-means-for-edtech

OECD. (2025). *OECD AI principles overview*. OECD.AI Policy Observatory. [Verify Date - Principles 2019/2024] Retrieved May 5, 2025, from https://oecd.ai/en/ai-principles

Owoc, M. L., Sawicka, A., & Weichbroth, P. (2021). Artificial intelligence technologies in education: Benefits, challenges and strategies of implementation. *arXiv preprint arXiv:2102.09365*. https://arxiv.org/abs/2102.09365

Temper, M., Tjoa, S., & David, L. (2025). Higher Education Act for AI (HEAT-AI): A framework to regulate the usage of AI in higher education institutions. *Frontiers in Education, 10*, 1505370. https://doi.org/10.3389/feduc.2025.1505370

U.S. Department of Education, Office of Educational Technology (OET). (2024). *Navigating Artificial Intelligence in postsecondary education: Building capacity for the road ahead*. https://portal.ct.gov/das/-/media/das/ctedtech/publications/2025/2025-used-oet-archive/os-24-002833-ai-in-postsecondary-ed.pdf

UNESCO. (2022). *Recommendation on the ethics of artificial intelligence*. UNESCO Publishing. https://unesdoc.unesco.org/ark:/48223/pf0000381137

Wu, C., Zhang, H., & Carroll, J. M. (2024). AI governance in higher education: Case studies of guidance at Big Ten universities. *Future Internet, 16*(10), 354. https://doi.org/10.3390/fi16100354

Wynants, S., Childers, G., De La Torre Roman, Y., Budar-Turner, D., & Vasquez, P. (2025, February). *ETHICAL principles AI framework for higher education*. CSU AI Commons. California State University, Fullerton. [Verify Date] https://genai.calstate.edu/communities/faculty/ethical-and-responsible-use-ai/ethical-principles-ai-framework-higher-education/

11

Phase 2: Implementing Human-Centric AI Initiatives

Introduction: Transitioning from Blueprint to Action

Following the strategic planning, readiness assessment, and pilot design undertaken in Phase 1 (Chapters 6, 7, 8, 9, and 10), this chapter delves into the core activities of Phase 2: Implementation and Piloting. This phase marks the transition from theoretical blueprints to tangible, real-world experimentation within the Higher Education Institution (HEI). It involves launching initial AI applications, often structured as Minimum Viable Products (MVPs) or prototypes, within controlled environments (Payne, 2025). This deliberate, iterative approach allows institutions to test feasibility, gather essential user feedback through mechanisms like Daily Feedback Logs, refine technical and ethical approaches based on practical experience, and validate alignment with institutional goals before committing to wider deployment (Behling & Law, 2020). This phase serves as a crucial bridge between strategy and operations, demanding agility, careful change management, and a commitment to learning from practice.

Successfully navigating this phase involves integrating planning insights with practical, adaptable implementation strategies. As HEIs operate within diverse sociocultural contexts, there is no single "correct" path for AI implementation (Carmody-Bubb, 2023). Therefore, the frameworks presented emphasize customization and leveraging AI itself as a strategic partner. Combining the iterative cycles of design thinking (Brown, 2019) with the structured approach of change management (Kotter et al., 2021) provides a robust methodology. This involves rapid prototyping based on user experience alongside change management tactics—such as mobilizing guiding coalitions (Smart Teams), communicating effectively, removing barriers, and celebrating wins—to navigate human dynamics and maintain momentum (Kotter et al., 2021; Kang et al., 2022).

Supplementary Information The online version contains supplementary material available at https://doi.org/10.1007/978-3-032-05205-6_11.

D. Woodring, *Advancing Higher Education with Human-Centric AI Partnerships*,
https://doi.org/10.1007/978-3-032-05205-6_11

Workbook Prep 1: Embracing Iteration and Context

Reflection: As you prepare to launch the pilot described in Chap. 10, why is adopting an iterative approach—launching an MVP, collecting feedback (like via Daily Logs), and refining—more suitable for complex AI implementations in HEIs than a large, fixed rollout? Consider alignment with Design Thinking's prototyping/testing and Change Management's navigation of uncertainty (Brown, 2019; Kotter et al., 2021).

UX Workbook Link: Begin with Chap. 11 UX Workbook "Preparatory Exercises" and Section 1 ("Planning: Finalizing Tools, Ethics, and Alignment for Pilot Implementation") activities to confirm the final pilot setup, considering context and iteration.

PALS Workbook Link & AI Partnership Guidance: Use the Chap. 11 PALS Workbook prompt for UX Section 1. Input pilot goals (Ch 10) and key context factors (Ch 8 readiness); ask your GenAI partner for a brief justification for an iterative pilot in this context, highlighting risk reduction and user-centered refinement (referencing HCASR Component 4).

Example Prompt: "Pilot Goal: [Goal]. Context: [Factor 1], [Factor 2]. Draft brief justification for iterative pilot in this context (HCASR C4)."

Human Oversight Reminder: AI articulates general benefits. Determining critical context factors and appropriate iteration scope/pace requires human judgment informed by institutional knowledge, resources, and stakeholder input (HCASR Component 4).

Core Frameworks: The frameworks mentioned below, such as HCASR, HCAS, RSA, EHCS, EAO, and AHIPs, are detailed in the companion *Human-GenAI Collaboration & Synergy (HAICS) Guide* and/or Part 1 of this textbook. Please refer to those sections for full definitions and explanations.

This chapter uses these frameworks, combined with the synergy between design thinking and change management (see HAICS Guide Part 4), to guide the transition from planning to action, emphasizing iterative development (HCASR Component 4), ethical implementation (HCASR Component 1, 4), dynamic feedback loops (HCASR Component 5), stakeholder empowerment (HCASR Component 4), and effective human-AI collaboration (HCAS, RSA) during the piloting phase.

The User Experience (UX) Workbook and the Personalized Adaptive Learning Sessions (PALS) Workbook for Chap. 11 provide practical activities and AI partnership guidance. These resources assist users in ethically and effectively using AI as a strategic partner for researching and implementing tools, solutions, resources, and User Outcomes & Deliverables (UODs) customized to their unique needs and contexts.

Pilot Objectives: Empowering Stakeholders and Enabling Continuous Assessment

The Implementation and Pilot phase (Phase 2) is driven by two primary, interconnected objectives building upon Phase 1 planning ([Table 11.1]):

1. **Empowering Stakeholders:** Actively engaging diverse users (students, faculty, staff, administrators) as participants and co-creators, not merely testers (Kang et al., 2022; Weil & Forrester, 2024). This requires transparent communication, clear roles, accessible training to build AI literacy and confidence, and fostering shared ownership (DerSimonian & Montagnino, 2025; Weil & Forrester, 2024; HCASR Component 4).
2. **Continuous Assessment & Feedback:** Establishing dynamic, purposeful feedback loops to gather real-time data on performance and ethics, driving iterative refinement (Digital Promise, 2023; Touchwork, 2025). Using mixed methods (surveys, focus groups, analytics, Daily Feedback Logs) ensures responsiveness to user experiences and allows rapid, evidence-based adjustments (Touchwork, 2025; Baker & Hadzovic, 2024; HCASR Component 5).

Table 11.1 Primary objectives of the AI implementation and pilot phase (phase 2)

Objective	Key elements/focus
1. Empowering stakeholders	Involve diverse users (students, faculty, staff, admin) as active participants and co-creators, not just testers; require transparent communication, defined roles, accessible training, fostering shared ownership and acceptance.
2. Continuous assessment and feedback	Establish and maintain dynamic, purposeful, bidirectional feedback loops; gather real-time data (quantitative & qualitative, incl. Daily Feedback Logs) on performance and ethics; drive iterative refinement based on user experience and evidence.

Note. These interconnected objectives ensure that AI pilots are both user-centered, fostering buy-in and relevance, and adaptive, allowing for continuous improvement based on real-world data

Workbook Prep 2: Defining Pilot Objectives

Reflection: Review pilot objectives (Chap. 10). How can they be refined for implementation (Chap. 11) to explicitly emphasize stakeholder empowerment (user skill development, co-design) and continuous assessment (specific feedback mechanisms like Daily Logs informing iteration)? Consider alignment with human-centric principles (HCAS).

UX Workbook Link: Use refined objectives when completing Chap. 11 UX Workbook, Section 1 (documenting ethical guidelines/alignment) and Section 2 (designing feedback collection strategies).

PALS Workbook Link & AI Partnership Guidance: Use the Chap. 11 PALS Workbook prompt for UX Section 2. Input draft pilot objectives; ask your GenAI partner to refine wording for SMARTIE criteria, explicitly incorporating stakeholder co-creation and continuous feedback mechanisms (like Daily Logs), referencing HCASR Component 1.

Example Prompt: "Refine draft objective: '[Draft]'. Ensure SMARTIE criteria, explicit stakeholder co-creation & Daily Feedback Log use (HCASR C1)."

Human Oversight Reminder: AI assists structuring goals. Ensuring objectives are relevant, achievable, inclusive, and drive meaningful assessment/empowerment requires strategic human insight and stakeholder validation (HCASR Component 1).

From Strategic Planning to Implementation: Building and Iterating

This section details translating approved pilot designs into operational reality, focusing on building initial versions and establishing processes for iterative refinement.

Building Minimum Viable Products (MVPs)

Implementation often starts with a Minimum Viable Product (MVP), the most basic, functional version needed to test core assumptions and gather initial user feedback efficiently (Lenarduzzi & Taibi, 2016; IES, 2019; FasterCapital, 2025a). This lean approach minimizes risk and focuses development on essential features addressing critical user needs identified in Phase 1 (Weil & Forrester, 2024; Srivastava, 2021). Human-centric

Table 11.2 Minimum viable product (MVP) principles for AI pilots

Aspect	Description/principle
Definition	The most basic, functional version of the AI tool/application needed to start the learning process.
Core purpose	Test core assumptions, validate the concept with actual users, and gather initial feedback efficiently with minimal upfront investment.
Feature selection	Prioritize only essential features addressing critical user needs/pain points; deliberately avoid complexity or "feature bloat."
Design focus	Must be human-centric, focusing on usability, positive user experience, intuitive interface, and delivering tangible value to stakeholders.
Feedback integration	Design must incorporate mechanisms for gathering user feedback early and often, enabling integration into the iterative development cycle.

Note. Using an MVP approach allows institutions to test AI concepts quickly, reduce risk, learn from users, and ensure development efforts remain focused on validated needs

design, focusing on usability and value, is key (Verhoeven & Hor, 2025). The MVP must include mechanisms for integrating user feedback to enable iteration (Dhruv, 2025; Verhoeven & Hor, 2025). Table 11.2 outlines key MVP principles.

Establishing the Iterative Piloting Process

Pilot implementation benefits from an iterative, agile methodology to navigate AI's complexities (Payne, 2025; Kotwel Technologies, 2025; Kang et al., 2022). This involves cycles of design, testing, feedback, and revision, common in educational research for ensuring feasibility and effectiveness (IES, 2019; IES, n.d.; Oudenampsen et al., 2024). The core iterative cycle ([Fig. 11.1]) typically involves deploying the MVP, collecting performance/user data (including Daily Logs), evaluating data against metrics, refining the tool/process, and redeploying (Payne, 2025; FasterCapital, 2025b). This continuous loop allows adaptation based on real-world use, minimizes risks, and enhances value delivery (Kotwel Technologies, 2025; Payne, 2025), aligning with design thinking (Brown, 2019) and HCASR Component 5.

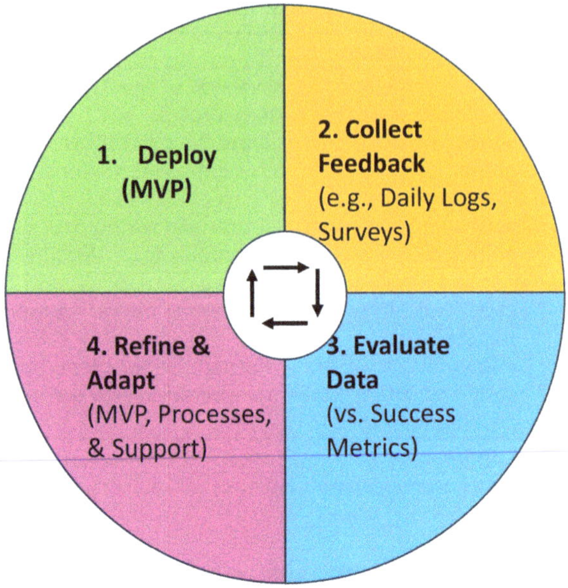

Fig. 11.1 Iterative AI pilot cycle

Gaining Practical Insights: Proprietary vs. Third-Party Solutions

Piloting either proprietary or third-party solutions provides crucial practical insights. Developing proprietary tools offers greater control but requires significant internal resources (O'Connell, 2024; Attridge, 2024; Paykamian, 2023). Integrating third-party solutions offers powerful platforms and faster deployment but requires scrutinizing contracts, data privacy, costs, integration, and support (Arizona State University, 2024; Todd, 2024; Rispens, 2024; Weil & Forrester, 2024). Pilot experience reveals real costs, technical challenges, data governance needs, and support requirements, providing evidence to validate or revise strategic sourcing decisions (Phase 1).

Workbook Prep 3: Defining the MVP and Iteration Plan

Reflection: For the specific pilot AI tool, what are the absolute minimum core MVP features for meaningful testing/feedback? Which features can be deferred? How will user feedback (esp. Daily Logs) be systematically collected/analyzed for the first iteration?

UX Workbook Link: Apply MVP decisions when working through Chap. 11 UX Workbook, Section 1 (Finalizing Tool Selection & MVP) and Section 2 (Mapping Workflows & Planning Iteration). Use MVP to inform usability testing plans (Section 3).

PALS Workbook Link & AI Partnership Guidance: Use the Chap. 11 PALS Workbook prompt for UX Section 3. Input pilot goal/target users; ask AI partner to draft MVP spec (core/excluded features, user stories) and outline first iteration cycle plan (feedback points, analysis, refinement areas), referencing HCASR Component 4.

Example Prompt: "Pilot: ['automating initial course advising'] for ['first-year students']. Draft MVP spec (3 core/2 excluded features). Outline 1st iteration cycle based on Daily Log feedback (HCASR C4)."

Human Oversight Reminder: AI structures specs/plans. Determining minimal viable features requires understanding user needs/technical feasibility. Designing effective iteration strategy depends on human judgment (feedback significance, resource allocation, strategic priorities) (HCASR Component 4).

AI Implementation Strategies: Integrating Change Management and Ethical Oversight

Successful pilot execution requires integrating effective change management, technical implementation, and continuous ethical oversight.

Balancing Leadership and Management: The Dual System in Action

Navigating AI pilots effectively requires balancing **leadership** (championing vision, building trust, supporting participants, celebrating wins) and **management** (coordinating resources, monitoring progress, managing technical details, ensuring protocol adherence, removing barriers), often termed a "dual

Table 11.3 Leadership versus management roles during AI pilot implementation

Role	Focus during pilot phase	Key actions during pilot phase
Leadership	The Why and the Who (Hearts & Heads)	Champion vision and strategic importance; build trust via transparent communication; visibly support teams/participants; foster psychological safety; identify and celebrate short-term wins; sustain momentum.
Management	The How and the What (Hands)	Ensure efficient operational execution; coordinate resources; monitor progress versus metrics/timelines; manage technical details; ensure protocol/ethical adherence; remove practical barriers; address operational issues.

Note. A balanced application of both leadership (inspiring change and maintaining vision) and management (ensuring effective execution and stability) is crucial for navigating the complexities of AI pilot implementation

system" (Kotter et al., 2021; DerSimonian & Montagnino, 2025; Prosci, 2025; Kang et al., 2022; Weil & Forrester, 2024). This ensures pilots remain strategically aligned while execution proceeds effectively. Table 11.3 summarizes roles during pilot implementation.

Establishing and Monitoring Ethical Guardrails

Defined ethical principles (Phase 1) must become operational practices via monitored guardrails. Key activities ([Table 11.4]) include regular bias audits,

Table 11.4 Key activities for monitoring ethical guardrails during AI pilots

Activity	Description/purpose
Regular bias audits	Systematically check AI outputs/performance data for potential biases (demographic, etc.) to ensure fairness and equity.
Data privacy enforcement	Rigorously apply data privacy protocols (minimization, anonymization/pseudonymization) and ensure adherence to regulations (FERPA, GDPR) and policies.
Ongoing monitoring and evaluation	Continuously monitor AI system performance, reliability, and adherence to ethical standards; track issues reported via feedback loops.
Transparency	Maintain clear communication with participants about AI capabilities, limitations, and data usage/protection throughout the pilot.
Reporting and recourse	Implement clear, accessible procedures for reporting ethical concerns, technical failures, or negative consequences, ensuring timely review and action.

Note. These activities translate ethical principles and risk management plans into operational practices, ensuring responsible conduct throughout the pilot phase

strict data privacy enforcement (FERPA/GDPR), ongoing monitoring of system performance/ethics, maintaining transparency with participants (data use, limitations), and clear procedures for reporting/addressing ethical concerns (including via Daily Feedback Logs) (Wynants et al., 2025; Harvard University Information Technology, 2025; Hannah, 2025; Touchwork, 2025). Referencing established frameworks (NIST AI RMF [Tabassi, 2023], UNESCO [2022], OECD [2024]) alongside institutional guidelines ensures responsible practices (Nguyen, 2024; Al-Omari et al., 2024).

Ensuring Pilot Alignment with Strategic Goals

Pilot activities require continuous monitoring for alignment with institutional mission and SMARTIE goals (Phase 1) (Weil & Forrester, 2024; Scott, 2024). Regular reviews should assess progress against KPIs and strategic value (Tobenkin, 2024). Significant divergence may necessitate intervention (redirection, pause, termination) (Weil & Forrester, 2024).

Implementing Dynamic Feedback Loops

Actively using dynamic feedback loops is crucial (Digital Promise, 2023; Touchwork, 2025). Use established mechanisms (surveys, Daily Feedback Logs, forums, analytics, check-ins) to systematically collect real-time user experience data (Touchwork, 2025). Promptly analyze data for themes/issues. Use insights directly for iterative refinement (Section 2.B), ensuring responsiveness to user needs/experience (Baker & Hadzovic, 2024; Weil & Forrester, 2024; HCASR Component 5). Figure 11.2 depicts this flow.

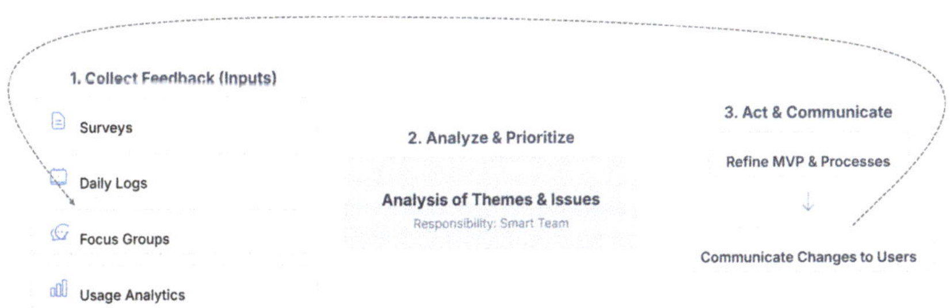

Fig. 11.2 Feedback loop data flow

Workbook Prep 4: Operationalizing Ethics and Feedback

Reflection: How will the pilot team ensure ethical guardrails (bias checks, privacy protocols from Ch 10) are actively monitored during implementation? What specific mechanism/role (weekly review of Daily Logs for ethical flags, designated ethics checker) operationalizes monitoring and ensures feedback informs ethical adjustments?

UX Workbook Link: Plan operational monitoring details when completing Chap. 11 UX Workbook, Section 3 (Planning Ethical AI Testing) and Section 4 (Verifying Pilot Readiness and Progress). Ensure alignment confirmed in Section 6 (Scaffolding).

PALS Workbook Link & AI Partnership Guidance: Use the Chap. 11 PALS Workbook prompt for UX Section 4. Input key ethical principle (e.g., "Fairness") and primary feedback mechanism (e.g., "Daily Logs"); ask AI partner to suggest 2–3 concrete weekly/bi-weekly actions/procedures to monitor adherence using specified feedback, referencing HCASR Component 4 & 5.

Example Prompt: "Principle: Fairness. Mechanism: Daily Logs. Suggest 2 actions team can take weekly to monitor fairness using log data (HCASR C4 & C5)."

Human Oversight Reminder: AI suggests activities. Designing practical, effective, non-burdensome monitoring requires human judgment (feasibility, meaningful indicators, accountability, planning actions based on findings) (HCASR Component 4 & 5).

AI Pilot Testing, User Experience, and Workflow Optimization

This section focuses on practical methods for AI pilot testing, using findings to refine User Experience (UX) and enable Workflow Optimization.

Comprehensive User Testing

Effective pilot testing assesses **Usability**, effectiveness, **Accessibility**, and overall UX quality. A mix of methods yields richer insights: **Usability Testing** (observing users performing tasks) identifies issues (Morris & Warman, 2015); **Think-Aloud Protocols** uncover user reasoning; **Scenario-Based Testing** evaluates effectiveness in context; **Accessibility Testing** verifies usability for diverse abilities (WCAG standards). Gathering **Subjective Feedback** (surveys, interviews) assesses satisfaction, perceived usefulness, ease of use, and trust (Weil & Forrester, 2024; Touchwork, 2025). The UX Workbook provides detailed planning activities for these methods.

Leveraging Stakeholder Input for Refinement

Data from testing and ongoing feedback (Daily Feedback Logs) must drive **Iterative Refinements** (Richards, 2024; Baker & Hadzovic, 2024). A consistent process: Analyze feedback -> Prioritize issues -> Brainstorm solutions -> Implement refinements -> Communicate changes back to users (Richards, 2024; Touchwork, 2025).

Task and Workflow Optimization Evaluation

Pilots often aim to optimize tasks/workflows (Kadiu, 2025; Moquin, 2024). Evaluation assesses if optimizations occur (faster process, fewer errors) and evaluates human-AI collaboration effectiveness within the workflow (Fragiadakis et al., 2024). Compare baseline metrics (Chap. 10) with pilot data (quantitative: task time, error rates; qualitative: satisfaction, workload perception from Daily Logs) (Fragiadakis et al., 2024; Touchwork, 2025). Frameworks (Fragiadakis et al., 2024; HCAS model) can analyze collaboration and inform redesigns (Fragiadakis et al., 2024).

Workbook Prep 5: Planning for Testing and Optimization

Reflection: For the pilot tool/workflow, which user testing method (think-aloud, scenario-based, accessibility check) is most critical? How will you specifically evaluate human-AI collaboration effectiveness/efficiency beyond just tool usability?

UX Workbook Link: Define testing methods and collaboration metrics when completing Chap. 11 UX Workbook, Section 3 (Defining Metrics, Planning Tests) and integrate findings into Section 5 (Benchmark Reports, Dashboards) and Section 7 (Pilot Report Template).

PALS Workbook Link & AI Partnership Guidance: Use the Chap. 11 PALS Workbook prompt for UX Section 5. Input primary pilot task/target users; ask AI partner to outline test plan (>=2 methods) and suggest 2–3 specific workflow optimization/human-AI synergy metrics (quant/qual), referencing HCAS/RSA principles and HCASR Component 4.

Example Prompt: "Pilot optimizes ['library search'] for ['grad students']. Outline test plan (Think-Aloud, Scenario Testing). Suggest 3 workflow metrics (speed, relevance, trust) (HCASR 4, HCAS)."

Human Oversight Reminder: AI suggests standard methods/metrics. Selecting appropriate methods/scenarios, setting realistic baselines, interpreting interaction nuances requires human expertise (UX research, evaluation design, domain knowledge) (HCASR Component 4).

Operationalizing Implementation: Documentation and Engagement

This section emphasizes robust documentation and continuous engagement during implementation.

Maintaining "Living Documents" and User Outcomes and Deliverables (UODs)

Maintaining comprehensive, accessible, and consistently updated documentation—referred to as "Living Documents"—is essential for management, transparency, and decision-making (Behling & Law, 2020; NARA, 2025). During active pilots, these documents (e.g., the Pilot Risk Register, MVP specifications, summaries of Daily Feedback Logs, meeting minutes, interim evaluation reports) must be actively maintained. This means regularly updating them with new data, decisions, and insights as the pilot progresses. For example, the Risk Register should be revisited weekly to reflect new issues identified through Daily Feedback Logs or testing. MVP specifications should be versioned as refinements are made. This active maintenance ensures that all stakeholders have access to the latest information, facilitates informed iteration, provides a clear audit trail, and builds a robust evidence base for the final pilot evaluation and decisions regarding broader scaling (Phase 3). All key outputs produced throughout the pilot, including these living documents and interim reports, constitute the UODs (user outcomes and deliverables).

Ensuring Continuous Stakeholder Engagement

Ongoing, inclusive engagement is vital for alignment, buy-in, identifying issues, achieving sustainable outcomes (Neto et al., 2023; Ochieng, 2024). It requires actively using feedback loops (Daily Logs, surveys, focus groups) (Digital Promise, 2023; Touchwork, 2025), transparent communication (progress, challenges, adaptations), managing expectations, soliciting diverse input, demonstrating responsiveness (Sito et al., 2023). Strong relationships and ensuring stakeholders feel heard foster shared ownership (Gillen, 2024; Forsythe et al., 2022; Neto et al., 2023).

Workbook Prep 6: Maintaining Documentation and Engagement

Reflection: What specific process/tool ensures key pilot documents (Risk Register, MVP spec, Feedback Summaries) remain "living," regularly updated? Beyond collecting feedback, how to ensure continuous two-way communication with participants about changes based on their input?

UX Workbook Link: Define processes when working through Chap. 11 UX Workbook, Section 5 (Living Documents: Maintaining Dynamic Pilot Records). Ensure communication plans updated (Section 7). Check role alignment (Section 9).

PALS Workbook Link & AI Partnership Guidance: Use the Chap. 11 PALS Workbook prompt for UX Section 5. Input key living documents/stakeholder group; ask AI partner to outline practical update schedule/process and suggest 2–3 ongoing two-way engagement methods, referencing HCASR Component 5 and stakeholder engagement principles.

Example Prompt: "Docs: Risk Register, Feedback Summary. Stakeholders: Faculty pilots. Outline monthly update process & suggest 2 ongoing faculty engagement methods (HCASR 5)."

Human Oversight Reminder: AI suggests standard schedules/tactics. Establishing realistic processes, choosing culturally appropriate engagement, ensuring follow-through require human planning, commitment, project management, communication skills (HCASR Component 4 & 5).

Role-Specific Implementation Strategies

Tailoring strategies leverages unique stakeholder contributions. During active pilot execution, each stakeholder group plays a distinct and vital role in the iterative development and assessment of the AI initiative. Understanding and clearly articulating these roles and their expected actions ensures comprehensive feedback and smooth operation. Key roles during implementation (Table 11.5):

- **Undergraduate Students:** As primary end-users, their actions involve actively using the AI tool as intended, providing timely and honest end-user feedback through mechanisms like Daily Feedback Logs and participation in testing sessions, and adhering to ethical usage guidelines (University of Michigan, 2025). Their contributions are crucial UODs for understanding real-world usability and impact.
- **Graduate Students:** Actions may include using the AI for more advanced tasks, offering more detailed technical or methodological feedback, and potentially assisting with data collection or preliminary analysis under supervision (Spring et al., 2023). Their contributions can enrich UODs with deeper insights.

Table 11.5 Role-specific contributions during AI pilot implementation

Stakeholder role	Key contribution during implementation phase	Example supporting citation
Undergraduate students	Provide end-user feedback on usability/effectiveness (e.g., via Daily Logs, testing); adhere to ethical use guidelines.	University of Michigan (2025)
Graduate students	Offer advanced feedback (e.g., on research/teaching tools); potentially assist in pilot data analysis or refinement.	Spring et al. (2023)
Faculty	Evaluate pedagogical impact; adapt teaching methods; model ethical use; provide feedback on workflow and student learning.	Weil and Forrester (2024)
Staff (support, IT, etc.)	Test tools in operational workflows; provide feedback on efficiency, integration, support needs; monitor technical performance.	Kadiu (2025); Moquin (2024)
Administrators	Provide strategic oversight; ensure alignment/resources; monitor compliance; remove high-level barriers; receive progress reports.	Tobenkin (2024); Glover (2023)
HEIs/Smart Teams	Coordinate activities; facilitate communication; collect/synthesize data; manage documentation (Living Docs); troubleshoot.	Jaskyte et al. (2023); Ellucian (2024)

Note. This table summarizes the distinct contributions expected from various stakeholder groups during the active implementation phase of AI pilots

- **Faculty:** Actions involve integrating the AI tool into their teaching or research workflows, evaluating its pedagogical or research impact, adapting teaching methods as needed, modeling ethical AI use for students, and providing structured feedback on the tool's effectiveness and alignment with academic goals (Weil & Forrester, 2024). Their UODs often include documented observations and curriculum adjustments.
- **Staff (e.g., IT, library, administrative):** Actions include testing the AI within operational workflows, providing feedback on efficiency gains or losses, integration challenges, technical support needs, and data security implications (Kadiu, 2025; Moquin, 2024). Their UODs contribute to understanding operational feasibility.
- **Administrators:** Actions focus on providing strategic oversight, ensuring the pilot remains aligned with institutional goals and policies, monitoring resource allocation and compliance, championing the initiative, and removing high-level barriers to progress (Tobenkin, 2024; Glover, 2023). They utilize UODs for strategic decision-making.
- **HEIs/Smart Teams:** Actions involve coordinating all pilot activities, managing communication channels, systematically collecting and synthesizing data from all stakeholder groups (including Daily Feedback Logs), maintaining Living Documents, troubleshooting issues, ensuring strategic and ethical adherence, and preparing interim UODs and recommendations for adjustments based on pilot findings (Jaskyte et al., 2023; Ellucian, 2024).

Note: Companion workbooks provide detailed role-specific activities.

Conclusion

Chapter 11 detailed the transition from planning to practical AI implementation and piloting (Phase 2), emphasizing MVPs and iterative cycles guided by design thinking and change management. Success hinges on empowering stakeholders, active feedback loops (Daily Logs), monitoring ethics (NIST AI RMF [Tabassi, 2023], UNESCO [2022]), user testing, and leveraging input for refinement. Maintaining "Living Documents" and continuous engagement ensures transparency/adaptability. Strategies must be tailored to context/roles. Phase 2 UODs provide the evidence base for data collection (Chap. 12) and Phase 3 evaluation/scaling (Chaps. 13, 14, and 15).

Looking Ahead & Engaging the Workbook

Chapter 11 provided strategies for implementing/piloting AI. The focus was practical, adaptive, human-centric, ethical implementation involving MVPs, iteration, feedback (Daily Logs), role contributions. The Chap. 11 UX Workbook offers structured activities/templates. Engaging produces UODs required for data collection (Chap. 12) and evaluation (Phase 3).

Long-Term Scaling, Sustainability & AI Governance: Learning from Proprietary TLARA Pilots

Chapter 11 has immersed you in the practicalities of implementing and managing a pilot for a proprietary GenAI system. This hands-on experience within a specific **TLARA (Teaching, Learning, Assessment, Research, Administration)** domain offers invaluable insights into user interaction, system performance, the efficacy of training and support, and the real-world ethical considerations that arise (Mollick, 2024). Documenting these experiences, from individual user feedback in specific **spaces** to system-**level** performance, is crucial for assessing the tool's potential for sustainable, long-term integration and for refining your institution's approach to AI governance over **time**. Both the Chap. 11 UX and PALS Workbooks are designed to help you ethically partner with AI to navigate and learn from these pilot implementations.

HEIs undertaking such pilots are essentially creating controlled experiments within their complex operational context (IDEAS, 2025). The PALS Workbook for Chap. 11, in its section "Long-Term Scaling, Sustainability & AI Governance: Iteratively Developing Your Custom TLARA-Informed AI Guidelines," guides you to connect these direct pilot experiences to your evolving framework. Using the "Topics for PALS TLARA Focus"—like "Ensuring Long-Term Ethical Governance for Proprietary AI Systems" and "Building Sustainable Internal Capacity"—your GenAI model will act as an advisor. It will help you distill lessons from the pilot to inform specific aspects of your TLARA guidelines, particularly regarding data governance, intellectual property, and the development of specialized AI literacy necessary for effectively using and potentially scaling proprietary tools (Ellucian, 2024).

References

Al-Omari, O., Alyousef, A., Fati, S., Shannaq, F., & Omari, A. (2024). Governance and ethical frameworks for AI integration in higher education: Enhancing personalized learning and legal compliance. *Journal of Ecohumanism, 4*(2), 80–86. https://doi.org/10.62754/joe.v4i2.5781

Arizona State University. (2024). *ASU and OpenAI collaboration*. https://ai.asu.edu/openAI

Attridge, M. (2024, April 24). Campus-GPT: How 2 University of California campuses are designing their own specialized AI tools. *BestColleges*. https://www.bestcolleges.com/news/uc-san-diego-irvine-design-specialized-ai-tools/

Baker, A., & Hadzovic, A. (2024, January 31). Collaborating for the future of teaching and learning with technology. *EdSurge*. (Sponsored by Digital Promise). https://www.edsurge.com/news/2024-01-31-collaborating-for-the-future-of-teaching-and-learning-with-technology

Behling, K., & Law, V. (2020). *Learning before going to scale: An introduction to conducting pilot studies*. Regional Educational Laboratory Appalachia / SRI International. https://ies.ed.gov/ncee/edlabs/regions/appalachia/pdf/REL_AP_Learning_Before_Going_to_Scale_508.pdf

Brown, T. (2019). *Change by design: How design thinking creates new alternatives for business and society* (Revised and Updated ed.). Harper Business.

Carmody-Bubb, M. (2023). *Cognition and decision making in complex adaptive systems: The human factor in organizational performance*. Springer.

DerSimonian, R., & Montagnino, C. (2025, March 26). Crafting thoughtful AI policy in higher education: A guide for institutional leaders. *Campus Technology*. https://campustechnology.com/articles/2025/03/26/crafting-thoughtful-ai-policy-in-higher-education-a-guide-for-institutional-leaders.aspx

Dhruv, S. (2025, March 4). How to build an AI MVP: Step by step process. *Aalpha Blog*. https://www.aalpha.net/blog/how-to-build-an-ai-mvp-step-by-step-process/

Digital Promise. (2023, December 7). *Principles for building effective feedback loops*. https://digitalpromise.org/2023/12/07/principles-for-building-effective-feedback-loops/

Ellucian. (2024, December 16). AI in higher education: Implementation strategies for institutions and tech partners. *Higher Ed Dive*. https://www.highereddive.com/spons/ai-in-higher-education-implementation-strategies-for-institutions-and-tech/735434/

FasterCapital. (2025a). *MVP pilot program*. Retrieved May 6, 2025, from https://fastercapital.com/content/MVP-Pilot-Program-Testing-Your-MVP-with-Real-Users.html

FasterCapital. (2025b). *MVP iteration cycle*. Retrieved May 6, 2025, from https://fastercapital.com/content/MVP-Iteration-Cycle-Refining-Your-MVP-Based-on-User-Feedback.html

Forsythe, L. P., Frank, L., Walker, K. O., Arnow, B. A., Urman, R., Workman, H. G., et al. (2022). Understanding the influence and impact of stakeholder engagement in patient-centered outcomes research: A qualitative study. *Journal of General Internal Medicine, 37*(Suppl 1), 73–83. https://doi.org/10.1007/s11606-021-07340-0

Fragiadakis, G., Diou, C., Kousiouris, G., & Nikolaidou, M. (2024). Evaluating Human-AI Collaboration: A review and methodological framework. *arXiv*. https://doi.org/10.48550/arXiv.2407.19098

Gillen, S. M. (2024). *Stakeholder engagement: Challenges, metrics, and tactical solutions in higher education strategic planning* [Doctoral dissertation, Murray State University]. Murray State's Digital Commons. https://digitalcommons.murraystate.edu/cgi/viewcontent.cgi?article=1466&context=etd

Glover, J. (2023, November). *Building an AI university: An administrator's guide.* University of Florida. https://www.ufl.edu/wp-content/uploads/sites/5/2023/11/Building-an-AI-university-An-administrators-guide.pdf

Hannah, A. (2025). Ethical considerations for AI in higher education: Ensuring fairness and transparency. *Liaison Resources Blog.* Retrieved May 6, 2025, from https://www.liaisonedu.com/resources/blog/ethical-considerations-for-ai-in-higher-education-ensuring-fairness-and-transparency/

Harvard University Information Technology. (2025). *Generative artificial intelligence (AI) guidelines.* Retrieved May 6, 2025, from https://huit.harvard.edu/ai/guidelines

IES. (2019, November 25). *Intelligent scaffolding for peer reviews of writing.* Institute of Education Sciences. https://ies.ed.gov/use-work/awards/intelligent-scaffolding-peer-reviews-writing

IES. (n.d.). *Developing and revising instructional activities to optimize cognitive engagement.* Institute of Education Sciences. Retrieved May 6, 2025, from https://ies.ed.gov/use-work/awards/developing-and-revising-instructional-activities-optimize-cognitive-engagement

Instructional Design, Engagement, and Support (IDEAS). (2025, January 27). *AI's impact on higher education through the eyes of faculty: 4 key trends.* University of Massachusetts Amherst. https://www.umass.edu/ideas/news/ais-impact-higher-education-through-eyes-faculty-4-key-trends

Jaskyte, K., Taylor, S., & Lichtenberg, A. (2023). Collaborative teams: A systematic review and framework for future research. *Group Decision and Negotiation, 32*(6), 1151–1189. https://doi.org/10.1007/s10726-023-09838-5

Kadiu, A. (2025, January 9). AI assistants and agents in US higher education: Beyond the classroom. *Element451.* https://element451.com/blog/ai-agents-in-us-higher-education

Kang, S. P., Chen, Y., Svihla, V., & Datye, A. K. (2022). Guiding change in higher education: An emergent, iterative application of Kotter's change model. *Studies in Higher Education, 47*(6), 1120–1136. https://doi.org/10.1080/03075079.2020.1741540

Kotter, J. P., Akhtar, V., & Gupta, G. (2021). *Change: How organizations achieve hard-to-imagine results in uncertain and volatile times.* Wiley.

Kotwel Technologies. (2025). *Continuous learning: Iterative improvement in AI development.* Retrieved May 6, 2025, from https://kotwel.com/continuous-learning-iterative-improvement-in-ai-development/

Lenarduzzi, V., & Taibi, D. (2016). MVP explained: A systematic mapping study on the definitions of minimal viable product. In *Proceedings of the 42nd Euromicro*

Conference on Software Engineering and Advanced Applications (SEAA) (pp. 112–119). IEEE. https://doi.org/10.1109/SEAA.2016.56

Mollick, E. (2024). *Co-intelligence: Living and working with AI*. Penguin Publishing Group.

Moquin, S. (2024, November 1). The role of AI in transforming higher education. *Enrollify*. https://www.enrollify.org/blog/the-role-of-ai-in-transforming-higher-education

Morris, H., & Warman, G. (2015, January 12). Using design thinking in higher education. *EDUCAUSE Review*. https://er.educause.edu/articles/2015/1/using-design-thinking-in-higher-education

NARA. (2025). *Guidance for proof of concept pilot*. U.S. national archives and records administration, records management policy. Retrieved May 6, 2025, from https://www.archives.gov/records-mgmt/policy/pilot-guidance.html

Neto, J. C. A., Mainardes, E. W., & Rodrigues, R. G. (2023). Stakeholder engagement in higher education: State of the art and research agenda. *Quality Assurance in Education, 31*(2), 241–257. https://doi.org/10.1108/QAE-10-2022-0105

Nguyen, N. (2024, February 12). From regulation to innovation: What the EU AI Act means for EdTech. *FeedbackFruits*. Retrieved May 6, 2025, from https://feedbackfruits.com/blog/from-regulation-to-innovation-what-the-eu-ai-act-means-for-edtech

O'Connell, A. J. (2024, February 7). How (and why) the University of Michigan built its own closed generative AI tools. *EDUCAUSE Review*. https://er.educause.edu/articles/2024/2/how-and-why-the-university-of-michigan-built-its-own-closed-generative-ai-tools

Ochieng, M. (2024). *Enhancing digital transformation success in education through effective stakeholder engagement strategies* [Doctoral dissertation, Walden University]. ScholarWorks. https://scholarworks.waldenu.edu/cgi/viewcontent.cgi?article=17311&context=dissertations

OECD. (2024). *OECD AI principles overview*. OECD.AI policy observatory. Retrieved May 6, 2025, from https://oecd.ai/en/ai-principles

Oudenampsen, J., Das, E., Blijlevens, N., & van de Pol, M. H. J. (2024). How to cross the line: Design principles for interdisciplinary education [version 2; peer review: 3 approved, 2 approved with reservations, 2 not approved]. *MedEdPublish, 14*(1), 35. https://doi.org/10.12688/mep.20016.2

Paykamian, B. (2023, November 16). Should universities build their own custom AI tools in-house? *Government Technology*. https://www.govtech.com/education/higher-ed/should-universities-build-their-own-custom-ai-tools-in-house

Payne, S. (2025, February 8). The art of iterative AI system development: A practical guide to evaluation-driven improvement. *Skylar Payne* (Wicked Data LLC). https://skylarbpayne.com/posts/real-life-evals/

Prosci. (2025). Kotter's change management theory explanation and applications. *Prosci Blog*. Retrieved May 6, 2025, from https://www.prosci.com/blog/kotters-change-management-theory

Richards, D. (2024, November 18). Navigating the human side of AI: A guide to stakeholder collaboration. *ATD Blog*. https://www.td.org/content/atd-blog/navigating-the-human-side-of-ai-a-guide-to-stakeholder-collaboration
Rispens, S. (2024, June 4). OpenAI builds ChatGPT system for colleges and universities. *EdScoop*. https://edscoop.com/openai-chatgpt-edu-colleges-universities-artificial-intelligence/
Scott, A. (2024, August 16). AI in higher education: Best practices for institutions and tech partners. *Medium*. https://medium.com/@aileenscott604/ai-in-higher-education-best-practices-for-institutions-and-tech-partners-16fb9c5066b7
Sito, L. B., Parsa, P., & Joksimović, S. (2023). Implementing educational technology in Higher Education Institutions: A review of technologies, stakeholder perceptions, frameworks and metrics. *Education and Information Technologies, 28*(12), 16403–16429. https://doi.org/10.1007/s10639-023-11846-x
Spring, E., Goel, A., & Johnston, B. (2023, September 21). UMich becomes first major university to develop its own generative AI. *The Michigan Daily*. https://www.michigandaily.com/news/administration/umich-becomes-first-major-university-to-develop-its-own-generative-ai/
Srivastava, R. (2021). Iterative minimum viable product approach to implementing AI, RPA, and BI solutions. *Westcliff International. Journal of Applied Research, 5*(1), 44–50. https://doi.org/10.47670/wuwijar202151RS
Tabassi, E. (2023, January 26). *Artificial intelligence risk management framework (AI RMF 1.0)* (NIST AI 100-1). National Institute of Standards and Technology. https://doi.org/10.6028/NIST.AI.100-1
Tobenkin, D. (2024). Artificial intelligence and the future of higher education, Part 2. *Trusteeship Magazine, 32*(3). Association of Governing Boards of Universities and Colleges (AGB). https://agb.org/trusteeship-article/artificial-intelligence-and-the-future-of-higher-education-part-2/
Todd, S. (2024, November 21). Rutgers Business School partners with Google to enhance teaching and classroom learning with Generative AI [Press release]. *PR Newswire*. https://www.business.rutgers.edu/news/rutgers-business-school-partners-google-enhance-teaching-and-classroom-learning-generative-ai
Touchwork. (2025). A guide to implementing real-time feedback systems in higher education institutions [Vendor White Paper]. Retrieved May 6, 2025, from https://touchwork.com/wp-content/uploads/2024/08/A-guide-to-implementing-real-time-feedback-systems-in-higher-education-institutions.pdf
UNESCO. (2023, November 30). *How generative AI is reshaping education in Asia-Pacific*. UNESCO. https://www.unesco.org/en/articles/how-generative-ai-reshaping-education-asia-pacific
University of Michigan. (2025). From theory to practice: How students are using AI at Michigan Ross. *Michigan Ross News*. Retrieved May 6, 2025, from https://

michiganross.umich.edu/news/theory-practice-how-students-are-using-ai-michigan-ross

Verhoeven, B., & Hor, T. (2025, February 12). A framework for human-centric AI-first teaching. *AACSB Insights*. https://www.aacsb.edu/insights/articles/2025/02/a-framework-for-human-centric-ai-first-teaching

Weil, D., & Forrester, J. (2024, October 17). A road map for leveraging AI at a smaller institution. *EDUCAUSE Review*. https://er.educause.edu/articles/2024/10/a-road-map-for-leveraging-ai-at-a-smaller-institution

Wynants, S., Childers, G., De La Torre Roman, Y., Budar-Turner, D., & Vasquez, P. (2025, February). *ETHICAL principles AI framework for higher education*. CSU AI Commons / California State University, Fullerton. https://www.google.com/search?q=https://genai.calstate.edu/communities/faculty/ethical-and-responsible-use-ai/ethical-principles-ai-framework-higher-education/

12

Collecting and Synthesizing Pilot Data (Phase 2)

Introduction: From Pilot Action to Data-Driven Insights

Following the active implementation and operation of initial Artificial Intelligence (AI) pilot projects (Chap. 11), Higher Education Institutions (HEIs) arrive at the concluding activities of Phase 2: systematically collecting data and feedback and conducting an initial pilot evaluation. This stage is vital for transitioning from experimentation to evidence-based understanding (Askew et al., 2024; Weil & Forrester, 2024). Before moving to the comprehensive monitoring, evaluation, and potential scaling decisions of Phase 3 (Chap. 13), institutions must rigorously gather, synthesize, and assess the data generated during pilot initiatives.

Effective data collection and initial evaluation align with core principles of Design Thinking and Change Management. The gathered data serves as vital "Inspiration," highlighting successes and challenges encountered during the pilot (Brown, 2019). Analyzing this data against predefined pilot goals reflects the "Test" phase of design thinking, assessing the effectiveness of the implemented solution (Brown, 2019). From a change management perspective, this evaluation provides concrete evidence of progress or identifies barriers, potentially documenting "Short-Term Wins" crucial for sustaining momentum and informing future actions (Kotter et al., 2021). This chapter details the methodologies for systematic data collection, analysis, synthesis, and initial evaluation, setting the stage for informed Phase 3 decisions.

Supplementary Information The online version contains supplementary material available at https://doi.org/10.1007/978-3-032-05205-6_12.

D. Woodring, *Advancing Higher Education with Human-Centric AI Partnerships*,
https://doi.org/10.1007/978-3-032-05205-6_12

Workbook Prep 1. Framing Pilot Data Collection Purpose

Reflection: As you conclude the active pilot phase detailed in Chap. 11, why is it essential to now systematically collect and synthesize all the generated data before making strategic decisions about scaling or discontinuation? How does this structured data collection process align with principles of evidence-based decision-making and responsible innovation within your HEI's context?

UX Workbook Link: Begin with Chap. 12 UX Workbook "Getting Started: Preparatory Exercises" (which prompts reflection on KPIs and lessons learned) and proceed to Section 1 ("Planning the Data Collection & Analysis") activities, which focus on finalizing the plan for gathering pilot evidence.

PALS Workbook Link & AI Partnership Guidance: Use the Chap. 12 PALS Workbook prompt corresponding to UX Workbook Section 1. Input your primary pilot goal and the key stakeholder groups involved; ask your GenAI partner to help draft a concise statement justifying the need for systematic data collection and initial evaluation before Phase 3 decisions are made, referencing HCASR Component 4 (Evaluation) & Component 5 (Accountability).

Example Prompt: "Pilot Goal: [Goal]. Stakeholders: [Groups]. Draft rationale justifying systematic data collection/eval before Phase 3 scaling (HCASR C4 & C5)."

Human Oversight Reminder: AI can articulate the general benefits of data collection. Defining the specific data crucial for your pilot's evaluation questions and ensuring the collection methods are ethical, appropriate, and feasible require human judgment informed by the pilot plan (Chap. 10) and institutional context (HCASR Component 1 & 4).

Systematic Pilot Data Collection Methodologies

Effective pilot evaluation relies on the systematic collection of comprehensive data, executing the plan developed earlier (Chaps. 10 and 11) to capture a mix of quantitative and qualitative evidence (Williams, 2025). Rigorous implementation ensures data quality and provides a reliable foundation for analysis. Transitioning from active piloting requires a deliberate shift to structured data gathering. Key data streams include:

Quantitative Performance Metrics Gathering objective data on predefined Key Performance Indicators (KPIs) is essential. This includes metrics such as operational efficiency (e.g., reduction in process time, error rate decreases), AI system performance (e.g., accuracy, uptime, response latency), user engagement (e.g., usage frequency, adoption rates across different groups, task completion rates), and potentially cost-related metrics compared to baseline (Acacia Advisors, 2023; Kanerika, 2024). Examples relevant to HEIs might include a reduction in student advising appointment wait times or improvements in pass rates on AI-supported formative quizzes.

Qualitative Stakeholder Feedback Capturing the subjective experiences, perceptions, and contextual insights of participants is equally crucial. Methods include post-pilot surveys (following best practices for design to ensure validity and reliability; Institute of Education Sciences [IES], 2021), structured analysis of Daily Feedback Logs (which provide continuous, in-the-moment, contextualized data on usability, challenges, and perceived value; Askew et al., 2024; Weil & Forrester, 2024), and targeted focus groups or interviews (to probe deeper into specific issues or gather nuanced perspectives; Askew et al., 2024; IES, 2024). Daily Logs are particularly valuable for capturing recurring usability frustrations or specific instances where the AI provided unexpected value or difficulty.

Observational Data Insights gleaned from direct observation of users interacting with the AI tool (potentially from usability testing conducted in Chap. 11) offer valuable data on actual usage patterns, navigational challenges, and deviations from intended workflows. Executing these methods ethically (e.g., ensuring informed consent, protecting anonymity where promised) and consistently across the pilot's conclusion yields a comprehensive dataset essential for robust analysis and evaluation (Brown, 2019).

Workbook Prep 2. Ensuring Comprehensive Data Collection

Reflection: Review the data collection plan finalized in the Chap. 10 pilot design (and potentially refined during Chap. 11). As you prepare for the final data gathering described in this chapter, are there any potential gaps? Does the plan adequately capture both quantitative performance metrics and rich qualitative feedback (especially contextual insights from Daily Feedback Logs) from all relevant stakeholder groups involved in the pilot?

UX Workbook Link: Execute the plan using Chap. 12 UX Workbook, Section 1 ("Planning the Data Collection & Analysis") to finalize methods and Section 2 ("Executing Data Collection & Initial Analysis") to guide the actual data gathering and monitoring.

PALS Workbook Link & AI Partnership Guidance: Use the Chap. 12 PALS Workbook prompt corresponding to UX Workbook Section 2. Input your final data collection plan summary; ask your GenAI partner to identify potential gaps based on the plan description and suggest one complementary method (quantitative or qualitative) to ensure a more holistic view of the pilot's impact, referencing HCASR Component 4 (Controlled Testing & Evaluation).

Example Prompt: "Review plan: [Paste Summary]. ID gaps & suggest 1 additional method for comprehensive quant/qual data capture (HCASR C4)."

Human Oversight Reminder: AI can suggest common data collection methods. Assessing the feasibility, ethical implications, resource requirements, and appropriateness of any additional methods requires human expertise and collaborative team consultation (HCASR Component 1 & 4). Ensure the final plan directly addresses the pilot's key evaluation questions.

Analyzing and Synthesizing Pilot Data

Raw data collected at the conclusion of the pilot requires systematic analysis and synthesis to transform it into meaningful insights and actionable knowledge (Acacia Advisors, 2023). The purpose of each analysis step within the pilot conclusion context is paramount, and readers are directed to the UX workbook activities and cited external resources for detailed methodological guidance.

Quantitative Data Analysis involves cleaning numerical data, calculating descriptive statistics, comparing pilot results against baselines and target KPIs (Acacia Advisors, 2023), and identifying trends, potentially segmenting data by group.

Qualitative Data Analysis requires organizing and coding feedback from sources like Daily Feedback Logs and open-ended responses. Thematic analysis helps identify recurring patterns, themes, and categories related to usability, benefits, challenges, ethical concerns, and suggestions (Askew et al., 2024; Ji et al., 2024), aiming to understand the "why" behind quantitative results.

Data Integration (Mixed Methods) combines quantitative and qualitative findings for a holistic view. Techniques include triangulation (comparing findings from different sources), explanation (using qualitative data to contextualize quantitative results, such as insights from Daily Logs explaining satisfaction scores), and complementarity (building a richer picture than either data type alone). This rigorous analysis, aided by data visualization, transforms raw data into actionable knowledge (Brown, 2019; The CASE HQ, 2023).

Workbook Prep 3. Planning Data Synthesis

Reflection: Once all pilot data is collected (Section Systematic Pilot Data Collection Methodologies), how will your team systematically synthesize the quantitative metrics (KPIs, usage data) and the diverse qualitative feedback (Daily Logs, surveys, interviews)? What specific process or framework will be used to identify key themes from the qualitative data and meaningfully relate them to the quantitative performance results or user experience scores?

UX Workbook Link: Implement this synthesis using Chap. 12 UX Workbook, Section 2 ("Executing Data Collection & Initial Analysis") for preliminary analysis and Section 3 ("Synthesizing Data & Identifying Themes") for detailed coding, triangulation, and integrated summary development.

PALS Workbook Link & AI Partnership Guidance: Use the Chap. 12 PALS Workbook prompt corresponding to UX Workbook Section 3. Input examples of the quantitative metrics being tracked and the types of qualitative data collected; ask your AI partner to outline a structured approach (mentioning techniques like triangulation or using qualitative data to explain quantitative trends) for synthesizing this mixed data to evaluate a specific pilot objective, referencing HCASR Component 5 (Data Analysis, Continuous Improvement) and mixed-methods research principles.

Example Prompt: "Objective: [Obj]. Data: [Quant Metrics], [Qual Data from Daily Logs & Surveys]. Outline steps to synthesize this data to evaluate objective (HCASR 5, mixed methods)."

Human Oversight Reminder: AI can suggest data synthesis structures. Interpreting nuanced qualitative feedback (especially from Daily Logs), applying analysis methods appropriately, ensuring thematic coding validity, and drawing unbiased conclusions require critical human judgment, methodological expertise, and collaborative team discussion (HCASR Component 5).

Initial Pilot Performance Evaluation

Following data synthesis, the initial evaluation assesses the pilot's performance against its objectives and success criteria defined in Chaps. 9 and 10. This preliminary assessment is based solely on pilot evidence. The focus of this initial evaluation is on the process and adherence to predefined metrics, as the detailed findings will be documented in the User Outcomes and Deliverables (UODs). Key dimensions include:

- Assessing performance against KPIs and benchmarks (Acacia Advisors, 2023; Kanerika, 2024)
- Evaluating user experience and usability (synthesizing qualitative feedback, especially from Daily Logs and usability tests)

- Reviewing ethical considerations in practice by documenting bias audit findings and reported concerns (Dion & Rattigan, 2025; Southern Regional Education Board [SREB], 2025)
- Assessing workflow integration effectiveness
- Identifying key successes and challenges

This objective, evidence-based evaluation informs immediate refinements and Phase 3 recommendations (Brown, 2019; Kotter et al., 2021). The UX workbook provides detailed activities and templates for this process.

Workbook Prep 4. Structuring the Initial Evaluation

Reflection: Consider the 2–3 most critical success criteria or KPIs defined for your pilot in Chap. 10. How can the structure of the initial pilot evaluation report (detailed in Chap. 12, Section Documenting Pilot Findings, Lessons Learned, and Preparing for Phase 3) most clearly and effectively communicate the pilot's performance against these specific, high-priority criteria, using the synthesized data from Section Analyzing and Synthesizing Pilot Data?

UX Workbook Link: Structure and conduct this evaluation using Chap. 12 UX Workbook, Section 4 ("Evaluating Against Benchmarks") which guides comparison to KPIs, and Section 5 ("Maintaining Living Documents for Pilot Evaluation") which involves updating benchmark reports and dashboards with final data.

PALS Workbook Link & AI Partnership Guidance: Use the Chap. 12 PALS Workbook prompt corresponding to UX Workbook Section 7 (Compiling Final Reports). Input the pilot's main objectives and key metrics; ask your AI partner to help draft a high-level outline for the "Initial Pilot Performance Evaluation Report," ensuring sections explicitly address performance against KPIs, overall user experience (informed by Daily Logs etc.), and ethical compliance findings, referencing HCASR Component 4 (Evaluation) & Component 5 (Accountability).

Example Prompt: "Objectives: [List 1–2], Metrics: [List 2–3 KPIs]. Draft outline for Initial Pilot Performance Eval Report (performance, UX, ethics, successes/challenges) (HCASR 4 & 5)."

Human Oversight Reminder: AI can generate report outlines. Determining critical evaluation criteria, interpreting findings' strategic significance, objectively assessing ethical performance, and formulating balanced conclusions require human analytical skills and team judgment (HCASR Component 4 & 5).

Refining Workflows & Risk Mitigation Based on Pilot Data

Pilot data actively informs immediate refinements to operational processes and risk management strategies. Documenting these data-driven adjustments is crucial.

Optimizing AI-Impacted Workflows Insights from pilot data (e.g., themes from feedback logs about confusion, efficiency metrics showing bottlenecks) should refine workflow maps. Adjustments might involve modifying human-AI collaboration protocols or reallocating tasks (applying HCAS principles). Table 12.1 outlines key activities. Figure 12.1 illustrates a comparative workflow map.

Table 12.1 Activities for optimizing workflows post-pilot implementation

Activity	Description
Process Mapping Analysis	Use pilot maps and feedback to identify bottlenecks
Human-AI Collaboration Adjust	Use feedback and performance data to refine task allocation and protocols

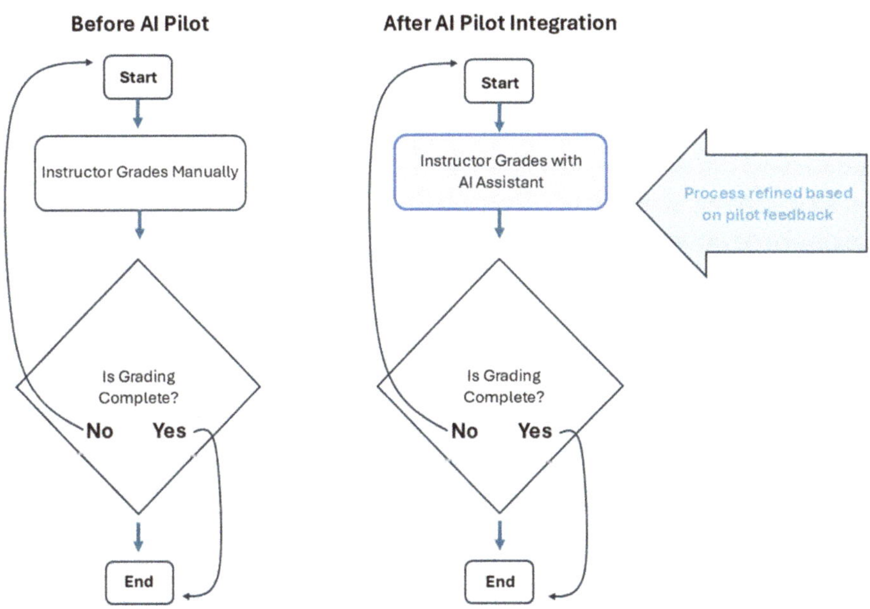

Fig. 12.1 Workflow for essay grading before and after AI pilot integration. (*Note*: Conceptual comparative flowchart illustrating a specific institutional workflow before and after AI pilot integration. The 'After' flow highlights steps involving AI and points where feedback or refinements occurred based on pilot data. (Requires final visual for accuracy))

Table 12.2 Activities for updating risk mitigation strategies post-pilot implementation

Activity	Description
Risk Identification & Assessment	Use logs and audits to find new risks
Refining Mitigation Measures	Use pilot data to adjust strategies
Ensuring Ethical Oversight Checks	Use compliance reports and feedback to verify mechanisms

Updating Risk Mitigation Strategies Real-world pilot data (e.g., error logs, user reports via Daily Logs, bias audit findings) provides input for managing risks, guided by frameworks like the NIST AI RMF (National Institute of Standards and Technology [NIST], 2023). This involves identifying new risks, reassessing known risks, refining or developing mitigations, and confirming ethical oversight mechanisms. Updates are documented in the pilot's Risk Register. Table 12.2 details these activities.

This documentation reflects iterative learning (Brown, 2019) and supports change management (Kotter et al., 2021).

Using AI as a Strategic Partner: Every HEI is situated in a unique sociohistorical context, which means there is no one-size-fits-all approach to AI integration in higher education (Carmody-Bubb, 2023). As institutions around the world integrate GenAI, contextual data from institutions can be ethically and effectively used to 1) guide custom tailored tools, resources, and solutions, and 2) mitigate Western (and other) biases that will be inherent within models as they are adopted globally. See the Human-GenAI Collaboration & Synergy (HAICS) Guide for tools and resources.

Using AI as a Strategic Partner: Every HEI is situated in a unique sociohistorical context, which means there is no one-size-fits-all approach to AI integration in higher education (Carmody-Bubb, 2023). As institutions around the world integrate GenAI, contextual data from institutions can be ethically and effectively used to 1) guide custom tailored tools, resources, and solutions, and 2) mitigate Western (and other) biases that will be inherent within models as they are adopted globally. See the Human-GenAI Collaboration & Synergy (HAICS) Guide for tools and resources.

Workbook Prep 5. Documenting Data-Driven Refinements

Reflection: Based on the pilot data analysis and initial evaluation (Sections Analyzing and Synthesizing Pilot Data & Initial Pilot Performance Evaluation), what was one specific workflow adjustment or update to a risk mitigation strategy that was clearly driven by the evidence gathered (e.g., a theme frequently mentioned in Daily Feedback Logs, a performance metric that missed its target, an unexpected technical issue)? How was the connection between the data/feedback and the resulting refinement documented to ensure learning is captured?

UX Workbook Link: Document these specific, data-driven updates using Chap. 12 UX Workbook, Section 6 ("Documenting Refinements & Risk Updates"), which involves updating process maps and finalizing the Pilot Risk Register.

PALS Workbook Link & AI Partnership Guidance: Use the Chap. 12 PALS Workbook prompt corresponding to UX Workbook Section 5 (Living Documents - adapt prompt focus to summarizing refinements). Input the specific refinement made and the data source that prompted it (e.g., analysis of Daily Logs, survey results); ask your AI partner to help draft a concise summary documenting this data-driven refinement and its rationale, suitable for inclusion in the final pilot report, referencing HCASR Component 4 (Iterative Improvement) & Component 5 (Documentation).

Example Prompt: "Refinement: [Describe]. Source: [Daily Log analysis]. Draft brief summary explaining data-driven refinement for pilot report (HCASR 4 & 5)."

Human Oversight Reminder: AI can assist in structuring documentation. Clearly identifying causal links between data (especially qualitative insights from logs) and actions, assessing refinement effectiveness, and accurately updating risk assessments require careful human analysis and validation (HCASR Component 4 & 5).

Documenting Pilot Findings, Lessons Learned, and Preparing for Phase 3

Key User Outcomes and Deliverables (UODs) ensure transparency and evidence-based decisions (Brown, 2019; Kotter et al., 2021). The purpose of each UOD is to provide a clear, evidence-based account of the pilot's activities, findings, and implications, collectively forming the foundation for Phase 3 decisions. These include:

A. AI Pilot Data Collection & Feedback Report: Compiles summaries of quantitative data (KPIs, usage metrics) and qualitative themes (from Daily Feedback Logs, surveys, etc.). Its purpose is to present all collected evidence systematically.
B. AI Pilot Performance Evaluation Report: Presents synthesized findings comparing pilot data against goals and KPIs, assessing performance, user

experience, and ethical compliance. Its purpose is to offer an objective assessment of the pilot's success against its defined objectives.

C. Updated Pilot Risk Register (Final Pilot Version): Documents the observed effectiveness of mitigations, new risks, actions taken, and residual risk status (NIST, 2023; SREB, 2025). Its purpose is to provide a final account of the risk landscape post-pilot.
D. Pilot Lessons Learned Document: Captures key technical, pedagogical/operational, ethical, and stakeholder engagement insights, drawing from Daily Feedback Logs and evaluation findings. Its purpose is to distill actionable knowledge from the pilot experience for future initiatives.
E. Pilot Conclusion Recommendation: The Smart Team's concise, evidence-based preliminary recommendation (e.g., Proceed to Phase 3 for Scaling/Refinement, or Discontinue) with rationale. Its purpose is to propose a data-driven next step for the initiative.

Workbook Prep 6. Compiling the Final Pilot Package

Reflection: As you prepare the final package of pilot reports (Data Collection Report, Performance Evaluation, Risk Register Update, Lessons Learned, Recommendation), how will the Smart Team ensure clear connections and traceability between the raw data summaries (Deliverable A), the objective evaluation against goals (B), the documented risk status (C), the actionable insights gained (D), and the final evidence-based recommendation for Phase 3 (E)? What review process will ensure consistency and coherence across these documents?

UX Workbook Link: Compile these final documents using Chap. 12 UX Workbook, Section 7 ("Compiling Final Pilot Reports") and formulate the final recommendation in Section 8 ("Formulating Recommendations for Phase 3").

PALS Workbook Link & AI Partnership Guidance: Use the Chap. 12 PALS Workbook prompt corresponding to UX Workbook Section 7. Input 2–3 key challenges or successes documented in the pilot evaluation report (Deliverable B); ask your AI partner to help formulate these as concise, actionable "Lessons Learned" statements suitable for Deliverable D, emphasizing insights applicable to future AI initiatives and referencing HCASR Component 5 (Continuous Improvement & Accountability). Example Prompt: "Finding: ['Users struggled with prompt specificity']. Formulate as actionable 'Lesson Learned' statement for future training/design (HCASR 5)."

Human Oversight Reminder: AI can assist in phrasing lessons. Ensuring final reports accurately reflect pilot data, objective evaluation, insightful lessons, and a logical recommendation requires human critical thinking, synthesis skills, and Smart Team consensus (HCASR Component 5).

Conclusion

Chapter 12 detailed the essential processes concluding Phase 2: the systematic collection, analysis, synthesis, and initial evaluation of AI pilot data. This stage transforms raw pilot experiences and ongoing feedback (including invaluable insights from Daily Feedback Logs) into structured, evidence-based insights. Methodologies for integrating quantitative metrics and qualitative feedback were outlined to provide a holistic understanding of the pilot's performance against goals, its impact on user experience, and its adherence to ethical considerations.

This chapter highlighted how pilot data directly informs immediate workflow and risk mitigation refinements, embodying the principles of iterative learning and adaptation. Compiling comprehensive documentation—data reports, evaluation summaries, updated risk registers, actionable lessons learned—creates the robust evidence package needed for the strategic evaluations and decisions of Phase 3 (Chap. 13), ensuring future AI integration efforts are data-driven, ethically grounded, and strategically aligned.

Looking Ahead & Engaging the Workbook

Chapter 12 established the framework for collecting, synthesizing, and conducting an initial evaluation of AI pilot data, concluding the active piloting phase. The Chap. 12 Companion UX Workbook provides the structured activities, templates, and guidance necessary to execute these critical tasks. Engaging with the workbook enables Smart Teams and stakeholders to systematically organize collected data, perform rigorous analyses, evaluate performance against defined criteria, document findings and refinements transparently, and formulate evidence-based recommendations. Completing these activities generates the key User Outcomes and Deliverables (UODs) that serve as the essential evidence base for the comprehensive evaluation, ethical reflection, adaptation, and strategic scaling decisions detailed in Phase 3 (Chaps. 13, 14, and 15).

Long-Term Scaling, Sustainability & AI Governance: Evidence-Informed TLARA Refinement

Chapter 12 concludes the active pilot phase by focusing on the systematic collection and synthesis of diverse data and feedback. This critical analysis of pilot performance against initial goals, stakeholder experiences across different levels and spaces, and documented lessons learned (including ethical observations) provides the evidence base for informed decision-making (U.S. Department of Education, OET, 2024). The insights gained during this period are fundamental for determining the future trajectory of AI initiatives within specific **TLARA (Teaching, Learning, Assessment, Research, Administration)** domains, particularly regarding sustainable scaling and effective long-term governance. Both the Chap. 12 UX and PALS Workbooks facilitate partnering with AI to manage and interpret this wealth of pilot data ethically.

The journey of AI integration within the complex sociocultural context of an HEI relies on such iterative cycles of action and reflection over time (DerSimonian & Montagnino, 2025). The PALS Workbook for Chap. 12, through "Long-Term Scaling, Sustainability & AI Governance: Iteratively Developing Your Custom TLARA-Informed AI Guidelines," prompts you to use this evidence to refine your framework. Engaging with "Topics for PALS TLARA Focus"—such as "Using Pilot Data to Inform Sustainable Scaling Decisions for TLARA" and "Lessons Learned for Refining Long-Term AI Governance"—your GenAI advisor will help you translate pilot outcomes into concrete adjustments and additions to your "Living Document." This ensures your TLARA guidelines become increasingly robust, evidence-backed, and prepared to support responsible, scaled AI deployment and enhance AI literacy (Ellucian, 2024).

References

Acacia Advisors. (2023). *Measuring success: Key metrics and KPIs for AI initiatives*. Retrieved May 2, 2025, from https://chooseacacia.com/measuring-success-key-metrics-and-kpis-for-ai-initiatives/

Askew, T., Mathew, R., Fishman, T., Kunkel, D., & Caron, B. (2024, September 25). How higher education can realize the potential of generative AI. *Deloitte Insights*. https://www2.deloitte.com/us/en/insights/industry/public-sector/generative-ai-higher-education.html

Brown, T. (2019). *Change by design: How design thinking transforms organizations and inspires innovation* (Rev. & Updated ed.). Harper Business.

Carmody-Bubb, M. (2023). *Cognition and decision making in complex adaptive systems: The human factor in organizational performance*. Springer.

DerSimonian, R., & Montagnino, C. (2025, March 26). Crafting thoughtful AI policy in higher education: A guide for institutional leaders. *Campus Technology*. https://campustechnology.com/articles/2025/03/26/crafting-thoughtful-ai-policy-in-higher-education-a-guide-for-institutional-leaders.aspx

Dion, K. E., & Rattigan, K. (2025, February 11). AI is transforming higher ed: What you need for compliance and governance. *University Business*. https://university-business.com/ai-is-transforming-higher-ed-what-you-need-for-compliance-and-governance/

Ellucian. (2024, December 16). AI in higher education: Implementation strategies for institutions and tech partners. *Higher Ed Dive*. https://www.highereddive.com/spons/ai-in-higher-education-implementation-strategies-for-institutions-and-tech/735434/

Institute of Education Sciences. (2021). *Creating effective surveys: Best practices in survey design*. U.S. Department of Education. Retrieved May 2, 2025, from https://ies.ed.gov/use-work/resource-library/resource/fact-sheetinfographicfaq/creating-effective-surveys-best-practices-survey-design

Institute of Education Sciences. (2024, July). *Considerations and best practices for focus group data collection*. U.S. Department of Education. Retrieved May 2, 2025, from https://ies.ed.gov/use-work/resource-library/resource/other-resource/ask-expert-considerations-and-best-practices-focus-group-data-collection

Ji, H., Suo, L., & Chen, H. (2024). AI performance assessment in blended learning: Mechanisms and effects on students' continuous learning motivation. *Frontiers in Psychology, 15*, 1447680. https://doi.org/10.3389/fpsyg.2024.1447680

Kanerika. (2024, January 9). *How to launch a successful AI pilot project: A comprehensive guide*. Retrieved May 2, 2025, from https://www.kanerika.com/blogs/ai-pilot/

Kotter, J. P., Akhtar, V., & Gupta, G. (2021). *Change: How organizations achieve hard-to-imagine results in uncertain and volatile times*. Wiley.

National Institute of Standards and Technology. (2023, January). *Artificial Intelligence Risk Management Framework (AI RMF 1.0)* (NIST AI 100-1). U.S. Department of Commerce. https://doi.org/10.6028/NIST.AI.100-1.

Southern Regional Education Board (SREB). (2025, February 18). *AI rec 7 risk management*. SREB. https://www.sreb.org/ai-rec-7-risk-management

The CASE HQ. (2023). *Transformative power of AI for assessment dashboards and performance reports*. Retrieved May 2, 2025, from https://thecasehq.com/transformative-power-of-ai-for-assessment-dashboards-and-performance-reports/.

U.S. Department of Education, Office of Educational Technology (OET). (2024). *Navigating Artificial Intelligence in postsecondary education: Building capacity for the road ahead*. https://portal.ct.gov/das/-/media/das/ctedtech/publications/2025/2025-used-oct-archive/os-24-002833-ai-in-postsecondary-ed.pdf

Weil, D., & Forrester, J. (2024, October 17). A road map for leveraging AI at a smaller institution. *EDUCAUSE Review*. https://er.educause.edu/articles/2024/10/a-road-map-for-leveraging-ai-at-a-smaller-institution

Williams, B. (2025). Evaluating a pilot program: Best practices. *Insight7*. https://insight7.io/evaluating-a-pilot-program-best-practices/

13

Phase 3: Monitoring and Evaluating Ongoing Initiatives

Introduction: Establishing Continuous Evaluation for Sustainable AI Integration

As Artificial Intelligence (AI) pilot programs conclude or mature within Higher Education Institutions (HEIs) and considerations for scaling emerge, the integration process enters the critical third phase: ongoing Monitoring and Evaluation. Transitioning from the initial implementation and data collection activities concluding Phase 2 (Chap. 12), this phase emphasizes the necessity of systematic, continuous assessment for AI initiatives considered "maturing" or "scaled." These terms refer to initiatives that have moved beyond initial, contained piloting and are in advanced testing, early stages of broader rollout, or fully operational within specific institutional contexts. This chapter focuses primarily on establishing the foundational concepts and frameworks for this ongoing monitoring and evaluation; subsequent chapters will delve into the specifics of scaling (Chap. 14) and adaptation (Chap. 15). This continuous evaluation ensures that AI initiatives remain effective, ethically sound, and consistently aligned with evolving institutional goals and stakeholder values, regardless of institutional context (Weil & Forrester, 2024).

Rigorous, continuous evaluation serves as a vital feedback mechanism for adaptation and sustainability. It provides essential data-driven insights required to refine strategies, optimize performance, proactively manage emerging risks (e.g., privacy, bias, transparency), and justify continued investment or strategic adjustments for AI applications (Dahlgren, 2025; SmartEvals, n.d.; Weil & Forrester, 2024). Without structured ongoing monitoring and evaluation, even initially successful AI implementations risk losing effectiveness or failing to

---

Supplementary Information The online version contains supplementary material available at https://doi.org/10.1007/978-3-032-05205-6_13.

D. Woodring, *Advancing Higher Education with Human-Centric AI Partnerships*,
https://doi.org/10.1007/978-3-032-05205-6_13

deliver sustainable value. This continuous evaluation process can be enhanced by applying Design Thinking principles, where performance data and user feedback ("Inspiration") fuel subsequent cycles of "Ideation" (identifying refinements) and "Implementation" (making adjustments) to maintain desirability, feasibility, and viability (Brown, 2019). It also aligns with Change Management concepts, particularly "Sustain Acceleration" and "Institute Change." Systematically monitoring progress and demonstrating value helps sustain momentum, while ongoing ethical oversight anchors changes within the institution's culture, reinforcing long-term sustainability and trustworthiness (Kotter et al., 2021; Robert & McCormack, 2024).

Workbook Prep 1. Rationale for Continuous Evaluation vs. Pilot Evaluation

Reflection: Why is ongoing monitoring/evaluation fundamentally different from initial pilot evaluation? What new challenges or opportunities arise with wider or longer AI implementation?

UX Workbook Link: See Chap. 13 UX Workbook "Preparatory Exercises" (focusing on long-term impact, ethics, and culture) & Section 1 ("Planning the Continuous Evaluation Framework").

PALS Workbook Link & AI Partnership Guidance: Use Chap. 13 PALS Workbook prompt for Section 1 ("Planning: Defining Evaluation Scope, Questions & Framework"). Input your scaling/operational AI initiative; ask your AI partner for rationale why continuous evaluation (versus just initial review) is essential for long-term success/ethics (HCASR C5) and help structuring the framework plan.

Example Prompt: "Initiative: [AI advising dashboard]. Help structure a 'Continuous Evaluation Framework Plan' and draft rationale justifying continuous evaluation for long-term success/adaptation (HCASR C5)."

Human Oversight Reminder: AI states general benefits and structures. Identifying specific risks/opportunities & designing feasible/meaningful continuous evaluation plans needs human strategic insight, context, and input (HCASR C5).

Core Frameworks Reference: The frameworks mentioned below, such as HCASR, HCAS, RSA, EHCS, EAO, and AHIPs, are detailed in the companion Human-GenAI Collaboration & Synergy (HAICS) Guide and/or Part 1 of this textbook. Please refer to those sections for full definitions and explanations.

This chapter uses these frameworks as they relate to evaluating the long-term impact and ethical integration of AI. The chapter conceptually links to the continuous application of Design Thinking (Brown, 2019) for ongoing refinement and Change Management (Kotter et al., 2021) principles for sustaining and institutionalizing AI initiatives effectively and ethically.

(The User Experience (UX) Workbook and Personalized Adaptive Learning Sessions (PALS) Workbook for Chap. 13, available via download, assist users in ethically and effectively partnering with AI to customize tools, solutions, resources, UODs (user outcomes and deliverables) for their unique needs and contexts.)

Goals of Ongoing Monitoring & Evaluation

The primary goals during the Monitor & Evaluate phase involve tracking performance against objectives, proactively addressing ethical challenges, and defining stakeholder roles to foster continuous refinement.

Tracking AI Performance Metrics & Governance Adherence

A central objective is the systematic, ongoing tracking of implemented AI tool performance against established Key Performance Indicators (KPIs), benchmarks, and institutional goals. Industry consultancies (e.g., Acacia Advisors, n.d.) and vendor analyses (e.g., Kanerika Editorial Team, 2025) often highlight operational and financial KPIs, such as efficiency gains, accuracy rates, user engagement, and cost savings. Analyzing these quantitative metrics alongside qualitative feedback provides crucial operational performance data (Acacia Advisors, n.d.; Kanerika Editorial Team, 2025).

However, in the higher education context, tracking performance must also encompass the development and application of user competencies, particularly AI literacy. This requires objective measurement tools distinct from potentially biased self-reports (Markus et al., n.d.; Jin et al., 2024; Li et al., n.d.). Continuous performance data, covering both operational effectiveness and verified user competencies, informs governance decisions regarding resources, policy adjustments, and strategic direction. Regularly assessing performance, including social impact and human-AI configuration effectiveness (NIST, 2023), supports change management by providing evidence of progress or areas needing action (Kotter et al., 2021) and aligns with design thinking's testing and inspiration cycles (Brown, 2019). Furthermore, tracking user competence development, particularly AI literacy assessed through objective measures (linking to the EHCS framework), is essential for evaluating the true impact of AI integration (Markus et al., n.d.; Jin et al., 2024; Li et al., n.d.).

Monitoring and Mitigating Ethical Challenges

A key goal is the continuous monitoring for, and proactive mitigation of, ethical challenges in scaled AI deployment, particularly as AI in education can be deemed "high-risk" (European Commission, 2024; Nguyen, 2025). This includes vigilance regarding algorithmic bias (Chapman University, n.d.; NIST, 2023), ongoing attention to transparency and explainability (NIST, 2023; Wynants et al., 2025), data privacy (Nguyen, 2025; Wynants et al., 2025), academic integrity, and accessibility (Wynants et al., 2025). Addressing this requires establishing processes for periodic ethical audits, verifying accuracy, analyzing user feedback, and maintaining awareness of evolving ethical best practices and regulations (NIST, 2023; Nguyen, 2025; Wynants et al., 2025). This commitment aligns with ensuring solutions remain ethically desirable (Brown, 2019) and anchoring ethical practices within the organizational culture (Kotter et al., 2021). This commitment to ongoing ethical vigilance operationalizes the core tenets of HCAS (ensuring human-centric values are upheld) and RSA (proactively regulating AI use to mitigate risks).

Defining Stakeholder Roles in Continuous Refinement

This phase requires clearly defined roles and responsibilities for diverse stakeholders (students, faculty, staff, administrators, technology leaders) in the ongoing evaluation ecosystem (Co-Action, Penn State University, n.d.; Wynants et al., 2025). Stakeholders impacted by AI should be involved in evaluation cycles (The Learning Accelerator, n.d.). Established channels for feedback and participation are necessary, along with clear communication about how input is considered (Co-Action, Penn State University, n.d.; The Learning Accelerator, n.d.). The aim is to cultivate an institutional culture where iterative refinement, driven by diverse feedback, is standard procedure. This embodies design thinking's iterative cycle (Brown, 2019) and mobilizes stakeholders to sustain change (Kotter et al., 2021). This collaborative approach reflects HCASR Component 4 (Stakeholder Engagement) and Component 5 (Collective Intelligence, Continuous Improvement), creating robust feedback loops.

Workbook Prep 2. Setting Specific Measurable Goals for Ongoing Monitoring

Reflection: For an AI tool scaling or in use, how do goals (Performance Tracking, Ethical Monitoring, Stakeholder Refinement Roles) apply to ensure long-term success and responsible use?

UX Workbook Link: See Chap. 13 UX Workbook, Section 1 activities (defining metrics, assembling governance teams) & Section 2 ("Executing Monitoring Activities & Audits").

PALS Workbook Link & AI Partnership Guidance: Use Chap. 13 PALS Workbook prompt for Section 2 ("Actions: Selecting Methods, Metrics & Planning Data Collection"). Input initiative/objective; ask AI to help draft 1–2 specific, measurable SMARTIE objectives for ongoing performance tracking (1.A) & ethical vigilance (1.B) (HCASR C5) and formulate audit findings into checklist items.

Example Prompt: "Initiative: [Name]. Objective: [Objective]. Draft 1 measurable SMARTIE sub-goal for ongoing performance tracking & 1 for ethical monitoring (HCASR 5). Also, formulate this audit finding '[finding]' as an actionable checklist item."

Human Oversight Reminder: AI helps formulate goals and checklist items. Defining meaningful, measurable, relevant monitoring goals (especially ethical ones) & appropriate actions needs human strategic judgment, expertise, and stakeholder consensus (HCASR C1 & 5).

AI Performance Evaluation & Benchmarking (Ongoing)

Evaluating the ongoing performance and impact of AI initiatives requires systematic assessment against predefined criteria, comparison against benchmarks, and consideration of long-term sustainability, adopting a performance-based approach focused on institutional goals (Whitmarsh, 2025).

Assessing Sustained Impact

Ongoing evaluation involves assessing AI's tangible effect on key institutional areas over time, considering both effectiveness (EAO framework) and the development of related user skills (EHCS framework). This requires tracking metrics related to student learning outcomes (Ghai, 2025; Verhoeven & Hor, 2025), including objective evidence showing how developing AI literacy relates to task performance. For instance, Jin et al. (2024) found performance-based GenAI literacy scores predict success in AI-assisted tasks. Evaluation also involves monitoring faculty workload and satisfaction (DerSimonian & Montagnino, 2025; TeacherMatic, n.d.), and administrative operations (Whitmarsh, 2025; Zimmermann, n.d.). Human validation of AI-generated insights remains crucial (Le Dinh et al., 2025; Verhoeven & Hor, 2025). Evaluating diverse impacts helps determine if AI solutiosns meet needs ("Desirability") and deliver value ("Viability") (Brown, 2019), providing evidence for continued investment ("Sustain Acceleration") (Kotter et al., 2021) (Table 13.1).

Evaluating Sustainability & Resource Management

Long-term success depends critically on sustainability. Evaluation must assess resource consumption relative to benefits, including operational costs, support time, and maintenance (Brown, 2019). This assessment should also consider the psychological factors influencing long-term, effective user engagement and adaptation, such as self-efficacy and self-management skills in using AI tools (Carolus et al., 2023). Monitoring the risk landscape is integral, requiring ongoing checks for bias creep, adherence to privacy and security measures (e.g., NIST AI RMF (NIST, 2023)), accessibility compliance, and system validity (SREB, 2025; World Economic Forum, 2024; Zimmermann, n.d.). Effective governance (DerSimonian & Montagnino, 2025; SREB, 2025) and ongoing training aligned with actual competency needs (Whitmarsh, 2025) are key to managing risks and fostering trust. Evaluating resource management ensures operational feasibility and ethical responsibility ("Institute Change") (Kotter et al., 2021). Monitoring progress can also be enhanced by utilizing a Continuous Quality Improvement Scorecard (see Fig. 13.1). Such a dashboard visually tracks key metrics related to AI performance, ethical compliance, and stakeholder satisfaction trends over time, potentially inspired by balanced scorecard approaches adapted for AI integration in HEIs (see Salmon & Psarras, 2024) (Table 13.2).

Utilizing Institutional Benchmarking

Comparing institutional AI performance and practices against relevant external benchmarks provides context. This can involve comparing KPIs against peer institutions (APRU, 2025), industry standards (e.g., EDUCAUSE), or established frameworks (e.g., SREB recommendations, NCEE framework, APRU's "CRAFT") (APRU, 2025; NCEE, 2024; SREB, 2025; Zimmermann,

Table 13.1 Key areas and example metrics for assessing sustained AI impact in higher education

Area of impact	Example metrics (quantitative)	Example metrics (qualitative)	Relevant framework(s)
Student Learning Outcomes	Course grades/GPA trends (in AI-affected courses/programs) Standardized test scores (if applicable) Retention/Persistence rates (by cohort/program) Time-to-completion rates Objective AI literacy scores (e.g., from GLAT, AICOS)	Student self-reports on learning gains (surveys, focus groups) Faculty observations of skill application Analysis of student work quality (rubrics, portfolio review) Employer feedback on graduate preparedness Analysis of student reflections on AI use	EAO, AHIPs EHCS, EAO
Faculty experience	Time spent on specific tasks (e.g., grading, prep) —pre-/post-AI Adoption rates of AI tools Professional development participation rates Faculty retention/satisfaction scores (departmental level)	Faculty surveys/interviews on workload perception Feedback on AI tool usability and effectiveness Reports on innovation in teaching practices using AI Anecdotal evidence of impact on research productivity	EHCS, RSA
Administrative operations	Process completion times (e.g., admissions, advising) Cost savings/resource reallocation Error rates in administrative tasks Staff time allocated to AI support/maintenance	Staff surveys/interviews on workflow efficiency Stakeholder satisfaction with AI-supported services (e.g., chatbot) Feedback on ease of access to information/services Qualitative assessment of improved decision-making	HCAS, RSA
Ethical compliance	Audit findings (bias, privacy incidents) Accessibility compliance rates Data breach incidents related to AI systems	Stakeholder reports/concerns regarding ethical issues User feedback on transparency/explainability Qualitative review of fairness perceptions	HCAS, RSA

Note. This table provides examples; specific metrics should be tailored to institutional goals, the specific AI initiative, and data availability. Metrics should be tracked over time to assess sustained impact. Framework references link to the HAICS Guide and incorporate concepts from sources such as Acacia Advisors (n.d.), Jin et al. (2024), Markus et al. (n.d.), DerSimonian and Montagnino (2025), Ghai (2025), and Zimmermann (n.d.)

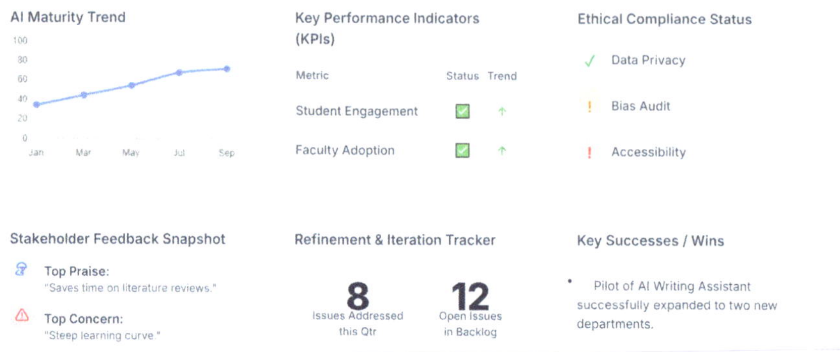

Fig. 13.1 Continuous quality improvement scorecard

Table 13.2 Key factors in evaluating AI sustainability

Factor	Key aspects to monitor/evaluate
Resource management	Ongoing operational costs (licensing, cloud fees, energy), staff support time, faculty/staff training needs, maintenance requirements, overall cost-benefit analysis ("Viability").
Ongoing risk management	Bias creep in algorithms, adherence to privacy regulations (FERPA) and data security protocols, cybersecurity vulnerability management, accessibility compliance, system validity/reliability.

Note. Evaluating resource consumption and proactively managing ongoing risks are crucial for ensuring AI initiatives remain operationally feasible, ethically responsible, and trustworthy over the long term ("Institute Change")

n.d.). Benchmarking helps understand relative position, identify best practices ("Inspiration"), validate successes ("Viability") (Brown, 2019), and maintain a drive for improvement ("Sense of Urgency") (Kotter et al., 2021) (Table 13.3).

AI Ethics & Literacy Audits & Policy Refinement

Continuous monitoring must extend to the ethical performance of AI systems and the effectiveness of AI literacy efforts, informing necessary policy refinements.

Table 13.3 Examples of benchmarks and frameworks for AI evaluation in higher education

Benchmark type	Example focus area	Potential benchmark sources	Relevant framework(s)
Peer institutions	AI adoption rates in specific functions	Consortia data, peer surveys/networking	HCAS (viability)
	Governance structures/policy approaches	Published case studies, conference presentations, inter-institutional collaborations	
	Student/faculty satisfaction with AI tools	Shared survey instruments/results (informal or formal networks)	
Industry standards (HEI/EdTech focused)	AI tool procurement vetting criteria	EDUCAUSE HECVAT (if AI modules added), industry best practice guides	RSA
	AI ethical guidelines/principles	EDUCAUSE resources, national/regional HE bodies (e.g., SREB), UNESCO guidelines	HCAS, RSA
	Accessibility standards for AI tools	WCAG applied to AI interfaces, HEI accessibility guidelines	HCAS (Equity)
	AI literacy competency frameworks	EDUCAUSE frameworks, UNESCO frameworks, published research/instruments	EHCS
	AI maturity models	Jisc AI Maturity Toolkit (2024), Asana/Anthropic model, Brady (2024) framework	DT-CM
General frameworks (applicable to HEI)	AI risk management practices	NIST AI RMF (2023), ISO standards related to AI risk/ethics	RSA, HCAS
	Data governance and privacy	NIST Privacy Framework, GDPR/FERPA compliance standards	RSA
	Responsible AI/trustworthiness principles	OECD AI Principles, EU AI Act guidelines, WEF AI Governance Alliance	HCAS
	Change management effectiveness	Kotter's framework success indicators, change readiness assessments	DT-CM

Note. Benchmarking involves comparing internal performance or practices against external references to identify areas for improvement and contextualize results. Sources should be selected based on relevance and credibility for the specific area being benchmarked. Framework references link to the HAICS Guide or Design Thinking/Change Management

Workbook Prep 3. Selecting Specific Metrics for Impact Assessment

Reflection: For an ongoing AI initiative, which impact area (Student Learning, Faculty Workload, Admin Ops, Sustainability/Risk) is most crucial to evaluate continuously? What specific metric offers insight?

UX Workbook Link: See Chap. 13 UX Workbook, Section 3 ("Ongoing UX & Ethical Assessment Strategies") activities (identifying metrics, planning benchmarking, reviewing ethics strategies).

PALS Workbook Link & AI Partnership Guidance: Use Chap. 13 PALS Workbook prompt for Section 3 ("UX Strategies: Stakeholder Input & Ethical Considerations in Evaluation"). Input initiative/impact area; ask AI for 2–3 specific, measurable KPIs to track impact over time (ref EAO framework) and help outline the ongoing ethical assessment strategy.

Example Prompt: "Initiative: [Name]. Impact Area: [Faculty Workload]. Suggest 2–3 measurable KPIs to track ongoing impact (EAO framework). Also, help outline key components for an 'Ongoing Ethical AI Assessment Strategy'."

Human Oversight Reminder: AI suggests KPIs & structure. Selecting meaningful, measurable, relevant, prioritized metrics, & designing ethical assessment methods needs human judgment, stakeholder agreement, & feasibility planning (HCASR C1 & 5).

Establishing Transparency & Accountability through Audits

Maintaining transparency and accountability requires periodic audits of AI systems and practices (Robert & McCormack, 2024). This involves established procedures, potentially overseen by cross-disciplinary governance committees (Robert & McCormack, 2024), to review usage, assess adherence to ethical guidelines (e.g., examining outputs for bias, reviewing data handling, assessing transparency), and investigate issues (DerSimonian & Montagnino, 2025; NTIA, 2024). Crucially, AI literacy audits gauge ongoing user understanding and application of AI principles, identifying training needs. Different approaches to AI literacy assessment exist: objective methods measure actual competencies via knowledge-based tests (e.g., AICOS; Markus et al., n.d.) or performance-based tasks (e.g., GLAT; Jin et al., 2024), while self-report scales assess perceived literacy and related psychological factors (e.g., MAILS; Carolus et al., 2023).

Given the potential disconnect between self-perceived and objective literacy (Markus et al., n.d.), incorporating objective measures into audits provides a more reliable assessment of actual user capabilities. The choice of instruments should align with the specific goals of the audit. Regular audits ensure ethics and literacy remain central (Brown, 2019) and become institutionalized (Kotter et al., 2021). Such audits are practical implementations of the regulatory aspects within the RSA framework and support the accountability goals of HCAS (Table 13.4).

Table 13.4 Types of audits for ongoing AI governance in higher education

Audit type	Purpose	Example focus areas/ questions	Frequency example	Relevant framework(s)
Ethical Compliance Audit	Verify adherence to institutional ethical AI principles, policies, and external regulations.	Is AI use disclosed transparently? Are bias mitigation steps documented? Is data handling compliant with privacy policies?	Annually	HCAS, RSA
Algorithmic Bias Audit	Assess AI model outputs and performance for systematic, unfair disparities across demographic groups.	Does the tool perform differently for various groups? Are fairness metrics tracked? Are there disproportionate negative impacts?	Annually/ biannually	HCAS (fairness)
Data Privacy Audit	Ensure compliance with data privacy regulations (e.g., FERPA) and institutional data handling policies.	Is PII appropriately anonymized/ protected? Are access controls effective? Is data retention policy followed?	Annually	RSA
Security Audit	Assess vulnerabilities, protection against unauthorized access, data breaches, or adversarial attacks.	Are security controls adequate? Is vulnerability scanning performed? Is incident response plan tested?	Annually/ biannually	RSA
AI Literacy Audit	Gauge ongoing user understanding and application of AI concepts, ethical use, capabilities, and limitations.	Do users understand institutional policies? Can users apply AI tools effectively for intended tasks? Are training needs being met?	Biannually/ ad hoc	EHCS, HCASR C5

(continued)

Table 13.4 (continued)

Audit type	Purpose	Example focus areas/ questions	Frequency example	Relevant framework(s)
Performance Audit	Evaluate if the AI system continues to meet its intended performance goals and KPIs over time.	Is accuracy degrading? Is efficiency maintained? Are outcomes aligned with original objectives?	Quarterly/ annually	EAO
Accessibility Audit	Ensure AI tool interfaces and outputs are accessible to users with diverse abilities.	Does the tool meet WCAG standards? Are alternative formats/methods available?	Annually/ad hoc	HCAS (equity)
Third-Party Vendor Audit	Review vendor compliance with contractual obligations regarding security, privacy, ethics, and SLAs.	Does vendor meet agreed standards? Are vendor audit reports available/ satisfactory?	Annually (contract)	RSA

Note. The specific types and frequency of audits should be determined based on institutional risk assessment, the nature of the AI application, regulatory requirements, and resource availability. Framework references link to the HAICS Guide or HCASR Components

Adapting Policies via Data-Driven Feedback

Institutional AI policies must evolve (Robert & McCormack, 2024). Data from ongoing monitoring (performance metrics, user feedback, audit findings, regulatory changes) provides the evidence for data-driven policy refinement (DerSimonian & Montagnino, 2025; Robert & McCormack, 2024). AI Governance bodies need transparent processes to review this data and determine necessary adjustments to existing policies (e.g., acceptable use, academic integrity, data privacy). This continuous adaptation aligns with design thinking's iterative cycle (Brown, 2019) and change management principles for sustaining and institutionalizing change (Kotter et al., 2021) (Fig. 13.2).

Using AI as a Strategic Partner: Every HEI is situated in a unique socio-historical context, which means there is no one-size-fits-all approach to AI integration in higher education (Carmody-Bubb, 2023). As institutions around the world integrate GenAI, contextual data from institutions can be ethically and effectively used to 1) guide custom-tailored tools, resources, and

Fig. 13.2 Policy refinement feedback loop

solutions, and 2) mitigate Western (and other) biases that will be inherent within models as they are adopted globally. See the Human-AI Collaboration & Synergy Guide for tools and resources.

Applying Evaluation for Continuous Refinement

Evaluation principles remain critical throughout Phase 3 for initiatives undergoing continuous improvement or staged rollouts, involving the use of frameworks and data for adjustments and ensuring institutional readiness. The insights gained from performance evaluation, benchmarking, ethical audits, and policy reviews must feed into a structured process for making adjustments and ensuring initiatives remain effective and aligned over time.

Workbook Prep 4. Planning Audit Cadence and Policy Review Processes

Reflection: How often should ethical audits and AI literacy assessments be conducted for ongoing AI initiatives? What data sources (metrics, logs, tickets) inform regular AI policy reviews?

UX Workbook Link: See Chap. 13 UX Workbook, Section 2 ("Executing Monitoring Activities & Audits"), Section 4 ("Ongoing Monitoring Checkpoints"), Section 6 ("Aligning with Institutional Frameworks").

PALS Workbook Link & AI Partnership Guidance: Use Chap. 13 PALS Workbook prompt for Section 4 ("Checkpoints: Validating Evaluation Plan Feasibility & Rigor"). Input AI application risk level; ask AI to help structure a checkpoint review agenda and draft a gap analysis based on sample findings, including linking findings to policy review needs (HCASR C5).

Example Prompt: "Structure a quarterly checkpoint review agenda for AI initiative '[Name]'. Also, draft brief gap analysis based on findings: '[Finding 1]', '[Finding 2]', linking to policy review needs (HCASR C5)."

Human Oversight Reminder: AI suggests agendas and analysis. Designing effective audits/literacy benchmarks, feasible cadences, meaningful consultation, & final policy decisions need human expertise (governance, ethics, assessment, context) (HCASR C5).

Utilizing Frameworks & KPIs for Continuous Assessment

Structured evaluation frameworks ensure consistent assessment. Institutions might adapt established tools (DuVarney & Anderson, 2024) or develop internal rubrics, potentially informed by vendor suggestions (e.g., SmartEvals, n.d.), to measure performance. Key Performance Indicators (KPIs) related to effectiveness, efficiency, user satisfaction, ethical compliance, and strategic alignment continue to be tracked, drawing examples from industry practice while adapting them to the HEI context (Acacia Advisors, n.d.; Kanerika Editorial Team, 2025). Critically, assessment must also include measures of user competence, such as AI literacy, using validated instruments (e.g., performance-based: Jin et al., 2024; objective knowledge: Markus et al., n.d.; self-report including meta-competencies: Carolus et al., 2023). Assessing institutional readiness for sustained use is also key (DuVarney & Anderson, 2024). Consistent framework application structures ongoing "Testing" (Brown, 2019) and aligns monitoring with priorities ("Institute Change") (Kotter et al., 2021). Applying frameworks consistently helps operationalize HCASR Component 5 (Continuous Improvement) and ensures assessments track progress towards enhancing outcomes (EAO) and skills (EHCS) responsibly (RSA).

Ensuring Data-Driven Decision Making

Synthesizing quantitative KPI data (operational and competency-based) and qualitative user feedback remains fundamental for informed decision-making (Deloitte, 2025; DerSimonian & Montagnino, 2025; Weil & Forrester, 2024). This evidence, including data on user AI literacy levels (Jin et al., 2024; Markus et al., n.d.; Carolus et al., 2023), informs resource allocation, expansion, strategy modification, training provision, policy revision, or phasing out initiatives. Managing expectations regarding return on investment (ROI) is important, recognizing that enterprise-focused ROI metrics (Deloitte, 2025) may need adaptation for the HEI context, where value also includes difficult-to-quantify impacts on student success and research innovation. This continuous cycle of data collection ("Inspiration"), analysis, decision-making, and action ("Implementation") (Brown, 2019) fuels refinement needed to sustain progress and ensure long-term value and ethical grounding (Kotter et al., 2021). This commitment to evidence-based adaptation is central to HCASR Component 5 (Continuous Improvement & Adaptation).

Workbook Prep 5. Developing Processes for Translating Evaluation Data into Action

Reflection: For potential findings (e.g., "User satisfaction declined," "Tool Z underperforming"), what process translates data to decision/action (investigate, refine, decommission)?

UX Workbook Link: See Chap. 13 UX Workbook, Section 5 ("Maintaining Ongoing Living Documents"), Section 7 ("Compiling Ongoing Evaluation Deliverables"), Section 8 ("Recognizing & Communicating Ongoing Success").

PALS Workbook Link & AI Partnership Guidance: Use Chap. 13 PALS Workbook prompt for Section 5 ("Living Documents: Comprehensive Evaluation Plan"). Input key findings for a living document update; ask AI to help structure the document template and draft a concise update summary highlighting changes and linking to potential actions/decisions (HCASR C5).

Example Prompt: "Help structure template for 'AI Literacy Report'. Also, draft update summary for intro based on new findings: '[Finding 1]', '[Finding 2]', linking to action needed (HCASR 5)."

Human Oversight Reminder: AI suggests structures and summaries. Deciding appropriate action & ensuring living docs effectively support decisions needs human judgment (severity, resources, priorities, impact, context) & clear governance processes (HCASR C5).

User Outcomes & Deliverables (UODs): Phase 3 Deliverables

The Monitor & Evaluate phase generates and regularly updates specific User Outcomes and Deliverables (UODs) to record performance, track compliance, communicate findings, and guide adjustments, ensuring transparency and accountability (NIST, 2023; NTIA, 2024). These are considered **living documents** because AI technology, institutional needs, and ethical understanding evolve, requiring these UODs to be continuously revisited and revised to remain relevant and effective guides for action during Phase 3. Key types include:

A. AI Implementation Roadmap (Updated Version): Reflects progress, incorporates monitoring findings, and outlines revised plans based on evaluation data (Brown, 2019; Kotter et al., 2021). Living because strategic priorities and technological feasibility change.
B. Ethics & Risk Management Monitoring Report (Periodic): Summarizes ethical audit findings, ongoing risk monitoring (e.g., NIST AI RMF (NIST, 2023)), stakeholder concerns, and incident responses (NIST, 2023; SREB, 2025; Zimmermann, n.d.). Living as new risks emerge and mitigation effectiveness is assessed.
C. AI Pilot Strategy Document (Archived/Final Evaluation Summary): For post-pilot initiatives, a final summary of goals, methods, findings, evaluation, and lessons learned, serving as institutional knowledge (APRU, 2025). Archived but informs future living documents.
D. AI Governance & Compliance Guide (Updated Version): The institution's repository of AI policies, ethical principles, usage guidelines, and compliance requirements, updated based on monitoring, audits, best practices, and regulatory changes (Robert & McCormack, 2024; SREB, 2025; Zimmermann, n.d.). Living due to evolving regulations, ethical understanding, and institutional learning.
E. AI Performance & Compliance Evaluation Report (Periodic): Synthesizes KPI trends, qualitative feedback, AI literacy results, compliance status, and overall impact assessment (Acacia Advisors, n.d.; Biagini et al., 2024; Deloitte, 2025; Kanerika Editorial Team, 2025; SREB, 2025; The Learning Accelerator, n.d.). Living as performance data is continually gathered and analyzed.
F. AI Continuous Improvement & Governance Strategy (Refined): Updated high-level strategy for monitoring, feedback integration, policy review, adaptation planning, and governance roles (NIST, 2023; Robert &

McCormack, 2024; The Learning Accelerator, n.d.). Living as institutional strategy and governance mechanisms adapt.

G. Public Transparency Report (Optional/Periodic): Outward-facing summaries of institutional AI usage, ethical commitments, risk management, and accountability, especially for high-impact systems (NTIA, 2024; SREB, 2025). Living as practices and public communication needs evolve.

These UODs help ensure monitoring informs transparent, accountable improvement.

Workbook Prep 6. Determining Audience and Key Insights for Specific Deliverables

Reflection: Which deliverables (A-G) are most critical for accountability & continuous improvement for ongoing AI in your context? Why?

UX Workbook Link: See Chap. 13 UX Workbook, Section 7 ("Compiling Ongoing Evaluation Deliverables") & Section 5 ("Maintaining Ongoing Living Documents" as these reports are updated).

PALS Workbook Link & AI Partnership Guidance: Use Chap. 13 PALS Workbook prompt for Section 7 ("Deliverables: Finalized Evaluation Plan Package"). Select deliverable (e.g., "Perf & Compliance Report"); ask AI to help structure the report based on key evaluation areas and suggest how to best present findings for accountability/improvement (HCASR C5).

Example Prompt: "Help structure 'Periodic Performance & Compliance Report' based on eval areas: [Area 1], [Area 2]. Suggest ways to present findings to drive improvement (HCASR 5)."

Human Oversight Reminder: AI suggests structures and presentation ideas. Determining key stakeholders, information needs, effective communication strategy, & ensuring reports drive accountability requires human understanding & strategic planning (HCASR C5).

Conclusion

Chapter 13 outlined the shift from initial pilot evaluations to continuous monitoring and evaluation for maturing AI initiatives (Phase 3). This ongoing process ensures AI tools remain effective, ethically aligned, sustainable, and responsive. Key goals include tracking performance, monitoring ethical challenges, and defining stakeholder roles in refinement. Strategies involve assessing sustained impact, evaluating sustainability and risk, benchmarking, conducting audits, and enabling data-driven policy refinement. The principles of Design Thinking (Brown, 2019) and Change Management (Kotter et al.,

2021) guide the required iterative, adaptive, and culturally sensitive approach. Phase 3 institutionalizes a cycle of vigilant observation, critical reflection, and purposeful adaptation for responsible AI integration.

Looking Ahead & Engaging the Workbook

Having established the importance and core components of ongoing monitoring and evaluation, the focus shifts to operationalizing these principles. This chapter outlined foundational strategies for continuous tracking, ethical oversight, impact assessment, benchmarking, and policy refinement. The Chap. 13 Companion UX and PALS Workbooks provide structured activities to build and implement these continuous monitoring frameworks, enabling Smart Teams and stakeholders to operationalize concepts, generate key Phase 3 outcomes and UODs, and strengthen the foundation for sustainable AI integration leading into Phase 4 (Chaps. 16 and 17).

Long-Term Scaling, Sustainability & AI Governance: Ensuring Enduring TLARA Effectiveness

Chapter 13 has underscored the importance of establishing robust frameworks for continuously monitoring and evaluating ongoing AI initiatives. This is not merely a compliance step but a strategic imperative for ensuring that AI applications within each **TLARA (Teaching, Learning, Assessment, Research, Administration)** domain remain effective, ethical, and aligned with institutional goals over **time** (U.S. Department of Education, OET, 2024). Regular evaluation across different **levels** and **spaces**, including periodic ethical audits and AI literacy assessments, provides the data needed for adaptive management and ensures that your AI ecosystem matures responsibly within its specific **sociocultural context**. Both the Chap. 13 UX and PALS Workbooks offer tools to ethically partner with AI in designing and implementing these crucial oversight processes.

HEIs, as dynamic learning organizations, must embed cycles of reflection and data-driven improvement to navigate the evolving AI landscape (IDEAS, 2025). The PALS Workbook for Chap. 13, through its section "Long-Term Scaling, Sustainability & AI Governance: Iteratively Developing Your Custom TLARA-Informed AI Guidelines," guides you to connect these M&E strategies to your framework. By engaging with "Topics for PALS TLARA Focus"—such as "Continuous Evaluation Frameworks for Long-Term AI Sustainability in TLARA" and "Ethical Audits as a Pillar of AI Governance"—your GenAI advisor will help you integrate principles of ongoing oversight and adaptive management into your "Living Document." This ensures your TLARA guidelines support not just initial deployment but also the sustained, ethical, and beneficial use of AI across the institution (Ellucian, 2024).

References

Acacia Advisors. (n.d.). *Measuring success: Key metrics and KPIs for AI initiatives.* Acacia Advisors. Retrieved May 2, 2025, from https://chooseacacia.com/measuring-success-key-metrics-and-kpis-for-ai-initiatives/

Association of Pacific Rim Universities (APRU). (2025, January). *Generative AI in higher education: Current practices and ways forward.* Retrieved May 2, 2025, from https://apru.org/news/transforming-higher-education-apru-publishes-generative-ai-whitepaper-and-project-report/

Biagini, G., Cuomo, S., & Ranieri, M. (2024, October). *Assessing AI Literacy: A framework-based approach* [Conference paper presentation]. University of Florence, Florence, Italy. Retrieved May 2, 2025, from https://www.researchgate.net/publication/384625476_Assessing_AI_Literacy_a_framework-based_approach

Brown, T. (2019). *Change by design: How design thinking transforms organizations and inspires innovation* (Rev. & updated ed.). Harper Business.

Carmody-Bubb, M. (2023). *Cognition and decision making in complex adaptive systems: The human factor in organizational performance.* Springer.

Carolus, A., Koch, M. J., Straka, S., Latoschik, M. E., & Wienrich, C. (2023). MAILS - Meta AI literacy scale: Development and testing of an AI literacy questionnaire based on well-founded competency models and psychological change- and meta-competencies. *Computers in Human Behavior: Artificial Humans, 1*(2), 100014. https://doi.org/10.1016/j.chbah.2023.100014

Chapman University. (n.d.). *Bias in AI.* Chapman University AI Hub. Retrieved May 2, 2025, from https://www.chapman.edu/ai/bias-in-ai.aspx

Co-Action, Penn State University. (n.d.). *Framework for stakeholder inclusion in the technology planning process.* Retrieved May 2, 2025, from https://coaction.psu.edu/inclusion-framework-phases/

Dahlgren, B. (2025, February 27). How university leaders can ethically and responsibly implement AI. *Campus Technology.* Retrieved May 2, 2025, from https://campustechnology.com/articles/2025/02/27/how-university-leaders-can-ethically-and-responsibly-implement-ai.aspx

Deloitte. (2025, January). *The state of generative AI in the enterprise: Now decides next (Quarter four report).* Deloitte.

DerSimonian, R., & Montagnino, C. (2025, March 26). Crafting thoughtful AI policy in higher education: A guide for institutional leaders. *Campus Technology.* Retrieved May 2, 2025, from https://campustechnology.com/articles/2025/03/26/crafting-thoughtful-ai-policy-in-higher-education-a-guide-for-institutional-leaders.aspx

DuVarney, D., & Anderson, J. (2024, May 3). *The art of possible: Intentional AI and data analytics in higher education.* Baker Tilly. Retrieved May 2, 2025, from https://www.bakertilly.com/insights/the-art-of-possible-intentional-ai-and-data-analytics

Ellucian. (2024, December 16). AI in higher education: Implementation strategies for institutions and tech partners. *Higher Ed Dive*. https://www.highereddive.com/spons/ai-in-higher-education-implementation-strategies-for-institutions-and-tech/735434/

European Commission. (2024, August). Artificial Intelligence Act. *Official Journal of the European Union*. Retrieved May 2, 2025, from https://digital-strategy.ec.europa.eu/en/policies/regulatory-framework-ai

Ghai, K. (2025, February 17). 73.5% of students saw improved learning with an AI course assistant. *Nectir Blog*. Retrieved May 2, 2025, from https://www.nectir.io/blog/73-5-of-students-saw-improved-learning-with-an-ai-course-assistant-nectir-ai-ccc-pilot-research-study-open-ai

Instructional Design, Engagement, and Support (IDEAS). (2025, January 27). *AI's impact on higher education through the eyes of faculty: 4 key trends*. University of Massachusetts Amherst. https://www.umass.edu/ideas/news/ais-impact-higher-education-through-eyes-faculty-4-key-trends

Jin, Y., Martinez-Maldonado, R., Gašević, D., & Yan, L. (2024). *GLAT: The Generative AI Literacy Assessment Test*. arXiv preprint arXiv:2411.00283. https://doi.org/10.48550/arXiv.2411.00283

Jisc. (2024, June 11). New toolkit for colleges and universities sets pathway for effective adoption of AI. *Jisc News*. https://www.jisc.ac.uk/news/all/new-toolkit-for-colleges-and-universities-sets-pathway-for-effective-adoption-of-ai

Kanerika Editorial Team. (2025, January 9). How to launch a successful AI pilot project: A comprehensive guide. *Kanerika Blogs*. Retrieved May 2, 2025, from https://kanerika.com/blogs/ai-pilot/

Kotter, J. P., Akhtar, V., & Gupta, G. (2021). *Change: How organizations achieve hard-to-imagine results in uncertain and volatile times*. Wiley.

Le Dinh, T., Le, T. D., Uwizeyemungu, S., & Pelletier, C. (2025). Human-centered artificial intelligence in higher education: A framework for systematic literature reviews. *Information, 16*(3), 240. https://doi.org/10.3390/info16030240

Li, N., Deng, W., & Chen, J. (n.d.). *From G-factor to A-factor: Establishing a psychometric framework for AI literacy* [Unpublished manuscript]. Department of Leadership and Organization Management, School of Economics and Management, Tsinghua University.

Markus, A., Carolus, A., & Wienrich, C. (n.d.). *Objective measurement of AI literacy: Development and validation of the AI competency objective scale (AICOS)* [Unpublished manuscript]. Institute Human-Computer-Media, Julius-Maximilians-University Wuerzburg.

National Center on Education and the Economy (NCEE). (2024, April 7). *Framework for AI-powered learning environments*. Retrieved May 2, 2025, from https://ncee.org/framework-for-ai-powered-learning-environments/

National Institute of Standards and Technology (NIST). (2023, January). *Artificial intelligence risk management framework (AI RMF 1.0)* (NIST AI 100-1). U.S. Department of Commerce. https://doi.org/10.6028/NIST.AI.100-1

National Telecommunications and Information Administration (NTIA). (2024, March 27). *AI accountability policy report: Overview*. Retrieved May 2, 2025, from https://www.ntia.gov/issues/artificial-intelligence/ai-accountability-policy-report/overview

Nguyen, N. (2025, February 12). From regulation to innovation: What the EU AI Act means for EdTech. *FeedbackFruits Blog*. Retrieved May 2, 2025, from https://feedbackfruits.com/blog/feedbackfruits-signs-ai-pact-voluntary-pledges-to-uphold-ethical-ai-standards

Robert, J., & McCormack, M. (2024, May 23). 2024 EDUCAUSE action plan: AI policies and guidelines. *EDUCAUSE*. Retrieved May 2, 2025, from https://www.educause.edu/research/2024/2024-educause-action-plan-ai-policies-and-guidelines

Salmon, I., & Psarras, A. (2024). Facilitating change management of higher education institutions by integrating artificial intelligence tools in their balanced scorecards. *Quest Journal of Research in Business and Management, 12*(10), 20–26. https://questjournals.org/jrbm/papers/vol12-issue10/12102026.pdf

SmartEvals. (n.d.). *How SmartEvals uses AI and data analytics to transform course evaluations*. Retrieved May 2, 2025, from https://info.smartevals.com/2738-2/

Southern Regional Education Board (SREB). (2025, February 18). *Risk management (Policy recommendation 7, Commission on AI in Education)*. Retrieved May 2, 2025, from https://www.sreb.org/ai-rec-7-risk-management

TeacherMatic. (n.d.). *TeacherMatic higher education pilot report*. Retrieved May 2, 2025, from https://teachermatic.com/case_study/jisc-higher-education-pilot-report/

The Learning Accelerator. (n.d.). Consistently engage stakeholders. In *EdTech Systems Guide: Equity-Driven selection, implementation, and evaluation*. Retrieved May 2, 2025, from https://practices.learningaccelerator.org/artifacts/consistently-engage-stakeholders-edtech-systems-guide

U.S. Department of Education, Office of Educational Technology (OET). (2024). *Navigating Artificial Intelligence in postsecondary education: Building capacity for the road ahead*. https://files.eric.ed.gov/fulltext/ED670768.pdf

Verhoeven, B., & Hor, T. (2025, February 12). A framework for human-centric AI-first teaching. *AACSB Insights*. Retrieved May 2, 2025, from https://www.aacsb.edu/insights/articles/2025/02/a-framework-for-human-centric-ai-first-teaching

Weil, D., & Forrester, J. (2024, October 17). A road map for leveraging AI at a smaller institution. *EDUCAUSE Review*. Retrieved May 2, 2025, from https://er.educause.edu/articles/2024/10/a-road-map-for-leveraging-ai-at-a-smaller-institution

Whitmarsh, L. (2025, February 21). AI-powered learning and human-centric development: A winning strategy. *Training Industry*. Retrieved May 2, 2025, from https://trainingindustry.com/articles/artificial-intelligence/ai-powered-learning-and-human-centric-development-a-winning-strategy/

World Economic Forum. (2024, January 23). *AI governance alliance: Briefing paper series*. Retrieved May 2, 2025, from https://www.weforum.org/publications/ai-governance-alliance-briefing-paper-series/

Wynants, S., Childers, G., De La Torre Roman, Y., Budar-Turner, D., & Vasquez, P. (2025). *ETHICAL principles AI framework for higher education*. California State University, Fullerton. Retrieved May 2, 2025, from https://genai.calstate.edu/communities/faculty/ethical-and-responsible-use-ai/ethical-principles-ai-framework-higher-education

Zimmermann, L. (n.d.). *Responsible AI for higher education and research*. Huron Consulting Group. Retrieved May 2, 2025, from https://www.huronconsultinggroup.com/insights/responsible-ai-higher-education

14

Phase 4: Scaling and Institutionalizing for Sustainability

Introduction: From Successful Pilots to Strategic Integration

This chapter introduces Phase 4 of AI integration for Higher Education Institutions (HEIs): Scaling and Institutionalization. Following the successful completion and rigorous evaluation of pilot projects (Phase 3), this phase marks a deliberate transition from contained experiments or limited AI usage to the broad, systematic, and sustainable integration of AI technologies into the core functions of the institution. Institutionalization, in this context, means embedding AI not just as a tool, but as a fundamental component of standard operating procedures, institutional culture, official policies, and core infrastructure, impacting teaching, learning, research, administration, and operations (Moquin, 2024; Tobenkin, 2024).

Building upon validated outcomes, established readiness, and compliance checks from earlier phases, this chapter guides HEIs in strategically embedding AI across diverse functions. This stage requires recognizing AI not merely as experimental, but as critical infrastructure and a core component of institutional strategy (Palmer, 2024). Demonstrating proven value from pilots and securing stakeholder buy-in are prerequisites for effectively institutionalizing the changes. Integrating Design Thinking (Brown, 2019) ensures scaled solutions achieve long-term "Viability." Concurrently, Change Management strategies (Kotter et al., 2021) guide this move beyond pilot successes to "Sustain Acceleration" and "Institute Change" by anchoring AI within the institution's culture and practices. It is also important to note that strategies for scaling and institutionalization may differ depending on whether the HEI is scaling its own proprietary AI systems or integrating third-party solutions, particularly concerning governance, data management, cost structures, and technical integration pathways.

Supplementary Information The online version contains supplementary material available at https://doi.org/10.1007/978-3-032-05205-6_14.

D. Woodring, *Advancing Higher Education with Human-Centric AI Partnerships*,
https://doi.org/10.1007/978-3-032-05205-6_14

Workbook Prep 1. Understanding the Scaling Imperative

Reflection: Why is moving beyond successful pilots to strategic scaling crucial for your institution's long-term mission and competitiveness? What are the biggest risks of not strategically scaling proven AI initiatives?

UX Workbook Link: See Chap. 14 UX Workbook, "Getting Started: Preparatory Exercises," focusing on articulating the strategic justification for Phase 4.

PALS Workbook Link & AI Partnership Guidance: Use Chap. 14 PALS Workbook prompts (Section 1: Aligning for Scaled Implementation) to explore the strategic rationale further.

Example Prompt: "Based on our institutional mission '[Insert Key Element]' and pilot evaluation findings showing [Brief Positive Outcome], help draft 2–3 points justifying the strategic imperative to scale this AI initiative institutionally, referencing HCASR Component 5 (Adaptive Strategy)."

Human Oversight Reminder: AI can help articulate strategic benefits. Grounding the rationale in specific institutional context, validated pilot data, and leadership vision requires human strategic analysis (HCASR Component 5).

Core Frameworks Reference: The frameworks mentioned below, such as HCASR, HCAS, RSA, EHCS, EAO, and AHIPs, are detailed in the companion Human-GenAI Collaboration & Synergy (HAICS) Guide and/or Part 1 of this textbook. Please refer to those sections for full definitions and explanations.

This chapter utilizes change management principles (Kotter et al., 2021)—particularly "Sustain Acceleration" and "Institute Change"—and design thinking's focus on achieving long-term "Viability" and "Feasibility" at scale (Brown, 2019). It applies HCASR principles related to Ethical Scaling & Oversight (Component 4) and Continuous Improvement, Lifelong Literacy, Strategic Policy Shifts, and Adaptive Strategy (Component 5). Frameworks like EHCS and EAO inform the focus on sustainable capacity building and competency development.

The User Experience (UX) Workbook and the Personalized Adaptive Learning Sessions (PALS) Workbook for Chap. 14 are available via download. These workbooks assist users in ethically and effectively using AI as a strategic partner for researching and implementing tools, solutions, and Users Outcomes and Deliverables (UODs) are customized and tailored to their unique needs and contexts.

The Rationale: Why Scaling and Institutionalizing AI is Essential

Moving AI from isolated pilots to institution-wide integration is driven by several rationales essential for long-term institutional function and relevance. First, it is a **Strategic Imperative** for AI investments to align with the institution's mission, values, and long-term goals (Palmer, 2024), contributing consistently to priorities like student success and maintaining competitiveness (Tobenkin, 2024). Institutions neglecting AI face challenges to their future relevance (Palmer, 2024). Second, **Breaking Down Silos** is required for transformation; AI strategy often occurs in isolated "pockets" (News Staff, 2025), hindering a holistic approach (U.S. Department of Education, Office of Educational Technology [OET], 2024). Scaling demands integration across departments and systems, promoting collaboration within a university-wide framework (U.S. Department of Education, OET, 2024). This aligns with Design Thinking's use of diverse teams (Brown, 2019) and Change Management's "Removing Barriers" (Kotter et al., 2021). Dedicated leadership can facilitate this integration (Rudra, 2024).

Third, scaling presents an **Ethical Necessity & Risk Mitigation** challenge. Widespread AI use necessitates robust, formalized governance to manage ethical considerations like bias, fairness, privacy, and data security (Hannah, 2024; Tobenkin, 2024), ensure equitable access (Ellucian, 2024), and mitigate risks institutionally. This reflects Change Management's "Instituting Change" (Kotter et al., 2021) and Design Thinking's focus on ethical "Desirability" and "Viability" (Brown, 2019). Finally, **Building Capacity** is fundamental. Effective and sustainable AI use depends on systematically developing AI literacy and competencies across all stakeholder groups (Palmer, 2024; U.S. Department of Education, OET, 2024), enabling Kotter et al.'s (2021) "Volunteer Army" and ensuring the change is sustainable (Table 14.1).

Establishing Permanent Ethical AI Governance and Leadership Structures

Institutionalizing AI requires transitioning from ad-hoc pilot oversight to permanent, formalized governance and leadership structures. This involves several key actions. **Formalizing Governance Bodies** means establishing or empowering permanent, cross-functional AI Councils or committees based on earlier frameworks (Chap. 9) (Ellucian, 2024; U.S. Department of

Table 14.1 Rationale for scaling and institutionalizing AI initiatives

Rationale	Core justification	Potential framework links
Strategic imperative	Align AI investments with mission/values; enhance student success; maintain institutional relevance and competitiveness.	HCASR C5 (Adaptive Strategy); Kotter (Vision)
Breaking down silos	Integrate AI efforts across departments/systems; promote holistic approach and collaboration.	DT (Diverse Teams); Kotter (Remove Barriers)
Ethical necessity and risk mitigation	"Establish robust, formalized governance; manage ethical risks (bias, privacy, equity) institutionally."	HCASR C4 (Ethical Scaling); DT (Desirability)
Building capacity	Systematically develop institution-wide AI literacy and competencies; empower stakeholders for sustainable adoption.	HCASR C5 (Lifelong Literacy); Kotter (Volunteer Army)

Note. This table summarizes the key drivers for moving AI initiatives beyond pilot stages to broader institutional integration, linking them to relevant conceptual frameworks. DT = Design Thinking; HCASR = Human-Centric AI Synergy & Regulation

Workbook Prep 2. Prioritizing Scaling Rationale

Reflection: Of the four rationales presented (Strategic Imperative, Breaking Silos, Ethical Necessity/Risk Mitigation, Building Capacity), which one presents the most compelling argument for scaling AI in your specific institutional context right now? Which presents the biggest challenge?

UX Workbook Link: See Chap. 14 UX Workbook, Activity 1.1 ("Align AI Scalability with Long-Term Leadership & Governance"), focusing on connecting scaling plans to the strategic vision.

PALS Workbook Link & AI Partnership Guidance: Use Chap. 14 PALS Workbook prompts (Section 1: Aligning for Scaled Implementation) to analyze these rationales.

Example Prompt: "Analyze the 'Breaking Down Silos' rationale for scaling AI at our institution, which has [Briefly describe departmental structure/culture]. Suggest one key challenge and one potential strategy for promoting cross-departmental collaboration during scaling, referencing Kotter's principles."

Human Oversight Reminder: AI can analyze frameworks. Understanding specific institutional politics, culture, and resource constraints related to each rationale requires human insight and stakeholder consultation (HCASR Component 4).

Education, OET, 2024). These bodies provide ongoing strategic direction, ethical review, and oversight (Tobenkin, 2024), thus "Instituting Change" (Kotter et al., 2021).

Concurrently, **Developing Comprehensive AI Policies** is necessary. These institution-wide policies, requiring regular updates aligned with the institutional mission, must address key areas including, but not limited to: data privacy and security, bias detection and mitigation, ethical use principles, academic integrity standards regarding AI assistance, responsible AI procurement guidelines, and digital accessibility requirements for AI tools (Klein, 2025; Palmer, 2024; U.S. Department of Education, OET, 2024). The policy development process should be iterative and inclusive (U.S. Department of Education, OET, 2024), helping to "Remove Barriers" (Kotter et al., 2021).

Defining Leadership Roles and Responsibilities is also important. Appointing a Chief AI Officer (CAIO) or similar senior role can provide crucial accountability and coordination, potentially overseeing functions like strategic AI alignment, cross-institutional initiative management, policy stewardship, ethical oversight coordination, and resource advocacy (Rudra, 2024). This strengthens the "Guiding Coalition" (Kotter et al., 2021). Lastly, **Ensuring Compliance and Continuous Improvement** necessitates processes for regular policy reviews, audits, ethical impact assessments, and feedback mechanisms for updates (Hannah, 2024; Klein, 2025). This mirrors Design Thinking's iterative approach applied to governance (Brown, 2019) (Table 14.2 and Fig. 14.1).

Building Sustainable AI Capacity: Competency Frameworks and Lifelong Learning

Embedding AI sustainably involves systematically building widespread AI capacity. This is essential for effective use and preparing students for the future workforce (Nikolopoulou, 2025; Howard et al., 2025). This process includes **Competency Mapping for All Roles**, defining detailed AI literacy, data literacy, and role-specific competencies (e.g., AI pedagogy for faculty, strategic oversight for administrators) needed for effective human-AI collaboration (Digital Education Council, 2025b; Howard et al., 2025; U.S. Department of Education, OET, 2024). This informs effective training design ("Empathize" - Brown, 2019).

Building on this, **Institution-Wide Lifelong Learning Strategies** are required. Institutions must offer diverse, ongoing professional development opportunities (workshops, online modules, peer learning) accessible to all

Table 14.2 Pillars of permanent ethical AI governance structures

Pillar	Key activities/focus
1. Formalizing governance bodies	"Establish or empower permanent, cross-functional AI Councils, Advisory Boards, or Ethics Committees to provide ongoing strategic direction, ethical review, oversight, and evaluation."
2. Developing comprehensive AI policies	"Create, maintain, and regularly update clear, institution-wide policies covering data privacy, bias mitigation, ethical use, academic integrity, procurement, accessibility, etc., using an iterative and inclusive process."
3. Defining leadership roles	"Define clear accountability and coordination roles, potentially including a Chief AI Officer (CAIO) or similar position, to champion strategy, secure resources, and facilitate cross-institutional efforts."
4. Ensuring compliance and continuous improvement	"Implement robust processes for regular policy reviews, audits (ethical, literacy, bias), ethical impact assessments, feedback mechanisms, transparency reporting, and human oversight to ensure ongoing responsibility and effectiveness."

Note. Establishing these permanent structures and processes is crucial for institutionalizing responsible AI use, moving beyond pilot phases to sustained, ethical integration aligned with institutional mission and values

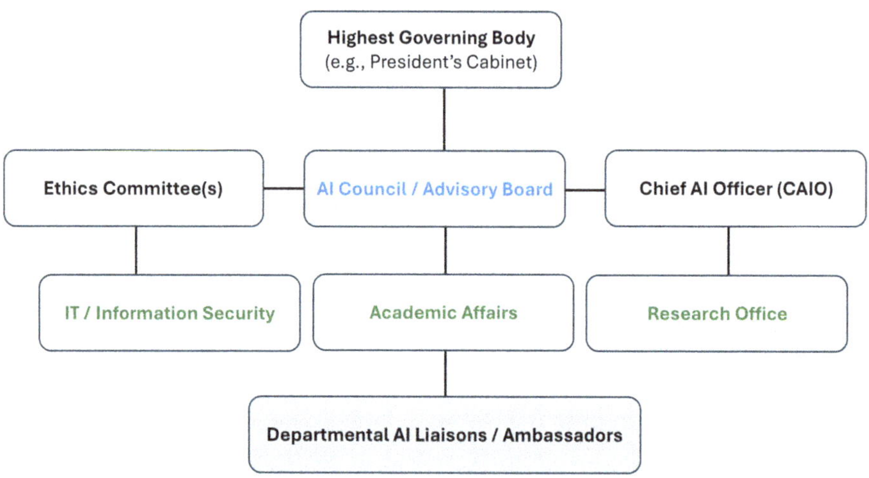

Fig. 14.1 Permanent AI governance structure chart. (*Note*: An organizational chart showing the relationship between the highest governing body (e.g., board/president), a central AI Council/Advisory Board, dedicated Ethics Committees, key leadership roles (e.g., Chief AI Officer), departmental AI liaisons/ambassadors, and operational units (IT, Academic Affairs, Research Office))

Workbook Prep 3. Designing Governance Structures for Scale

Reflection: How does your institution's current (or planned permanent) AI governance structure need to adapt to effectively oversee scaled AI initiatives across multiple departments? Are the roles (e.g., AI Council, CAIO) sufficiently empowered for institution-wide oversight?

UX Workbook Link: See Chap. 14 UX Workbook, Activity 2.1 ("AI Sustainability Planning") which includes identifying future leadership/governance needs, and Activity 1.1.

PALS Workbook Link & AI Partnership Guidance: Use Chap. 14 PALS Workbook prompts (Section 1, related to governance) to explore governance models.

Example Prompt: "Based on best practices for institutional AI governance (e.g., OET report, Ellucian), suggest 2–3 specific responsibilities for a permanent AI Council focused on overseeing scaled implementations, ensuring ethical compliance and continuous improvement (HCASR Component 5)."

Human Oversight Reminder: AI can suggest roles and responsibilities. Designing governance structures that are effective, efficient, representative, and fit the institutional culture requires careful human deliberation, leadership buy-in, and clear chartering (HCASR Component 4 & 5).

stakeholders, recognizing that faculty unfamiliarity can be a barrier (Digital Education Council, 2025a; Howard et al., 2025; U.S. Department of Education, OET, 2024). Success requires leadership buy-in and incentives (Digital Education Council, 2025a). This ensures the workforce continually adapts ("Sustain Acceleration" - Kotter et al., 2021).

Finally, **Fostering Communities of Practice (CoPs)** focused on AI accelerates learning, innovation, peer support, and sharing best practices (Digital Education Council, 2025a, 2025b; Ellucian, 2024). For instance, institutional support for CoPs might include providing dedicated online collaboration spaces, offering small grants for CoP-led projects or speakers, and formally recognizing CoP leadership contributions. These CoPs strengthen collaborative ideation (Brown, 2019) and informal networks essential for sustaining change (Kotter et al., 2021). Establishing these robust competency frameworks, lifelong learning strategies, and supportive communities of practice provides the essential human foundation required to successfully undertake the systematic process of scaling AI integration across the institution (Table 14.3 and Fig. 14.2).

Table 14.3 Strategies for building sustainable AI capacity

Strategy	Key activities/focus
1. Competency mapping for all roles	"Define detailed AI literacy, data literacy, adaptive teaching skills, and specific competencies required for different roles (students, faculty, staff, admin) to enable effective and critical human-AI collaboration."
2. Institution-wide lifelong learning	"Offer a diverse portfolio of ongoing, accessible AI-related professional development (workshops, modules, micro-credentials, peer learning); requires leadership buy-in and incentives (e.g., grants, promotion recognition)."
3. Fostering communities of practice (CoPs)	"Create and support groups focused on AI (by role, discipline, tool) to accelerate learning, innovation, peer support, and the sharing of best practices and challenges."

Note. Systematically developing AI competencies and fostering a culture of continuous learning are essential for embedding AI effectively, ethically, and sustainably across the institution

Fig. 14.2 AI Competency framework matrix (advanced). (*Note*: A matrix listing key stakeholder roles (Undergrad, Grad, Faculty, Staff, Admin) as rows and advanced/role-specific AI competencies (e.g., Ethical AI Auditing, AI Pedagogy Design, AI Workflow Automation, Strategic AI Planning, Advanced Prompt Engineering) as columns. Cells could indicate required proficiency levels (e.g., Foundational, Intermediate, Advanced) or link to specific training resources/modules)

Workbook Prep 4. Designing Capacity Building Strategies

Reflection: What are the most critical advanced AI competencies needed for faculty and staff in your unit/department to effectively support scaled AI use? What is the biggest challenge to implementing a sustainable lifelong learning strategy for AI at your institution?

UX Workbook Link: See Chap. 14 UX Workbook, Activity 2.2 ("Develop AI Competency Frameworks") and Activity 2.3 ("Design Lifelong Learning AI Strategies").

PALS Workbook Link & AI Partnership Guidance: Use Chap. 14 PALS Workbook prompts (Section 2: Building Capacity and Sustainability).

Example Prompt: "For the role of '[Specific Staff Role, e.g., Academic Advisor]', brainstorm 2–3 advanced AI competencies needed for scaled use of our new AI advising tool. Suggest one lifelong learning activity (e.g., advanced workshop, CoP participation) to develop one of these competencies (HCASR Component 5)."

Human Oversight Reminder: AI can help brainstorm competencies and learning activities. Defining meaningful, role-specific competencies and designing engaging, effective, and accessible lifelong learning programs require human expertise in job analysis, instructional design, and resource management (HCASR Component 5).

The Process of Scaling AI Integration Across the Institution

Scaling successful AI pilots uses a systematic, phased approach guided by established governance and capacity-building (Complete College America, 2025; U.S. Department of Education, OET, 2024). First, **Solidify Governance and Change Frameworks** to ensure AI adoption aligns with clear policies, ethical guidelines, institutional strategy, and change management principles (Complete College America, 2025; Kotter et al., 2021). Second, **Build Sustainability and Competency Models** by implementing lifelong learning strategies and supporting CoPs to develop the human capacity required for widespread adoption (Complete College America, 2025; Kotter et al., 2021; U.S. Department of Education, OET, 2024).

Third, **Expand Integration Systematically** by embedding proven AI tools across targeted areas. This requires planning for technical integration, robust data governance, scalable infrastructure, resource management, vendor vetting, and user support (Brown, 2019; Complete College America, 2025; U.S. Department of Education, OET, 2024). Planning for this systematic expansion must carefully consider whether the AI source is proprietary or third-party, as this significantly impacts technical integration complexity, data governance protocols, vendor management needs, and security vetting processes. This structured process operationalizes Kotter et al.'s (2021) "Sustain Acceleration" and "Institute Change" by ensuring necessary structures and skills are in place before embedding AI into standard operations.

Using AI as a Strategic Partner: Every HEI is situated in a unique socio-historical context, which means there is no one-size-fits-all approach to AI integration in higher education (Carmody-Bubb, 2023). As institutions around the world integrate GenAI, contextual data from institutions can be ethically and effectively used to 1) guide custom-tailored tools, resources, and solutions, and 2) mitigate Western (and other) biases that will be inherent within models as they are adopted globally. See the Human-AI Collaboration & Synergy Guide for tools and resources.

Workbook Prep 5. Planning the Scaling Sequence

Reflection: Considering the 3-step scaling process (Solidify Governance -> Build Capacity -> Expand Systematically), where does your institution currently stand? What is the most immediate priority for moving to the next step for a specific AI initiative ready for scaling?

UX Workbook Link: See Chap. 14 UX Workbook, Activity 1.2 ("Prepare for Multi-Department or Full-Institution Rollout"), focusing on prioritizing and sequencing.

PALS Workbook Link & AI Partnership Guidance: Use Chap. 14 PALS Workbook prompts (Section 1, related to rollout planning).

Example Prompt: "We plan to scale [Specific AI Tool] based on successful pilot results. Using the 3-step scaling process, help outline key considerations for the 'Build Sustainability and Competency Models' step specifically for this tool before we move to systematic expansion. Reference HCASR C5."

Human Oversight Reminder: AI can help structure planning based on models. Determining the actual readiness to proceed through scaling stages, managing dependencies, and adapting the sequence based on real-time institutional context requires human project management and strategic oversight (HCASR Component 5).

Expanding AI Applications Across University Functions

With institutional capacity built and governance established, proven AI applications can be strategically expanded across core university functions, tailored to specific needs (Kadiu, 2025; Nikolopoulou, 2025).

Curriculum and Co-curricular Activities: Scaling can involve deploying institution-wide AI literacy modules within first-year experiences, providing faculty with AI-powered tools integrated into the Learning Management System (LMS) for generating personalized practice sets or offering preliminary writing feedback, making AI tutors available for foundational courses, or using AI to enhance accessibility features in digital learning materials (e.g., improved automated captioning for lectures) (Klein, 2025; Kadiu, 2025; Nikolopoulou, 2025).

Research and Collaboration: Expansion may include providing researchers campus-wide access to licensed AI platforms for complex data analysis or modeling, integrating AI literature review tools with library databases, offering AI-assisted grant proposal writing workshops, or establishing secure platforms that facilitate interdisciplinary research collaborations using shared AI tools and datasets (Palmer, 2024; Tobenkin, 2024; Nikolopoulou, 2025; U.S. Department of Education, OET, 2024).

Administration and Operations: Scaling commonly focuses on implementing AI chatbots for student services (admissions, financial aid, IT help desk) trained on institutional data, automating routine administrative tasks like transcript processing or initial vendor screening, using AI analytics dashboards for enrollment forecasting or resource allocation planning, or deploying AI-driven cybersecurity tools to monitor network traffic (Kadiu, 2025; Nikolopoulou, 2025; Risk Management Partners, 2024).

Applying Design Thinking (Brown, 2019) ensures these scaled solutions are tailored effectively for each functional area, while successful expansion provides tangible evidence of "Instituting Change" (Kotter et al., 2021).

Workbook Prep 6. Identifying Expansion Opportunities

Reflection: Based on successful pilots or identified needs, which university function (Curriculum, Research, Admin/Ops) offers the most strategic opportunity for scaled AI integration at your institution in the near term? What is one specific application within that function?

UX Workbook Link: See Chap. 14 UX Workbook, Activity 1.1 (Gathering departmental scaling input) and Activity 1.2 (Prioritizing initiatives).

PALS Workbook Link & AI Partnership Guidance: Use Chap. 14 PALS Workbook prompts (relevant sections) to explore potential applications.

Example Prompt: "Brainstorm 3 specific ways GenAI could be scaled within 'Research and Collaboration' functions at a research-intensive university, focusing on enhancing efficiency while upholding research integrity. Suggest one potential benefit and one ethical risk for each, referencing HCASR C4."

Human Oversight Reminder: AI can generate application ideas. Assessing the strategic fit, feasibility, resource implications, and ethical readiness for scaling specific applications requires domain expertise, stakeholder consultation, and alignment with institutional priorities (HCASR Component 4).

Strategic Implementation: Alignment and Actionable Steps

Formalizing the scaling phase requires alignment with the overall AI plan and institutional mission, positioning AI as integral to governance and strategy (Complete College America, 2025). This involves defining concrete implementation actions (News Staff, 2025). Key actions include: Developing and maintaining an AI Long-Term Roadmap (Palmer, 2024); Producing periodic AI Scalability and Compliance Reports (Ellucian, 2024; Risk Management Partners, 2024); Establishing an AI Ethics and Sustainability Strategy (Complete College America, 2025; Hannah, 2024); Formalizing AI Governance Structures; and Planning for Integration into Core Institutional Infrastructure (LMS, SIS, etc.), requiring robust data governance and security (Complete College America, 2025; U.S. Department of Education, OET, 2024). These represent steps within Kotter et al.'s (2021) "Sustain Acceleration" and "Institute Change," reflecting Brown's (2019) "Implementation" focused on long-term "Viability" and "Feasibility."

Workbook Prep 7. Planning Implementation Actions

Reflection: Consider one key UOD mentioned (e.g., AI Long-Term Roadmap, Ethics & Sustainability Strategy). What is the most critical first actionable step your institution needs to take to develop or formalize this deliverable for the scaling phase? Who needs to be involved?

UX Workbook Link: See Chap. 14 UX Workbook, Section 7 ("Deliverables: Finalizing Scaling & Institutionalization Documents"), which consolidates the planning into final documents. Also links to specific activities like Section 1 (Roadmap outline), Section 2 (Sustainability outline), Section 4 (Checkpoints informing reports).

PALS Workbook Link & AI Partnership Guidance: Use Chap. 14 PALS Workbook prompts (e.g., Section 7) to outline UODs.

Example Prompt: "Help outline the key sections for an 'AI Ethics & Sustainability Strategy' document intended to guide scaled AI deployment. For each section, suggest one key question it should address, referencing HCASR Component 5 (Strategic Policy, Adaptation)."

Human Oversight Reminder: AI can help structure document outlines. Defining the specific content, metrics, implementation details, and accountabilities within these strategic documents requires collaborative human effort, leadership input, and governance approval (HCASR Component 5). With permanent governance structures and policies in place, the focus shifts to ensuring the institution has the human capacity required to utilize and sustain these scaled AI initiatives effectively.

Key Deliverables for Scaling & Institutionalization

This phase culminates in key User Outcomes and Deliverables (UODs) solidifying AI's integration. These are often living documents, refined over time:

1. **AI Long-Term Roadmap (Updated)**: A revised strategic plan with updated timelines, resource allocations, and priorities based on scaling experience.
2. **AI Scalability and Compliance Report(s)**: Periodic reports summarizing scaling progress, outcomes versus KPIs, and compliance with policies and ethical guidelines.
3. **AI Ethics and Sustainability Strategy**: A formal document outlining long-term ethical commitments, risk mitigation approaches, and plans for sustainable AI use and development.
4. **Formalized AI Governance Structures and Processes Documentation**: Updated charters, policy manuals, and procedural documents for permanent AI oversight bodies and roles.
5. **AI Infrastructure Integration Plan Updates**: Documentation of completed integrations and evolving plans for future infrastructure needs to support scaled AI.
6. **Institutional AI Literacy & Competency Framework**: A comprehensive, living framework defining AI-related skills and knowledge required across different roles, guiding training and development.
7. **Community of Practice (CoP) Frameworks/Charters**: Documents outlining the structure, purpose, and operational guidelines for AI-focused CoPs.
8. **AI-Integrated Curriculum & Policy Playbook**: A collection of resources, guidelines, and examples for integrating AI into curricula and applying relevant institutional policies.

These UODs provide the structure for ongoing, responsible AI use and align with terms used in the UX workbook.

Conclusion: Anchoring AI in the Institution's Future

Successfully scaling and institutionalizing AI transitions it from experimental technology to integral infrastructure (Palmer, 2024). This phase solidifies AI's role in achieving the institution's vision but demands significant responsibility (Hannah, 2024; U.S. Department of Education, OET, 2024). By formalizing governance, building sustainable capacity through lifelong learning (Nikolopoulou, 2025; U.S. Department of Education, OET, 2024), embedding AI applications thoughtfully across functions (Klein, 2025), and continuously monitoring impact and ethics (Ellucian, 2024; Hannah, 2024), HEIs ensure enduring adoption. This achieves Design Thinking's trifecta ("Desirability, Feasibility, and Viability") institutionally (Brown, 2019). It represents the culmination of Change Management steps, where change is fully "Instituted" within the culture (Kotter et al., 2021). This responsible institutionalization allows HEIs to utilize AI's potential while remaining vigilant and adaptive (Palmer, 2024; Tobenkin, 2024).

Looking Ahead & Engaging the Workbook

Chapter 14 outlined the strategic imperatives and processes for scaling successful AI initiatives and institutionalizing them. It established requirements for permanent governance, sustainable capacity, systematic expansion, and alignment with change management and design principles. The Chap. 14 Companion UX and PALS Workbooks provide structured activities and templates to execute this phase effectively, enabling stakeholders to operationalize these concepts, produce key UODs, and prepare for ongoing Adaptation & Refinement (Chap. 15).

The PALS Workbook for Chap. 14, particularly its section "Long-Term Scaling, Sustainability & AI Governance: Iteratively Developing Your Custom TLARA-Informed AI Guidelines," guides the development of a "Living Document." This ensures TLARA (Teaching, Learning, Assessment, Research, Administration) guidelines actively support institution-wide capacity building, permanent AI governance, and a pervasive AI literacy culture (IDEAS, 2025), all critical for embedding AI deeply and sustainably.

Long-Term Scaling, Sustainability & AI Governance: Embedding TLARA for Institutional Impact

Chapter 14 focuses on the critical transition from successful pilots to strategic scaling and institutionalization of AI initiatives. This phase requires careful planning to ensure that AI's benefits are realized equitably and sustainably across all levels and spaces of the institution, aligning with the long-term vision and governance structures (DerSimonian & Montagnino, 2025). Successfully embedding AI means moving beyond isolated projects to integrate AI capabilities into the core **TLARA (Teaching, Learning, Assessment, Research, Administration)** functions, supported by robust resources, well-defined competencies, and a culture of lifelong learning that adapts over time. Both the Chap. 14 UX and PALS Workbooks are designed to help you ethically partner with AI in strategizing for this significant institutional evolution.

For HEIs to achieve lasting transformation, AI integration must be woven into their operational and sociocultural context, not merely bolted on (Ellucian, 2024). The PALS Workbook for Chap. 14, in its section "Long-Term Scaling, Sustainability & AI Governance: Iteratively Developing Your Custom TLARA-Informed AI Guidelines," prompts you to use your GenAI partner to develop strategies for this deep embedding. By exploring "Topics for PALS TLARA Focus"—like "Strategic Scaling for Sustainable, Institution-Wide TLARA Enhancement," and "Cultivating a Deep and Sustainable AI Literacy Culture Across TLARA"—you will refine your "Living Document." This ensures your TLARA guidelines actively support institution-wide capacity building, the formalization of permanent AI governance, and the cultivation of a pervasive AI literacy culture (IDEAS, 2025).

References

Brown, T. (2019). *Change by design: How design thinking transforms organizations and inspires innovation* (Rev. & updated ed.). Harper Business.

Carmody-Bubb, M. (2023). *Cognition and decision making in complex adaptive systems: The human factor in organizational performance*. Springer.

Complete College America. (2025, January 14). *Generating College Completion: Charting a Path to Institutional AI Adoption for Student Success in Higher Education*. https://completecollege.org/resource/generating-college-completion-charting-a-path-to-institutional-ai-adoption-for-student-success-in-higher-education/

DerSimonian, R., & Montagnino, C. (2025, March 26). Crafting thoughtful AI policy in higher education: A guide for institutional leaders. *Campus Technology*. https://campustechnology.com/articles/2025/03/26/crafting-thoughtful-ai-policy-in-higher-education-a-guide-for-institutional-leaders.aspx

Digital Education Council. (2025a). *Breaking faculty barriers to AI literacy: Adoption, incentives, and scaling strategies*. Retrieved April 25, 2025, from https://www.digitaleducationcouncil.com/post/breaking-faculty-barriers-to-ai-literacy-adoption-incentives-and-scaling-strategies

Digital Education Council. (2025b). *Empowering faculty to experiment, share and learn: Advancing AI literacy in higher education*. Retrieved April 25, 2025, from https://www.digitaleducationcouncil.com/post/empowering-faculty-to-experiment-share-and-learn-advancing-ai-literacy-in-higher-education

Ellucian. (2024, December 16). AI in higher education: Implementation strategies for institutions and tech partners. *Higher Ed Dive*. https://www.highereddive.com/spons/ai-in-higher-education-implementation-strategies-for-institutions-and-tech/735434/

Hannah, A. (2024, May 9). Ethical considerations for AI in higher education: Ensuring fairness and transparency. *Liaison Resources Blog*. https://www.liaison-edu.com/resources/blog/ethical-considerations-for-ai-in-higher-education-ensuring-fairness-and-transparency/

Howard, C., Rainie, L., & Dircksen, A. (2025, January 22). Higher education leaders navigate AI disruption. *AAC&U Newsroom*. https://www.aacu.org/newsroom/higher-education-leaders-navigate-ai-disruption

Instructional Design, Engagement, and Support (IDEAS). (2025, January 27). *AI's impact on higher education through the eyes of faculty: 4 key trends*. University of Massachusetts Amherst. https://www.umass.edu/ideas/news/ais-impact-higher-education-through-eyes-faculty-4-key-trends

Kadiu, A. (2025, January 9). AI assistants and agents in US higher education: Beyond the classroom. *Element451*. https://element451.com/blog/ai-agents-in-us-higher-education

Klein, E. (2025, January 15). How artificial intelligence is transforming higher education. *ACE News & Insights*. American Council on Education. https://ace.edu/news/how-artificial-intelligence-is-transforming-higher-education/

Kotter, J. P., Akhtar, V., & Gupta, G. (2021). *Change: How organizations achieve hard-to-imagine results in uncertain and volatile times*. Wiley.

Moquin, S. (2024, November 1). The role of AI in transforming higher education. *Enrollify*. https://www.enrollify.org/blog/the-role-of-ai-in-transforming-higher-education

News Staff. (2025, February 25). Survey: Higher Ed AI adoption faces financial, policy hurdles. *Government Technology*. https://www.govtech.com/education/higher-ed/survey-higher-ed-ai-adoption-faces-financial-policy-hurdles

Nikolopoulou, K. (2025). Generative artificial intelligence and sustainable higher education: Mapping the potential. *Journal of Digital. Educational Technology, 5*(1), ep2506. https://doi.org/10.30935/jdet/15860

Palmer, K. (2024, December 19). How will AI influence higher ed in 2025? *Inside Higher Ed*. https://www.insidehighered.com/news/tech-innovation/artificial-intelligence/2024/12/19/how-will-ai-influence-higher-ed-2025

Risk Management Partners. (2024, June). *Risk control: Artificial intelligence in higher education*. https://rmpartners.co.uk/wp-content/uploads/2024/06/RMP-Risk-Control-Artificial-Intelligence-in-HE.pdfs

Rudra, S. (2024, February 6). What is a chief AI officer, and should your university appoint one? *EdTech Magazine*. https://edtechmagazine.com/higher/article/2024/02/what-chief-ai-officer-and-should-your-university-appoint-one-perfcon

Tobenkin, D. (2024, January/February). Artificial intelligence and the future of higher education, Part 1. *Trusteeship Magazine, 32*(1). https://agb.org/trusteeship-article/artificial-intelligence-and-the-future-of-higher-education-part-1/

U.S. Department of Education, Office of Educational Technology (OET). (2024, January). *Navigating artificial intelligence in postsecondary education: Building capacity for the road ahead.* https://portal.ct.gov/das/-/media/das/ctedtech/publications/2025/2025-used-oet-archive/os-24-002833-ai-in-postsecondary-ed.pdf

15

Ongoing Adaptation and Refinement (Phase 4)

Introduction: Navigating Continuous AI Adaptation

This chapter introduces the ongoing Adaptation and Refinement phase necessary after the initial scaling of Generative Artificial Intelligence (GenAI) in Higher Education Institutions (HEIs). As AI tools become embedded in daily operations, this phase requires continuous adjustments at cultural and procedural levels, often moving at the speed of organizational change rather than technology alone (Deloitte, 2025). It requires embedding AI thoughtfully, ethically, and sustainably across the institution (Cornell University Task Force on GenAI in Academic Research, 2023; Maneval, 2025; UNESCO, 2023). This signifies a shift towards a mature state where AI integration is constantly evaluated, improved, and strategically aligned to create demonstrable value (Relyea et al., 2024). The distinction between this phase and the preceding Scaling phase (Chap. 14) is crucial; while scaling focuses on broadening the reach of proven AI, adaptation centers on sustaining effectiveness, optimizing existing AI use, and evolving practices in response to ongoing learning and environmental shifts (see Table 15.1).

Key activities in the Adaptation and Refinement phase include iteratively updating policies and governance based on real-world use and feedback (Cornell University Task Force on GenAI in Academic Research, 2023; Maneval, 2025; UNESCO, 2023), providing ongoing competency development (Cornell University Task Force on GenAI in Academic Research, 2023; WEF, 2025), and maintaining dynamic stakeholder engagement (Maneval,

---

Supplementary Information The online version contains supplementary material available at https://doi.org/10.1007/978-3-032-05205-6_15.

D. Woodring, *Advancing Higher Education with Human-Centric AI Partnerships*,
https://doi.org/10.1007/978-3-032-05205-6_15

Table 15.1 Shifting focus from scaling phase (Chap. 14) to adaptation phase (Chap. 15)

Aspect	Scaling phase (Chap. 14) emphasis	Adaptation phase (Chap. 15) emphasis
Primary goal	Broaden reach of proven AI; embed into standard operations.	Sustain effectiveness; improve/optimize existing AI use; evolve practices.
Governance focus	Formalize permanent structures; establish baseline policies.	Implement adaptive governance; iterate policies based on real-world use.
Policy focus	Develop comprehensive, institution-wide baseline policies.	Refine/update policies as "living documents"; address emerging issues.
Training focus	Build widespread capacity; develop advanced role competencies.	Provide ongoing lifelong learning; address evolving skill gaps; update training.
Stakeholder role	Adopt scaled tools; contribute to CoPs; provide initial feedback.	Provide continuous feedback; participate in refinement; adapt practices.
Key challenge	Managing large-scale rollout; ensuring initial adoption/buy-in.	Sustaining momentum; managing cultural change; keeping pace with AI evolution.
Guiding framework focus	Kotter (Sustain Acceleration, Institute Change); DT (Viability)	HCASR C5 (Continuous Improvement); DT (Iteration); CM (Sustaining Change)

Note. This table contrasts the primary focus areas and activities between the Scaling and Institutionalization phase (Chap. 14) and the ongoing Adaptation and Refinement phase (Chap. 15). DT = Design Thinking; CM = Change Management; HCASR = Human-Centric AI Synergy & Regulation

2025; UNESCO, 2023). Three interconnected themes guide this phase: Ongoing Change Management, Iterative Policy Development and Governance, and Dynamic Stakeholder Coordination. This phase applies Design Thinking's (Brown, 2019) continuous iterative cycle ("Inspiration" -> "Ideation" -> "Implementation") and Change Management's (Kotter et al., 2021) principles for sustaining change ("Sustain Acceleration," "Institute Change") by ensuring human experience and organizational capacity evolve together.

Workbook Prep 1. Understanding Ongoing Adaptation Needs

Reflection: Now that AI tools are becoming embedded, what is the single biggest challenge your institution (or unit) faces in continuously adapting to them? Is it keeping policies current, ensuring skills keep pace, managing cultural shifts, or something else?

UX Workbook Link: See Chap. 15 UX Workbook, "Getting Started: Preparatory Exercises," focusing on initial reflections about adaptation challenges and needs.

PALS Workbook Link & AI Partnership Guidance: Use Chap. 15 PALS Workbook prompts (Introduction/Section 1) to explore the landscape of ongoing adaptation.

Example Prompt: "Based on the concept of continuous AI adaptation in HEIs (Chap. 15 Introduction), help brainstorm 3 key areas where our institution might need ongoing refinement after scaling our [Specific AI Initiative]. Reference HCASR Component 5 (Continuous Improvement)."

Human Oversight Reminder: AI can synthesize common adaptation challenges. Identifying the most pressing and context-specific needs for your institution requires human insight, stakeholder consultation, and analysis of local data (HCASR Component 4 & 5).

Core Frameworks Reference: The frameworks mentioned below, such as HCASR, HCAS, RSA, EHCS, EAO, and AHIPs, are detailed in the companion Human-GenAI Collaboration & Synergy (HAICS) Guide and/or Part 1 of this textbook. Please refer to those sections for full definitions and explanations.

This chapter utilizes principles from both Design Thinking and Change Management (integrating Brown, 2019 and Kotter et al., 2021). It operationalizes Human-Centric AI Synergy & Regulation (HCASR) Component 5 (Continuous Improvement & Advanced Human–AI Synergy), particularly regarding feedback loops, iterative policy development, and lifelong AI literacy. Principles from HCAS (Human-Centric AI Synergy), RSA (Regulation Synergy & Augmentation), EHCS (Enhanced Human-Centric Skills), and EAO (Enhanced Academic Outcomes) inform strategies for ongoing skill development, ethical oversight, and ensuring AI continues to enhance institutional goals.

The User Experience (UX) Workbook and the Personalized Adaptive Learning Sessions (PALS) Workbook for Chap. 15 are available via download. These workbooks assist users in ethically and effectively using AI as a strategic partner for researching and implementing tools, solutions, resources, and Users Outcomes & Deliverables (UODs) that are customized and tailored to their unique needs and contexts.

Fostering Adaptive Culture and Managing Change

Successfully embedding GenAI requires proactive change management focused on cultivating an adaptable, AI-literate culture (Relyea et al., 2024). A lack of expertise and literacy among stakeholders can impede effective adaptation and trust (Gouseti et al., 2025; Hassan et al., 2024). Therefore, the first crucial element is **Building a Culture of AI Literacy** through ongoing competency development pathways. This involves Upskilling Faculty, Staff, and Students through continuous professional development and curricular integration (PwC, 2024; Relyea et al., 2024). Training must extend beyond basic usage to include critical evaluation, ethical application (privacy, bias, fairness, transparency), prompt design, and understanding data implications (Cornell University Task Force on GenAI in Academic Research, 2023; Gouseti et al., 2025; Nathan & Welburn, 2024; UNESCO, 2023).

Simultaneously, efforts must focus on **Reducing Resistance and Encouraging Adoption**. Resistance may stem from fear, uncertainty, lack of trust, or perceived threats to roles and values (Gouseti et al., 2025; Hassan et al., 2024; Mills et al., 2025). Effective strategies include clear communication about the AI strategy, benefits, limitations, and risks (Cornell University Task Force on GenAI in Academic Research, 2023; Mills et al., 2025; Relyea et al., 2024), targeted training (Gouseti et al., 2025; Relyea et al., 2024), and demonstrating tangible value to build confidence and momentum (Relyea et al., 2024). Understanding the institution's current stage of AI adoption and cultural readiness, potentially through AI maturity models, can help tailor these change management strategies. AI maturity models typically assess an institution's progress across several dimensions such as AI strategy, data governance, technological infrastructure, workforce skills, and ethical considerations, providing a benchmark for targeted improvements. These are explored further in the UX Workbook. Managing these cultural and change aspects is fundamental before delving into the specifics of policy and governance adaptation.

Using AI as a Strategic Partner: Every HEI is situated in a unique socio-historical context, which means there is no one-size-fits-all approach to AI integration in higher education (Carmody-Bubb, 2023). As institutions around the world integrate GenAI, contextual data from institutions can be ethically and effectively used to 1) guide custom-tailored tools, resources, and solutions, and 2) mitigate Western (and other) biases that will be inherent within models as they are adopted globally. See the Human-AI Collaboration & Synergy Guide for tools and resources.

Workbook Prep 2. Reflecting on Cultural Adaptation Needs

Reflection: Consider the strategies for reducing resistance (clear communication, targeted training, demonstrating value). Which strategy do you think would be most effective for fostering a more adaptive AI culture within your specific team or department? Why? What specific action could you take?

UX Workbook Link: See Chap. 15 UX Workbook, Activity 1.2 ("Planning Organizational Adaptation") and Activity 1.4 ("Designing Ongoing Competency Development Pathways").

PALS Workbook Link & AI Partnership Guidance: Use Chap. 15 PALS Workbook prompts (Section 1: Planning Adaptation).

Example Prompt: "Analyze potential sources of resistance to ongoing AI adaptation among [Specific Stakeholder Group, e.g., mid-career faculty]. Suggest one communication strategy and one training approach to address these concerns, referencing principles from Kotter or HCASR Component 5."

Human Oversight Reminder: AI can suggest change management tactics. Understanding the specific cultural dynamics, fears, and motivations within your context requires human empathy, active listening, and tailored communication strategies (HCASR Component 4).

Iterative Policy Development and Adaptive Governance

With a focus on cultural adaptation underway, the next critical area is ensuring that policies and governance structures evolve. The dynamic nature of GenAI requires **Iterative Policy Development and Adaptive Governance**. Static policies become outdated quickly (UNESCO, 2023). Institutions require holistic and adaptive AI governance frameworks, treating policies as "living documents"—that is, documents designed for regular, evidence-based updates rather than static archiving upon initial completion (Maneval, 2025). This involves **Regular Review Cycles** (e.g., semi-annually) of AI policies, ethical guidelines, and risk protocols (Cornell University Task Force on GenAI in Academic Research, 2023; Maneval, 2025; NIST, 2023).

The process must incorporate **Structured Feedback Integration**, using formal channels for diverse stakeholders to provide input based on practical experience (Cornell University Task Force on GenAI in Academic Research, 2023; Hassan et al., 2024; Heintz et al., 2024; Mills et al., 2025; NIST, 2023). Furthermore, ongoing **Monitoring and Data Analysis** of usage, performance, risk assessments, user feedback, and external shifts inform timely

Table 15.2 Components of iterative policy development and adaptive governance for AI

Component	Description/purpose
1. Regular review cycles	Establish planned, periodic reviews (e.g., semi-annual, annual) of AI policies, ethical guidelines, acceptable use standards, risk protocols, and the governance structure itself to ensure they remain relevant and effective.
2. Structured feedback integration	Maintain formal, accessible channels for diverse stakeholders to provide input on policies and practices based on their real-world experiences, concerns, and evolving needs.
3. Ongoing monitoring and data analysis	Continuously collect and analyze data on AI usage, performance, risk assessments (bias, security, accuracy), user feedback, and external shifts (regulatory, technological) to inform necessary policy updates.

Note. Treating AI governance as an ongoing, adaptive process informed by data and feedback, rather than a static endpoint, is crucial for navigating the dynamic AI landscape responsibly

policy updates and ensure accountability (Maneval, 2025; Mills et al., 2025; NIST, 2023). This iterative process reflects Design Thinking (Brown, 2019) applied to governance (Nathan & Welburn, 2024) (Table 15.2).

Workbook Prep 3. Planning for Policy Iteration

Reflection: How effective are your current institutional mechanisms for gathering feedback on existing policies (AI-related or otherwise)? How could these mechanisms be improved or adapted to support the regular, structured feedback needed for iterative AI policy development?

UX Workbook Link: See Chap. 15 UX Workbook, Activity 2.1 ("Gathering Real-World Feedback"), Activity 2.2 ("Drafting Policy Updates"), and Activity 2.3 ("Proposing Enhancements to Oversight Structures").

PALS Workbook Link & AI Partnership Guidance: Use Chap. 15 PALS Workbook prompts (Section 2: Refining Policies).

Example Prompt: "Outline a 4-step process for incorporating structured stakeholder feedback into our planned semi-annual review cycle for the institutional AI Acceptable Use Policy, ensuring transparency and alignment with HCASR Component 5 principles."

Human Oversight Reminder: AI can help structure processes. Designing effective feedback mechanisms, facilitating inclusive policy review discussions, navigating institutional bureaucracy, and making final policy decisions require human leadership, collaboration, and governance expertise (HCASR Component 4 & 5).

Sustaining Dynamic Stakeholder Coordination

Building upon iterative policy development, sustaining effective integration requires **Continuous Stakeholder Engagement** and dynamic coordination (Appaya & Ng, 2024; Maneval, 2025; WEF, 2025). Strategies include maintaining **Ongoing Dialogue Forums** (workshops, town halls, committee meetings) where diverse stakeholders discuss issues, share practices, and provide input (Heintz et al., 2024). Ensuring **Accessible Feedback Channels** remain open and responsive for submitting suggestions or reporting concerns is vital (Heintz et al., 2024; Mills et al., 2025).

Collaborative Refinement should directly involve representative stakeholders in reviewing feedback and co-creating adjustments to policies, training, and tools (Heintz et al., 2024). This participatory approach builds trust and shared ownership (Heintz et al., 2024; WEF, 2025). Design Thinking principles (Brown, 2019) are applied through continuous feedback ("Empathize," "Test"). Change Management principles (Kotter et al., 2021) are operationalized by equipping stakeholders ("Volunteer Army"), sustaining momentum ("Sustain Acceleration"), and embedding changes ("Institute Change") through inclusive processes (Table 15.3 and Fig. 15.1).

Workbook Prep 4. Coordinating Ongoing Stakeholder Input

Reflection: Consider the stakeholder coordination strategies (Dialogue Forums, Feedback Channels, Collaborative Refinement). Which strategy is currently strongest at your institution regarding AI? Which needs the most development to ensure dynamic, ongoing input during this adaptation phase?

UX Workbook Link: Integrates with activities across the Chap. 15 UX Workbook, particularly those involving feedback collection (e.g., Activity 2.1) and refinement (e.g., Activity 3.4).

PALS Workbook Link & AI Partnership Guidance: Use Chap. 15 PALS Workbook prompts (relevant sections, e.g., Section 9 on roles).

Example Prompt: "Suggest 2 specific methods for establishing more accessible and responsive feedback channels for staff using a newly scaled administrative AI tool, focusing on capturing input for continuous refinement (HCASR Component 5)."

Human Oversight Reminder: AI can suggest coordination methods. Building trust, fostering genuine dialogue, facilitating effective collaboration across diverse groups, and ensuring feedback leads to action require strong human interpersonal, facilitation, and leadership skills (HCASR Component 4).

Table 15.3 Strategies for sustaining dynamic stakeholder coordination

Strategy	Key activities/focus
1. Ongoing dialogue forums	Maintain regular platforms (workshops, town halls, committee meetings, online forums) for diverse stakeholders to discuss emerging issues, share practices, voice concerns, and provide input.
2. Accessible feedback channels	Ensure channels for submitting suggestions, reporting concerns (errors, biases), or escalating issues related to AI remain open, known, and responsive.
3. Collaborative refinement	Directly involve representative stakeholders (faculty, staff, students, admin) in reviewing feedback/data and co-creating adjustments to policies, training, support, and AI tools.

Note. Continuous engagement and co-creation build trust, ensure solutions remain relevant, and foster shared ownership, which are vital for sustaining successful AI integration

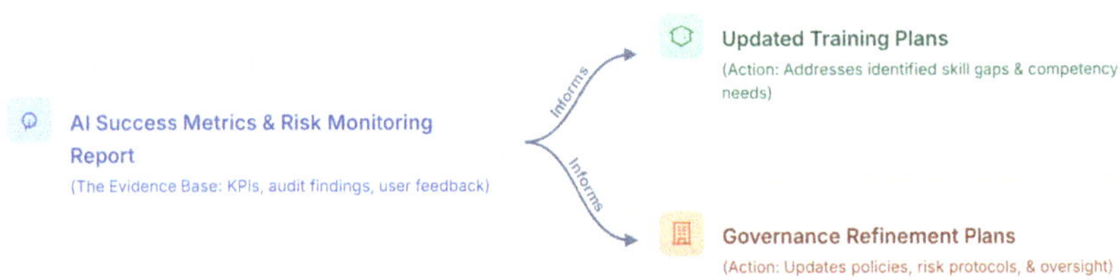

Fig. 15.1 Iterative AI adaptation cycle. (*Note*: A circular diagram illustrating the AI adaptation and refinement cycle, moving from monitoring/feedback collection through analysis, ideation, implementation, and impact evaluation, with stakeholder engagement inputs)

User Outcomes & Deliverables (UODs) for Adaptation and Refinement

This ongoing phase relies on specific User Outcomes and Deliverables (UODs) to structure the process and guide data-driven decisions (Nathan & Welburn, 2024). Each deliverable serves a clear purpose in guiding the institution's adaptation journey:

1. **Updated Faculty, Staff, and Student Training Plans:** These UODs detail current training modules, resources, and competency pathways. Their purpose is to ensure all stakeholders possess the necessary, up-to-date skills for ethical and effective AI use, informed by ongoing needs assessments and feedback.

2. **Periodic AI Success Metrics and Risk Monitoring Reports:** These UODs track Key Performance Indicators (KPIs)—such as AI adoption rates, user satisfaction, efficiency gains, impact on learning outcomes, and Return on Investment (ROI)—alongside critical risk data, including bias audits, security incidents, and ethical concerns (Mills et al., 2025; NIST, 2023). Their purpose is to provide quantitative and qualitative evidence of AI's performance and risk profile, enabling informed adjustments.
3. **Governance Refinement Plans:** These UODs formally document the schedule and process for reviewing and updating AI policies, guidelines, and governance structures. Their purpose is to ensure that oversight mechanisms remain adaptive, responsive to new challenges, and aligned with institutional values.

These UODs work interdependently: metrics reports provide the data foundation, governance plans ensure structured and timely updates to policies and oversight, and training plans address the evolving skill requirements identified through this adaptive process, thereby facilitating continuous institutional improvement (Table 15.4 and Fig. 15.2).

Workbook Prep 5. Detailing Adaptation Deliverables

Reflection: Review the key deliverables listed (Updated Training Plans, Metrics/Risk Reports, Governance Refinement Plans). Which of these deliverables do you think will be most critical for guiding evidence-based decision-making during the adaptation phase at your institution? Why?

UX Workbook Link: See Chap. 15 UX Workbook, Section 5 ("Managing Living Documents") and Section 7 ("Finalizing Adaptation & Refinement Plans"), which focus on creating and maintaining these UODs.

PALS Workbook Link & AI Partnership Guidance: Use Chap. 15 PALS Workbook prompts (e.g., Section 7).

Example Prompt: "Help outline the key sections and data types needed for our periodic 'AI Success Metrics and Risk Monitoring Report' deliverable, ensuring it provides actionable insights for governance refinement (HCASR Component 5)."

Human Oversight Reminder: AI can help structure deliverables. Determining the most relevant metrics, defining appropriate content, ensuring data accuracy, and using the deliverables effectively for strategic decision-making require human expertise, institutional knowledge, and leadership engagement (HCASR Component 5).

Table 15.4 Key deliverables for ongoing AI adaptation and refinement

Deliverable	Purpose/content	Key inputs/triggers
1. Updated Training Plans	Dynamic documents detailing current and upcoming training modules, resources (e.g., competency frameworks), and pathways to address evolving skill needs and AI capabilities.	Regular needs assessments, analysis of performance metrics and user feedback (from Metrics Reports), identified skill gaps.
2. AI Success Metrics & Risk Monitoring Report	Periodic reports (e.g., quarterly, annually) tracking agreed-upon KPIs (adoption, satisfaction, efficiency, impact, ROI) and critical risk monitoring data (bias audits, security, performance, ethical concerns).	Ongoing data collection (KPIs, risk monitoring), scheduled reporting cycle.
3. Governance Refinement Plans	Formal documents outlining the schedule, structured process (incl. stakeholder consultation, feedback integration), and approval workflows for regular review/updates of AI policies, guidelines, and governance structures.	Scheduled review cycle, findings from Metrics & Risk Reports, external regulatory changes, stakeholder feedback.

Note. These deliverables work together, using data from monitoring reports to inform structured updates to training and governance, thus facilitating continuous, data-driven improvement and adaptation of the institution's AI ecosystem

Conclusion

The Adaptation and Refinement phase embodies the ongoing commitment required for successful, ethical AI integration. It moves beyond initial deployment to embed AI thoughtfully through continuous cycles of monitoring, feedback, adjustment, and learning. Navigating this phase relies on fostering an adaptable culture, maintaining dynamic governance and stakeholder engagement, and using key UODs to guide decisions. By applying principles of iterative refinement informed by Design Thinking (Brown, 2019) and Change Management (Kotter et al., 2021), HEIs ensure AI use evolves responsibly, aligns with strategic goals, and enhances outcomes. This sustained effort prepares the institution for long-term strategic considerations.

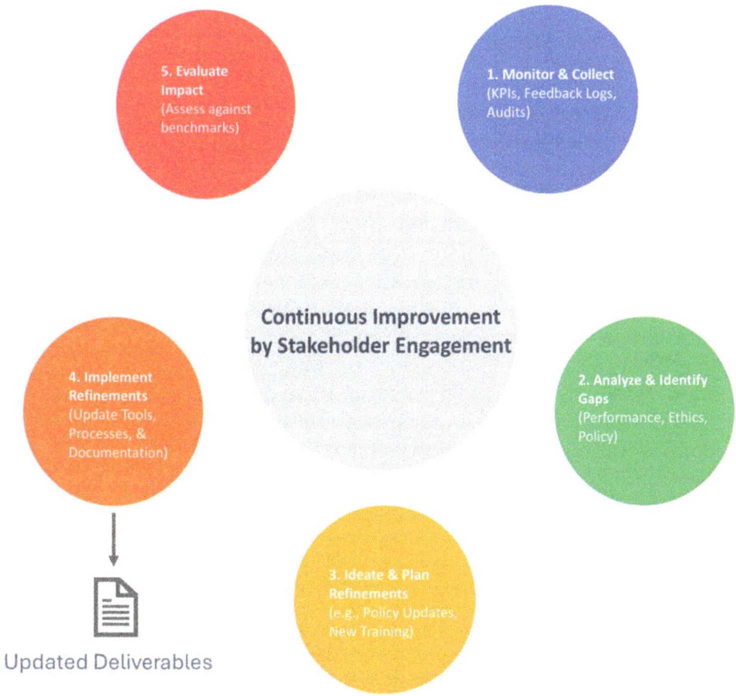

Fig. 15.2 Deliverables integration for continuous improvement. (*Note*: A flowchart depicting how key deliverables integrate: Success Metrics and Risk Reports inform both Updated Training Plans and Governance Refinement Plans, supporting continuous improvement)

Looking Ahead & Engaging the Workbook

Chapter 15 detailed the ongoing processes of adaptation and refinement after AI initiatives are scaled. We explored strategies for cultural adaptation, sustained AI literacy, iterative policy development, dynamic stakeholder coordination, and key UODs. The next chapter, Chap. 16, addresses long-term strategic considerations, focusing on customized AI ethics frameworks and sustainable governance. The Chap. 15 Companion UX and PALS Workbooks provide structured activities to operationalize the adaptation and refinement processes discussed here, enabling teams to manage continuous improvement effectively.

Long-Term Scaling, Sustainability & AI Governance: Ensuring TLARA Frameworks Evolve

Chapter 15 addresses the necessity of ongoing adaptation and refinement as AI technologies and their applications within Higher Education continuously evolve. This final chapter of Phase 3 emphasizes that institutional AI strategies, governance frameworks, and the specific guidelines within your custom **TLARA (Teaching, Learning, Assessment, Research, Administration)** structure must be dynamic and responsive to change over **time** (Wynants et al., 2025). Cultivating organizational agility, ensuring evolving skill development, and establishing iterative refinement cycles for AI tools and policies across all institutional levels and spaces are key to maintaining ethical alignment and effectiveness in a rapidly changing sociocultural context. Both the Chapter 15 UX and PALS Workbooks help you partner with AI to build these adaptive capacities.

The long-term success of AI integration in HEIs depends on this ability to learn and adapt, ensuring that initial investments lead to sustained value (Ellucian, 2024; IDEAS, 2025). The PALS Workbook for Chap. 15, via "Long-Term Scaling, Sustainability & AI Governance: Iteratively Developing Your Custom TLARA-Informed AI Guidelines," guides you to consider these dynamics. Engaging with "Topics for PALS TLARA Focus"—such as "Adaptive AI Governance for Continuous TLARA Alignment," and "Iterative Refinement for Sustained Value in TLARA Applications"—your GenAI advisor will assist in incorporating principles of continuous improvement and flexibility into your "Living Document." This prepares your TLARA guidelines and overall AI governance approach to remain resilient, relevant, and supportive of a thriving AI literacy culture well into the future (DerSimonian & Montagnino, 2025).

References

Appaya, S., & Ng, J. (2024). *Global trends in AI governance: Evolving country approaches*. World Bank Group.

Brown, T. (2019). *Change by design: How design thinking transforms organizations and inspires innovation* (Rev. & updated ed.). Harper Business.

Carmody-Bubb, M. (2023). *Cognition and decision making in complex adaptive systems: The human factor in organizational performance*. Springer.

Cornell University Task Force on GenAI in Academic Research. (2023). *Generative AI in academic research: Perspectives and cultural norms*. Cornell University Office of the Vice President for Research and Innovation. https://research-and-innovation.cornell.edu/generative-ai-in-academic-research/

Deloitte. (2025, January). *The state of generative AI in the enterprise: Now decides next (Quarter four report)*. https://www2.deloitte.com/content/dam/Deloitte/bo/Documents/consultoria/2025/state-of-gen-ai-report-wave-4.pdf

DerSimonian, R., & Montagnino, C. (2025, March 26). Crafting thoughtful AI policy in higher education: A guide for institutional leaders. *Campus Technology*. https://campustechnology.com/articles/2025/03/26/crafting-thoughtful-ai-policy-in-higher-education-a-guide-for-institutional-leaders.aspx

Ellucian. (2024, December 16). AI in higher education: Implementation strategies for institutions and tech partners. *Higher Ed Dive*. https://www.highereddive.com/spons/ai-in-higher-education-implementation-strategies-for-institutions-and-tech/735434/

Gouseti, A., James, F., Fallin, L., & Burden, K. (2025). The ethics of using AI in K-12 education: A systematic literature review. *Technology, Pedagogy and Education, 34*(2), 161–182. https://doi.org/10.1080/1475939X.2024.2428601

Hassan, M., Kushniruk, A., & Borycki, E. (2024). Barriers to and facilitators of artificial intelligence adoption in health care: Scoping review. *JMIR Human Factors, 11*, e48633. https://doi.org/10.2196/48633

Heintz, F., Linden, K., & Geurts, J. (2024, February). Actionable recommendations and support for updating the SRIDA (Deliverable D1.5). *Adra-e*. https://adra-e.eu/sites/default/files/2024-02/Adra-e_D1.5_Actionable%20recommendations%20and%20support%20for%20updating%20the%20SRIDA.pdf

Instructional Design, Engagement, and Support (IDEAS). (2025, January 27). *AI's impact on higher education through the eyes of faculty: 4 key trends*. University of Massachusetts Amherst. https://www.umass.edu/ideas/news/ais-impact-higher-education-through-eyes-faculty-4-key-trends

Kotter, J. P., Akhtar, V., & Gupta, G. (2021). *Change: How organizations achieve hard-to-imagine results in uncertain and volatile times*. Wiley.

Maneval, M. (2025, April 2). Ensuring ethical and responsible AI: Tools and tips for establishing AI governance. *LogicGate*. https://www.logicgate.com/blog/ensuring-ethical-and-responsible-ai-tools-and-tips-for-establishing-ai-governance/

Mills, S., Broestl, N., & Kleppe, A. (2025, March 27). *You won't get GenAI right if you get human oversight wrong*. Boston Consulting Group. https://www.bcg.com/publications/2025/wont-get-gen-ai-right-if-human-oversight-wrong

Nathan, L., & Welburn, W. C. (Eds.). (2024). *New horizons in artificial intelligence in libraries*. De Gruyter Saur. https://library.oapen.org/bitstream/handle/20.500.12657/98734/9783111336435.pdf?sequence=1&isAllowed=y

National Institute of Standards and Technology. (2023, January). *Artificial intelligence risk management framework (AI RMF 1.0)* (NIST AI 100-1). U.S. Department of Commerce. https://nvlpubs.nist.gov/nistpubs/ai/NIST.AI.100-1.pdf

PwC. (2024). *2025 AI business predictions*. https://www.pwc.com/us/en/tech-effect/ai-analytics/ai-predictions.html

Relyea, C., Maor, D., Durth, S., & Bouly, J. (2024, August 7). *Gen AI's next inflection point: From employee experimentation to organizational transformation*. McKinsey & Company. https://www.mckinsey.com/capabilities/people-and-organizational-performance/our-insights/gen-ais-next-inflection-point-from-employee-experimentation-to-organizational-transformation

United Nations Educational, Scientific and Cultural Organization. (2023). *Guidance for generative AI in education and research*. UNESCO. https://unesdoc.unesco.org/ark:/48223/pf0000386693

World Economic Forum. (2025, January). *AI in action: Beyond experimentation to transform industry*. https://reports.weforum.org/docs/WEF_AI_in_Action_Beyond_Experimentation_to_Transform_Industry_2025.pdf

Wynants, S., Childers, G., De La Torre Roman, Y., Budar-Turner, D., & Vasquez, P. (2025). *ETHICAL principles AI framework for higher education*. California State University, Fullerton. Retrieved June 30, 2025, from https://genai.calstate.edu/communities/faculty/ethical-and-responsible-use-ai/ethical-principles-ai-framework-higher-education

16

Custom AI Ethics and Literacy Frameworks (Phase 4)

Introduction: Planning for Enduring Ethical AI Integration

As Artificial Intelligence (AI), particularly Generative AI (GenAI), increasingly integrates within Higher Education Institutions (HEIs) worldwide, developing comprehensive AI literacy (Chap. 17) and robust ethical frameworks is paramount for long-term sustainability and effective AI Governance. This chapter provides the foundational understanding for these endeavors, ensuring technologies are harnessed responsibly and equitably. Entering Phase 4 of the AI integration journey—Long-Term Scaling, Sustainability & AI Governance—this chapter focuses on enduring ethical coherence and operational sustainability over five to ten years. A central aim is to empower HEIs globally to use LLMs/GenAI as strategic partners. This involves institutions tailoring timelines and strategies to their unique sociocultural contexts (Carmody-Bubb, 2023), engaging diverse stakeholders to identify appropriate tools and outcomes aligned with their mission and values. This aligns with this book's philosophy of leveraging human-centric frameworks (see HAICS Guide) and GenAI as strategic partners for tailored, not prescriptive, solutions. The Workbook Prep sections throughout this chapter are integral for directing users to the PALS and UX workbooks, where they can ethically and effectively search, find, explore, and create tools, solutions, and UODs that align with their unique needs and sociocultural contexts.

This chapter addresses the development and implementation of such a long-range plan, laying the groundwork for implementation, piloting, evaluation, and scaling. This foresight is essential for navigating AI's rapid evolution, ensuring alignment with the institution's mission and values, meeting dynamic accreditation standards (e.g., ACCJC, 2020, 2024), and adapting to the evolving global

Supplementary Information The online version contains supplementary material available at https://doi.org/10.1007/978-3-032-05205-6_16.

D. Woodring, *Advancing Higher Education with Human-Centric AI Partnerships*,
https://doi.org/10.1007/978-3-032-05205-6_16

regulatory landscape (Appaya & Ng, 2024; OECD, n.d.-a; The World Bank, 2024; State University of New York [SUNY], n.d.). It moves beyond reactive measures to strategically embed ethical considerations into the institution's long-range vision, fostering responsible innovation (EDUCAUSE, 2024; Higher Education Strategy Associates [HESA], 2024).

A foundational element of this strategy is institutionalizing a dynamic process for stakeholders to collaboratively develop, apply, and refine customized AI ethics and literacy guidelines. This chapter introduces the TLARA conceptual structure—an adaptable organizational scaffold guiding institutions in building their own ethical frameworks specific to Teaching, Learning, Assessment, Research, and Administration. Crucially, TLARA is not a predefined set of rules but an organizational tool. HEIs can utilize it, leveraging research, stakeholder engagement (Chap. 7), and strategic AI partnership (guided by frameworks like HCASR from the HAICS Guide), to develop specific ethical principles, governance procedures, and usage policies with rigorous human oversight. This customization, initiated conceptually in Chap. 5, enables institutions to build upon initial explorations of core frameworks (HCAS, RSA, EHCS, EAO, AHIPs) and tailor them to their distinct environments. A customizable approach addresses barriers like lack of strategy and understanding (EDUCAUSE, 2024). This process, guided by the TLARA structure, supports a holistic, integrated approach, embedding responsible, equitable, and human-centric AI governance across core functions. Such a human-centric approach, as emphasized by the HCAS framework, ensures that AI augments human capabilities rather than merely replacing roles, empowering institutions to tailor ethical considerations to unique challenges within each operational area (Wynants et al., 2025; UC Davis AI Council, 2025).

Workbook Prep 1: Long-Term Ethical Foresight Needs

Reflection Question: What is the most significant AI-related ethical challenge or opportunity your institution will face in the next 5–10 years, and are current governance structures prepared for this long-range planning? How can your institution proactively prepare for the ethical dimensions of next-generation AI, ensuring alignment with its mission, values, and evolving external standards, potentially using GenAI as a strategic partner for research and brainstorming?

UX Workbook Link: Engage with Chap. 16 UX Workbook, Section 1: "Planning: Establishing Permanent Governance & Framework Integration." This is where you will apply your reflections.

PALS Workbook Link & AI Partnership Guidance: Utilize the Chap. 16 PALS Workbook prompts corresponding to UX Workbook Section 1 to leverage GenAI in developing your plans.

Example PALS Prompt: "Input your institutional mission's core tenets and current AI adoption status. Ask your GenAI partner to help brainstorm three potential elements for a 10-year institutional vision for ethical AI, and identify one potential long-term risk related to emerging AI trends that current governance might not address, referencing HCASR Component 1 (Ethical & Effective Goal Setting) & Component 3 (Rapid Research & Review)."

Human Oversight Reminder: Strategic foresight requires human leadership. Use AI for brainstorming and trend analysis (as a strategic partner), but ensure the final vision reflects realistic capacity and core values, validated through diverse stakeholder engagement and governance processes (HCASR C1 & C5).

Strategic Imperatives for a 5 to 10-Year Ethical AI Horizon

A comprehensive long-term plan must address key strategic areas for effective and sustainable ethical AI integration. This involves proactive planning and structural adaptation, acknowledging AI's evolving nature demands adaptive governance (UNESCO, 2023b). These imperatives provide the foundations for action.

Long-Term Ethical AI Planning and Foresight

The rapid pace of AI development necessitates proactive foresight. Institutions should anticipate future ethical dilemmas from next-generation AI, consider societal impacts, and plan adaptive governance. This includes ongoing environmental scanning of technological advancements (Deloitte, 2025), global regulatory shifts (Appaya & Ng, 2024; OECD, n.d.-a; The World Bank, 2024), evolving societal expectations, and building a foundation for continuous AI literacy education (Chap. 17). Understanding future trends, including Artificial General Intelligence (AGI), is crucial (EDUCAUSE, 2024). Strategic foresight activities, potentially AI-supported (guided by HCASR Component 3 as a strategic partner), should inform regular reviews and updates of the institution's AI vision and strategic plan (Brown, 2019).

Integrated and Sustained Governance Structures

Effective long-term governance requires permanent, integrated structures. Establishing bodies like a central AI Council, supported by domain-specific ethics subcommittees (potentially aligning with TLARA domains), provides enduring oversight and strategic direction (DerSimonian & Montagnino, 2025; EDUCAUSE, 2024; UC Davis AI Council, 2025). These multi-stakeholder committees need clear mandates for developing and overseeing institution-specific policies (informed by custom TLARA-informed frameworks), compliance monitoring, and resolving ethical concerns. Such integration embeds AI governance within the institution's operational fabric (Kotter et al., 2021). The RSA framework's emphasis on regulatory synergy ensures these structures are robust and adaptable, supporting HEIs in creating solutions tailored to their unique contexts.

Sustainable Adoption and Continuous Improvement Cycles

Long-term success depends on embedding continuous improvement. This involves regular ethical audits, systematic stakeholder feedback, periodic reviews of AI policies and custom ethical guidelines (developed using the TLARA structure), ongoing AI literacy programs (Chap. 17), and iterative refinement of AI applications. Treating policies as "living documents" (guided by HCASR Component 5) reflects this commitment. This cycle ensures AI integration remains effective, ethical, and aligned with evolving needs, aligning with design thinking (Brown, 2019) and change management (Kotter et al., 2021) (Table 16.1).

Workbook Prep 2: Planning Sustainable Governance

Reflection Question: What mechanisms are needed for regular review of AI policies and incorporation of new ethical/technological developments? What is the role of ongoing AI literacy in maintaining governance responsiveness, especially when aiming for solutions tailored to your HEI's unique global context?

Table 16.1 Select governance frameworks emphasizing continuous improvement cycles

Framework	Relevant component/ principle	Emphasis on continuous improvement/ adaptation
NIST AI RMF	"Govern, Map, Measure, Manage functions (especially Measure & Manage)"	"Continuous monitoring, risk management adaptation, iterative improvements based on ongoing assessment."
COBIT	"Monitor, Evaluate, and Assess (MEA) Domain"	"Regular performance assessments, governance evaluations, identifying areas for improvement, adaptation to needs."
ISO/IEC 42001	Plan-Do-Check-Act (PDCA) Cycle for AI Management System	"Requires establishing, implementing, maintaining, and continually improving the AI management system; regular reviews."

Note. Frameworks like the National Institute of Standards and Technology (NIST) AI Risk Management Framework (RMF) (NIST, 2023), ISACA's COBIT (ISACA, 2025), and the ISO/IEC 42001 standard for AI management systems all incorporate iterative cycles

UX Workbook Link: See Chap. 16 UX Workbook, Section 2: "Actions: Developing Strategies & Defining Leadership Roles," specifically Activity 2.1 (Develop Expanded Governance Strategies based on TLARA), which addresses anticipating future challenges. Also relevant is Section 4: "Checkpoints," particularly Activity 4.3 on integrating feedback for continuous improvement.

PALS Workbook Link & AI Partnership Guidance: Use Chap. 16 PALS Workbook prompts corresponding to UX Workbook Sections 2 & 4.

Example PALS Prompt: "Input your institution's proposed AI governance model (e.g., Central AI Council with TLARA subcommittees). Ask your GenAI partner to suggest three specific mechanisms that would ensure this governance structure remains adaptive over a 10-year period, considering how to integrate stakeholder feedback, incorporate findings from strategic foresight activities, and align with Kotter's principles for anchoring change, referencing HCASR Component 5 (Continuous Improvement)."

Human Oversight Reminder: AI can suggest standard mechanisms for adaptability. Designing robust, truly adaptive governance requires human understanding of institutional culture, bureaucratic realities, and resource constraints. Ensure processes facilitate meaningful input and are not merely performative, focusing on genuine responsiveness and building stakeholder trust (HCASR C4 & C5).

Core Rationale for Utilizing the AI Ethics TLARA Structure for Custom Framework Development

Focusing on Teaching, Learning, Assessment, Research, and Administration (TLARA) through a multilevel systems approach (Carmody-Bubb, 2023; Bronfenbrenner & Morris, 2006) provides a practical heuristic for HEIs to address AI integration comprehensively. This holistic approach is essential for institutions to develop their own specific, actionable ethical frameworks and responsibly navigate long-term AI integration, using GenAI as a strategic partner in this development.

The Imperative for Domain-Specific Ethical Oversight

Widespread AI deployment introduces context-specific risks in HEIs (Coffey, 2024; Sain et al., 2025; OECD, 2023b). General AI principles may not adequately address these nuances, necessitating context-aware, domain-specific ethical frameworks (HSD Author Collective, 2025; UNESCO, 2023b). The TLARA structure prompts institutions to develop specific ethical lenses for each core domain, enabling proactive risk mitigation through their own collaboratively developed guidelines, tailored to their unique sociocultural environment.

Ensuring Alignment with Institutional Mission, Values, and Evolving Standards

Developing ethical guidelines using TLARA ensures AI integration aligns with the HEI's mission, values, external standards (e.g., ACCJC, 2024), and sociocultural norms. Institutions should prioritize augmenting human capabilities, a core tenet of the HCAS framework. The TLARA structure reinforces this human-centric stance by prompting critical consideration of AI's appropriate role within each domain, ensuring AI applications adhere to data privacy laws, academic freedom, and student rights (UNESCO, 2021, 2023a).

The Need for a Unified yet Customizable Approach via the TLARA Structure

TLARA helps prevent fragmented AI practices by prompting ethical considerations and policy development within each domain:

- Teaching: e.g., AI-assisted curriculum development, ethical AI-supported grading (Coffey, 2024; Verhoeven & Hor, 2025).
- Learning: e.g., ethical use of AI tutors, ensuring equitable access (Coffey, 2024; Xie et al., 2024).
- Assessment: e.g., policies for AI proctoring, maintaining academic integrity (Coffey, 2024; OECD, 2023b).
- Research: e.g., ethical guidelines for AI in data analysis, ensuring research integrity (Le Dinh et al., 2025; Cornell University Task Force on GenAI, 2023).
- Administration: e.g., ethical AI in admissions, responsible AI in automation (Coffey, 2024).

Developing guidelines using this structure, with active stakeholder participation and GenAI as a potential research partner, ensures systematic application of ethical principles, fostering institutional coherence and addressing diverse needs (Brown, 2019) (Table 16.2 and Figs. 16.1 and 16.2).

Workbook Prep 3: Identifying Domain-Specific AI Challenges for Guideline Development

Reflection Question: Within your institutional context, which TLARA domain currently faces the most pressing ethical challenges due to AI, thus needing urgent attention for developing clear, custom guidelines? Why is a unified organizing structure like TLARA crucial for developing policies that are both coherent across the institution and effectively address these unique domain-specific issues, especially when aiming for global applicability and cultural sensitivity?

UX Workbook Link: Engage with Chap. 16 UX Workbook, Section 1, Activity 1.2 ("Integrating Foundational Frameworks into TLARA Domains") and Section 3, Activity 3.2 ("Policy Mapping for Long-Term Coherence").

PALS Workbook Link & AI Partnership Guidance: Utilize the Chap. 16 PALS Workbook prompts corresponding to UX Workbook Sections 1 & 3 to use GenAI as a strategic partner in analyzing these challenges.

Example PALS Prompt: "Select one TLARA domain (e.g., 'Assessment'). Ask your GenAI partner to elaborate on two specific ethical risks GenAI poses within this domain in a higher education context. Then, prompt it to explain how using the TLARA structure helps an institution to proactively develop targeted guidelines to mitigate these risks, referencing HCASR Component 3 (Rapid Research & Review) for supporting evidence or examples."

Table 16.2 TLARA structure: domain examples and ethical considerations for institutional guideline development

TLARA domain	Example AI applications	Key ethical considerations for custom guidelines (examples)
Teaching	AI-assisted course design tools AI grading support/ feedback tools	"Ensuring pedagogical soundness, transparency in AI use, instructor oversight" "Maintaining academic standards, addressing bias in AI feedback, data privacy"
Learning	Personalized AI tutors/ learning pathways Adaptive learning systems	"Equity of access, impact on critical thinking, data security" "Algorithmic transparency, student agency/choice, potential for over-reliance"
Assessment	AI-based proctoring/ academic integrity tools AI-driven feedback on assignments	"Student privacy, algorithmic bias, accuracy/reliability, clear use policies" "Fairness, quality of feedback, definition of plagiarism/authorship"
Research	AI tools for literature review/data analysis AI in peer-review processes	"Research integrity, proper citation/ attribution of AI use, data confidentiality" "Bias in review, confidentiality, accountability for AI-assisted conclusions"
Administration	AI for admissions screening/student support routing AI for workflow automation/resource allocation	"Algorithmic fairness/bias, transparency of decision processes, data governance" "Impact on staff roles, accountability for automated decisions, data security"

Note. This table provides illustrative examples within the Teaching, Learning, Assessment, Research, and Administration (TLARA) domains. Institutions should use this structure to develop their own detailed ethical guidelines reflecting their specific context, values, risks, and stakeholder needs

Human Oversight Reminder: Ensure the rationale for developing custom guidelines within the TLARA structure is clear, compelling, and addresses institutional context. AI-generated elaborations on risks and benefits must be validated by human experts for accuracy, relevance, and ethical soundness (HCASR C3 & C4).

Using AI as a Strategic Partner in Diverse Global Contexts for Long-Term Governance

Every HEI's unique sociohistorical context necessitates tailored AI integration (Carmody-Bubb, 2023). For long-term governance in diverse global settings,

Fig. 16.1 TLARA-informed implementation process flowchart. (*Note*: A flowchart with three main sequential stages labeled: (1) Establish Permanent AI Ethics Governance. (2) Implement Comprehensive Stakeholder Training. (3) Develop & Scale AI Integration. A feedback loop arrow from Stage 3 points back to Stage 1, indicating continuous improvement)

GenAI tools, used strategically and critically (guided by frameworks like HCASR from the HAICS Guide), can be invaluable. Institutions can ethically partner with AI using contextual data to: 1) co-create custom governance tools and ethical solutions resonating with local needs, values, and sociocultural nuances, and 2) identify and mitigate dominant cultural biases (often Western-centric) within AI models that might impact framework development (Kim et al., 2025; Tao et al., 2024). While not replacing human expertise in ethical deliberation or understanding cultural norms, AI's ability to process vast information on global regulatory trends, diverse ethical philosophies, or cross-cultural communication strategies makes it a vital strategic partner in planning for adaptable and globally relevant long-term AI governance. This supports the development of frameworks that are ethically sound, culturally sensitive, and practically applicable across varied international contexts, empowering HEIs worldwide. See the Human-GenAI Collaboration & Synergy (HAICS) Guide for resources on ethical AI partnership.

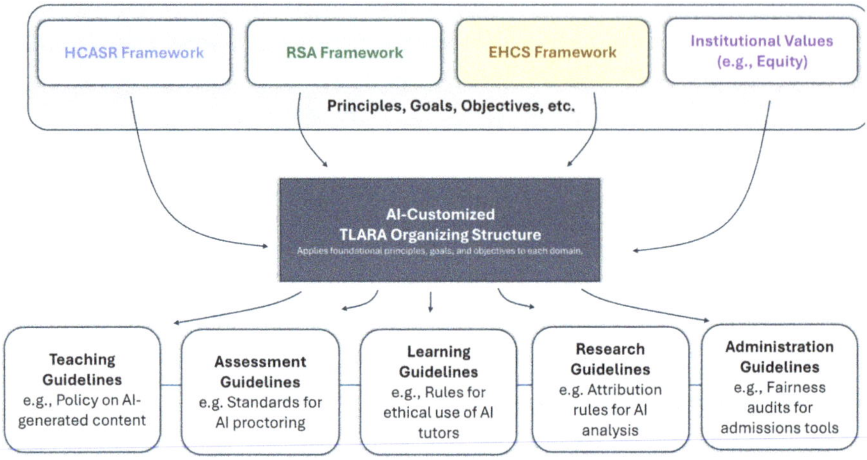

Fig. 16.2 Structuring framework application with TLARA. (*Note*: A diagram illustrating how the TLARA structure organizes the application of foundational institutional frameworks. It shows principles from core frameworks (e.g., HCASR, RSA) flowing into and being organized by the five TLARA domain containers (Teaching, Learning, Assessment, Research, Administration)

Institutionalizing the TLARA-Informed Framework Development Process

Institutionalizing the development and application of AI Ethics & Literacy guidelines involves a structured process integrated into the institution's long-term strategy. This provides the foundation for embedding ethical considerations systematically. The following adaptable steps are based on research and emerging best practices (EDUCAUSE, 2024; UC Davis AI Council, 2025).

Step 1: Establish Permanent, Empowered AI Ethics Governance Structures

Implement permanent, equitable, transparent, and accountable AI oversight structures. These guide collaborative development, implementation, and refinement of policies addressing TLARA domains, tailored to the HEI's context. This typically involves formalizing a cross-departmental AI Governance Council (EDUCAUSE, 2024; UC Davis AI Council, 2025). This council and its subcommittees develop, endorse, and update institutional policies, using the institution's custom TLARA-informed framework (Ellucian, 2024; Husch Blackwell, 2024), establishing the "Guiding Coalition" (Kotter et al., 2021).

Step 2: Implement Comprehensive Stakeholder Training in Human-AI Collaboration (HAIC) and AI Literacy

Foster institution-wide AI literacy and promote effective, ethical human-AI interaction based on HAIC principles (Part 1; HAICS Guide) and the institution's custom, TLARA-informed guidelines. Deploy comprehensive, ongoing, role-specific AI literacy programs covering AI concepts, institutional ethical principles, bias awareness, data privacy, responsible use, and HAIC strategies (EDUCAUSE, 2024; Schroeder, 2025). This empowers the community through user-centric training (Brown, 2019; Kotter et al., 2021).

Step 3: Develop and Scale AI-Integrated Programs and Processes Guided by Custom Frameworks

Systematically embed approved AI tools and custom TLARA-informed ethical principles into curricula, research, and administrative workflows. Embed institution-specific AI Ethics & Fair Use policies into syllabi, research protocols, administrative procedures, and technology adoption guidelines (Wynants et al., 2025). Foster cross-disciplinary AI education and collaborative research adhering to custom guidelines (AAMC, n.d.; Jumadinova, n.d.; Xie et al., 2024), accounting for institutional practicalities (EDUCAUSE, 2024) (Table 16.3).

Workbook Prep 4: Evaluating Governance Structure Readiness

Reflection Question: Does your institution's permanent AI Governance Council (or equivalent) have sufficient representation from all TLARA domains, clearly defined authority, and adequate resources to effectively oversee the development, implementation, and continuous refinement of policies across these areas, tailored to your HEI's needs? Is there a transparent and planned process for domain-specific policy creation that uses the TLARA structure as an organizational guide?

Table 16.3 Process for institutionalizing TLARA-informed framework development

Step	Step name	Key activities/focus
1	Establish Permanent AI Ethics Governance Structures	"Formalize/empower permanent, representative AI Governance Council & subcommittees; Task Council with developing, enforcing, and regularly updating detailed institutional policies using the TLARA structure as a guide for domain coverage (T, L, A, R, A)."
2	Implement Comprehensive Stakeholder Training	"Deploy ongoing, comprehensive AI literacy and Human-AI Collaboration (HAIC) training for all stakeholders covering foundational concepts, institution-specific TLARA-informed ethics, bias, privacy, responsible use, and role-specific HAIC strategies based on custom guidelines."
3	Develop & Scale AI-Integrated Programs & Processes	"Systematically embed custom TLARA-informed policies into curricula (syllabi), research protocols (IRB), administrative procedures, and tech adoption guidelines; Foster cross-disciplinary AI education and research adhering to these institutional guidelines."

Note. This structured process aims to systematically embed ethical considerations based on institutionally developed guidelines using the TLARA structure into governance, training, and operations

UX Workbook Link: Refer to Chap. 16 UX Workbook, Section 1: "Planning," Activity 1.1 ("Establish AI Ethics Governance as a Permanent Fixture"), which finalizes the governance structure. Also relevant is Section 2: "Actions," Activity 2.1 ("Develop Expanded Governance Strategies"), which involves planning domain-specific approaches.

PALS Workbook Link & AI Partnership Guidance: Use Chap. 16 PALS Workbook prompts corresponding to UX Workbook Sections 1 & 2.

Example PALS Prompt: "Input the five TLARA domains. Ask your GenAI partner to suggest three potential subcommittee structures or working group models that an AI Governance Council might establish to effectively manage domain-specific policy development and oversight, referencing established governance models (e.g., from UC Davis AI Council report or general HEI governance literature) and HCASR Component 4 (Participatory Design)."

Human Oversight Reminder: Proposed governance structures must be practical and have clear mandates. Effective policy development utilizing the TLARA structure requires domain experts and diverse stakeholders to ensure feasibility, buy-in, and contextual relevance (HCASR C4).

TLARA-Informed Guideline Development Across AI Rollout Phases

Developing custom AI ethics guidelines using the TLARA structure is an evolving process throughout the four-phase AI rollout (Research & Planning; Implementation & Piloting; Evaluation & Adaptation; Scaling & Institutionalization), maturing with institutional experience and stakeholder dialogue.

- **Phase 1 (Research, Engagement, Planning):** Initial ethical principles are defined. The need for domain-specific guidelines (using TLARA as a scaffold) is assessed (Chap. 8), and stakeholders begin conceptualizing custom approaches (Chap. 5).
- **Phase 2 (Implementation & Piloting):** Draft guidelines, organized by TLARA, provide preliminary ethical parameters for pilot projects (Chaps. 10 and 11). Adherence is monitored, and feedback collected (Chap. 12).
- **Phase 3 (Ethical Evaluation & Refinement):** Pilot outcomes are evaluated against developing TLARA-informed principles (Chap. 13). Guidelines are significantly refined based on evaluations and practical application (Chap. 15).
- **Phase 4 (Long-Term Scaling, Culture, Governance, Sustainability):** Mature, refined institutional guidelines, collaboratively developed using TLARA, are fully institutionalized within policies, training (Chap. 17), workflows, and permanent governance structures (this Chap. 16), becoming the standard for ethical AI practice.

This lifecycle demonstrates iterative ethical framework evolution, informed by experience, stakeholder input, and continuous improvement, mirroring design thinking (Brown, 2019) and organizational change stages (Kotter et al., 2021).

Workbook Prep 5: Aligning TLARA-Informed Development with Current Phase

Reflection Question: Considering your institution's current phase in the AI integration journey, how are (or how could) custom, TLARA-informed ethical guidelines being applied, developed, or considered? Are iterative dynamics—such as learning from pilot experiences (Phase 2/3) or responding to new

AI capabilities (Phase 4)—evident in this process? How does the work in the current phase build upon previous efforts and prepare for the next stage of AI governance and ethical practice, ensuring solutions are tailored to your HEI's unique context?

UX Workbook Link: This is an overarching contextual reflection relevant to all UX Workbook activities. Specific phase alignment activities are embedded within each chapter's workbook. Section 6 of the Chap. 16 UX workbook explicitly addresses scaffolding from previous phases.

PALS Workbook Link & AI Partnership Guidance: Utilize the Chap. 16 PALS Workbook prompts corresponding to UX Workbook Section 6 ("Scaffolding & Ensuring Continuity").

Example PALS Prompt: "Describe your institution's current phase in the AI rollout (e.g., 'Early Phase 3 - Evaluating initial pilots') and a specific challenge being faced (e.g., 'Faculty uncertainty regarding AI use in student assessment'). Ask your GenAI partner how applying the developing TLARA-informed Assessment principles, alongside frameworks like Brown's Design Thinking (for empathy and iteration) or Kotter's Change Management (for communication and removing barriers), could help address this challenge, referencing HCASR Component 4 (Human-Centric Implementation) and Component 5 (Continuous Improvement)."

Human Oversight Reminder: Framework alignment is conceptual. Real-world guideline implementation and problem-solving require flexible adaptation guided by human leadership and proactive, empathetic stakeholder engagement (HCASR C5).

Critical Importance of Developing Customized Multilevel Frameworks: Using the TLARA Structure for HEIs

Institutionalizing a stakeholder-driven process for developing domain-specific ethical guidelines, using TLARA as an organizational aid and potentially GenAI as a research and drafting partner, is vital for long-term success and responsible AI integration. Its importance spans risk prevention, compliance, community empowerment, and fostering responsible innovation adaptable to global contexts.

Prevention of Ethical Risks and Failures

A TLARA-structured, collaborative process enables proactive identification and management of AI-related risks (American Bar Association, 2024; ISACA Now, 2024; Xie et al., 2024). Prompting ethical checks within TLARA domains helps mitigate risks, aligning with DEIA commitments (AAMC, n.d.; Fairfax, 2025; Wargo & Anderson, 2024).

Assurance of Compliance and Standards Alignment

Consistently applying institutionally developed guidelines, organized via TLARA, provides a transparent approach to meeting external and internal standards (ACCJC, 2020, 2024; UNESCO, 2021, 2023a). This underpins AI literacy, fostering informed compliance (Coffey, 2024; Stone, 2025). Institutions must also proactively manage bureaucratic hurdles (EDUCAUSE, 2024).

Empowerment of the Institutional Community and Global Adaptability

A well-implemented process for developing TLARA-informed guidelines promotes Human-AI Collaboration (HAIC) that emphasizes human agency and meaningful oversight, empowering users (Holzinger et al., 2025b; Jourdan et al., 2024). Co-created, domain-specific guidelines enable confident, critical, and responsible AI use. This empowerment aligns with human-centric principles (HAICS Guide) and community engagement (Kotter et al., 2021). Because TLARA is a scaffold for institutions to build their own frameworks, it inherently supports global adaptability, allowing HEIs worldwide to define ethical AI use aligned with their unique values, sociocultural contexts, and regulatory environments (Appaya & Ng, 2024; UNESCO, 2023b).

Workbook Prep 6: Communicating the Value of Developing TLARA-Informed Guidelines

Reflection Question: How would you explain the strategic importance of your institution investing time and resources in collaboratively developing its own custom AI ethical guidelines (using the TLARA structure) to a skeptical member of the Board of Trustees or a faculty senate? What are the key benefits for risk prevention, compliance, innovation, and overall mission alignment, particularly for an HEI operating in a specific global context? How does this process specifically empower individuals within your own role or unit?

UX Workbook Link: The practical importance is demonstrated through completing Chap. 16 UX Workbook activities (e.g., governance planning, policy mapping). Section 8 specifically focuses on developing the Recognition & Excellence Framework, which involves communicating value.

PALS Workbook Link & AI Partnership Guidance: Use Chap. 16 PALS Workbook prompts corresponding to UX Workbook Section 8 to leverage GenAI for drafting these communications.

Example PALS Prompt: "Ask your GenAI partner to draft three key talking points for presentation to the Board of Trustees, outlining the value proposition of the institution investing in the collaborative development of custom AI ethical guidelines using the TLARA structure. Emphasize benefits such as risk mitigation, ensuring compliance with evolving standards (e.g., ACCJC, global trends), supporting mission alignment, and fostering responsible innovation, referencing HCASR Component 1 (Ethical Goal Setting) and Component 5 (Strategic Policy)."

Human Oversight Reminder: Talking points must be tailored to the specific audience's priorities and concerns. Support arguments with concrete examples or data where possible, and ensure delivery is by trusted leaders who can speak with clarity and conviction (HCASR C5).

Strategic Recommendations for Long-Term Governance Based on Custom Frameworks Using the TLARA Structure

HEIs should consider these adaptable long-term strategic recommendations as part of their change management strategy, derived from principles guiding custom framework development with the TLARA structure. These provide a foundation for sustained ethical AI integration.

Governance

Collaboratively develop, implement, enforce, and continuously update comprehensive AI governance guidelines derived through their process utilizing the TLARA structure, tailored to their unique institutional and sociocultural context. Establish clear policies overseen by permanent, representative governance structures (DerSimonian & Montagnino, 2025; Dion & Rattigan, 2025; RNL, n.d.; Wu et al., 2024; UC Davis AI Council, 2025).

Training

Implement mandatory, role-specific, ongoing AI literacy, ethics (based on custom TLARA-informed guidelines), and fair use training. Training must evolve, covering tool usage, critical evaluation, institutional ethical implications, HAIC principles, and policies (Digital Education Council, 2025; Higher Ed Dive, 2024; Schroeder, 2025; U.S. Department of Education, OET, 2024).

Integration

Systematically embed approved AI tools and custom TLARA-informed ethical guidelines into curricula, research, administrative processes, and procurement. Foster cross-disciplinary collaboration and responsible innovation guided by the institution's unique framework (Fairfax, 2025; Higher Ed Dive, 2024; Wargo & Anderson, 2024; Wynants et al., 2025).

Oversight

Prioritize AI for augmenting human judgment, not solely automating high-consequence decisions. Maintain clear human accountability in critical TLARA-domain processes. Implement robust mechanisms for human review and intervention in high-stakes AI use, ensuring meaningful oversight (Huron Consulting Group, 2025; Stone, 2025; U.S. Department of Education, OET, 2024; Wynants et al., 2025).

Adopting and adapting these recommendations helps institutionalize ethical AI, representing actionable steps for "Instituting Long-Term Cultural Change" (Kotter et al., 2021) and achieving feasible, desirable, and viable AI integration (Brown, 2019).

Workbook Prep 7: Implementing Strategic Recommendations

Reflection Question: Regarding the four strategic recommendations, which presents the most significant implementation challenge for your institution over the next 5–10 years if it were to fully adopt your custom, TLARA-informed guidelines tailored to its unique context? What concrete first step could your specific unit or department take to begin addressing the "Integration" recommendation in line with these guidelines, potentially using GenAI as a strategic partner for planning or resource identification?

UX Workbook Link: The Chap. 16 UX Workbook provides the structure for planning the implementation of these recommendations through activities related to governance (Sect. 1), framework integration (Sect. 1), leadership (Sect. 2), policy (Sect. 3), training (Sect. 5), and sustainability (Sect. 5).

PALS Workbook Link & AI Partnership Guidance: Utilize the Chap. 16 PALS Workbook prompts corresponding to relevant UX Workbook sections (especially Section 5 on the HAIC Playbook and Sustainability Roadmap) to explore implementation strategies with GenAI.

Example PALS Prompt: "Select the 'Integration' recommendation. For the [e.g., Undergraduate Business Curriculum] context, ask your GenAI partner to suggest two practical examples of how faculty could begin to embed ethical AI use, based on your institution's developing TLARA-informed guidelines (focusing on principles like bias awareness and data privacy for student projects), referencing HCASR Component 4 (Human-Centric Implementation)."

Human Oversight Reminder: Implementation of strategic recommendations requires sustained commitment, allocation of resources, and broad collaboration. Use AI for brainstorming initial ideas, but detailed planning, resource negotiation, and practical application guided by institutional guidelines require human judgment and leadership from relevant committees and departments (HCASR C4 & C5).

Users Outcomes & Deliverables (UODs) for Long-Term AI Governance Planning

Strategic planning in this chapter, operationalized via the integral Chap. 16 UX Workbook (and supported by the PALS workbook), should yield key User Outcomes and Deliverables (UODs). These co-created documents formalize the institution's long-term ethical AI approach, reflecting custom guidelines developed using the TLARA structure and human-centric frameworks. Core UODs (see Chap. 16 UX Workbook, Section 7, Table 16.12 for details) include:

- Long-Term AI Governance & Ethical Integration Plan (V1)
- Framework Integration Document (TLARA Focus - V1)
- AI Roles & Competencies Document (Updated)
- Master AI Plan & Certified GenAI Ethics & Literacy Training Modules (V1)
- Leadership & Policy UX Summary (V1)
- Policy Mapping Matrix (V1)
- AI Sustainability Roadmap (e.g., 5 to 10-Year Horizon - V1)
- Institutional AI Maturity Model (V1)
- HAIC TLA & Research Playbook (V1)
- Recognition & Excellence Framework (V1)

These "living documents," subject to regular review, provide structure for sustained, ethical AI integration and cultivating an AI-literate campus culture (Chap. 17). They are the tangible outcomes of HEIs partnering with AI strategically to create solutions aligned with their unique needs.

Long-Term Scaling, Sustainability & AI Governance: Institutionalizing Your Custom TLARA Frameworks

Chapter 16 has guided you into the long-term strategic horizon, emphasizing the institutionalization of customized AI ethics frameworks and robust, permanent AI governance. This advanced stage moves beyond initial planning and piloting to embed ethical AI practices deeply within the operational and sociocultural context of your institution, ensuring that your approach to AI is both strategic and sustainable (EDUCAUSE, 2023; Mollick & Tambe, 2024). The development of TLARA-informed guidelines, as explored throughout this book, culminates here in establishing enduring structures and processes that promote responsible innovation and human oversight (Shneiderman, 2022). Both the Chap. 16 UX and PALS Workbooks are integral tools designed to help you ethically partner with AI to solidity these long-range plans, creating dellverables tailored to your HEI's unique context.

HEIs are evolving ecosystems, and their AI governance must be designed for continuous adaptation and foresight (IDEAS, 2025). The PALS Workbook for Chap. 16, through its section "Long-Term Scaling, Sustainability & AI Governance: Iteratively Developing Your Custom TLARA-Informed AI Guidelines," provides a customizable prompt. This tool positions your GenAI as an expert strategic partner to help you translate your learning into concrete, long-term ethical frameworks and governance models for each TLARA domain. Engaging with "Topics for PALS TLARA Focus"—like "Institutionalizing TLARA-Informed Ethical Frameworks" and "Establishing Permanent and Adaptive AI Governance"—ensures your "Living Document" transforms into a cornerstone of your institution's lasting commitment to ethical AI, fostering a mature AI literacy culture (DerSimonian & Montagnino, 2025) and empowering your HEI to create solutions truly aligned with its global and sociocultural needs.

References

Accrediting Commission for Community and Junior Colleges (ACCJC). (2020). *ACCJC 2030 Strategic Plan*. ACCJC. https://accjc.org/wp-content/uploads/ACCJC-2030-Strategic-Plan.pdf

Accrediting Commission for Community and Junior Colleges (ACCJC). (2024). *2024 Accreditation Standards*. ACCJC.

American Bar Association Standing Committee on Ethics and Professional Responsibility. (2024, July 29). *Formal opinion 512: Generative artificial intelligence tools*. American Bar Association. https://www.americanbar.org/groups/professional_responsibility/publications/formal_opinions/512/

Appaya, S., & Ng, J. (2024). *Global trends in AI governance: Evolving country approaches*. World Bank Group.

Association of American Medical Colleges. (n.d.). Artificial intelligence in academic medicine: Principles and guidelines. *AAMC*. Retrieved April 25, 2025, from https://www.aamc.org/what-we-do/mission-areas/medical-research/artificial-intelligence

Bobula, M. (2024). Generative artificial intelligence (AI) in higher education: A comprehensive review of challenges, opportunities, and implications. *Journal of Learning Development in Higher Education, 30*. https://doi.org/10.47408/jldhe.vi30.1137

Bronfenbrenner, U., & Morris, P. A. (2006). The bioecological model of human development. In R. M. Lerner & W. Damon (Eds.), *Handbook of child psychology: Theoretical models of human development* (Vol. 1, 6th ed., pp. 793–828). John Wiley & Sons.

Brown, T. (2019). *Change by design: How design thinking transforms organizations and inspires innovation* (Revised & Updated ed.). Harper Business.

Carmody-Bubb, M. (2023). *Cognition and decision making in complex adaptive systems: The human factor in organizational performance*. Springer.

Coffey, L. (2024, June 25). A new digital divide: Student AI use surges, leaving faculty behind. *Inside Higher Ed*. https://www.insidehighered.com/news/tech-innovation/artificial-intelligence/2024/06/25/digital-divide-students-surge-ahead

Cornell University Task Force on GenAI in Academic Research. (2023). *Generative AI in academic research: Perspectives and cultural norms*. Cornell University Office of the Vice President for Research and Innovation.

Deloitte. (2025, January). *The state of generative AI in the enterprise: Now decides next (Quarter four report)*. Deloitte.

DerSimonian, R., & Montagnino, C. (2025, March 26). Crafting thoughtful AI policy in higher education: A guide for institutional leaders. *Campus Technology*. https://campustechnology.com/articles/2025/03/26/crafting-thoughtful-ai-policy-in-higher-education-a-guide-for-institutional-leaders.aspx

Digital Education Council. (2025). *Empowering faculty to experiment, share and learn: Advancing AI literacy in higher education. Digital Education Council Blog*.

Dion, K. E., & Rattigan, K. (2025, February 11). AI is transforming higher ed: What you need for compliance and governance. *University Business*. https://universitybusiness.com/ai-is-transforming-higher-ed-what-you-need-for-compliance-and-governance/

EDUCAUSE. (2023, October 23). *EDUCAUSE AI landscape analysis: Current and future states of AI in higher education*. EDUCAUSE.

EDUCAUSE. (2024). *2024 EDUCAUSE Horizon Report: Teaching and learning edition*. https://library.educause.edu/resources/2024/5/2024-educause-horizon-report-teaching-and-learning-edition

Ellucian. (2024, December 16). AI in higher education: Implementation strategies for institutions and tech partners. *Higher Ed Dive*. https://www.highereddive.com/spons/ai-in-higher-education-implementation-strategies-for-institutions-and-tech/735434/

Fairfax, A. (2025). Responsible innovation and cross-disciplinary collaboration in the age of AI. *Higher Education Quarterly, 39*(1), 45–58. https://doi.org/10.1111/hequ.12345

Francis, N. J., Jones, S., & Smith, D. P. (2025). Generative AI in higher education: Balancing innovation and integrity. *British Journal of Biomedical Science, 81*, 14048. https://doi.org/10.3389/bjbs.2024.14048

Guidry, K. R. (n.d.). *AI literacy literature summary*. CSU AI Commons. Retrieved April 25, 2025, from https://genai.calstate.edu/communities/faculty/ai-literacy-literature-summary

Higher Ed Dive. (2024, March 15). Strategies for scaling AI literacy across campus. *Industry Dive*. https://www.highereddive.com/news/strategies-scaling-ai-literacy-campus/2024/

Higher Education Strategy Associates (HESA). (2024, October 18). *HESA's AI observatory: What's new in higher education*. Higher Education Strategy Associates. https://higheredstrategy.com/hesas-ai-observatory-whats-new-in-higher-education-october-18-2024/

Holzinger, A., Keiblinger, K., Holub, P., Zatloukal, M., & Simonic, M. (2025a). The explainability (XAI) pillar: A survey of the foundations of XAI. In *Trustworthy AI* (pp. 1–103). Springer.

Holzinger, A., Zatloukal, K., & Müller, H. (2025b). Is human oversight to AI systems still possible? *New Biotechnology, 80*, 59–62. https://doi.org/10.1016/j.nbt.2024.12.003

HSD Author Collective. (2025). AI ethics: Contextual frameworks and domain-specific concerns. *Humanities & Social Sciences Discussions, 7*(1), 1–14. https://hsd.ardascience.com/index.php/journal/article/view/1224

Huron Consulting Group. (2025). *Responsible AI for higher education and research.* https://www.huronconsultinggroup.com/insights/responsible-ai-higher-education

Husch Blackwell. (2024, January 5). How to approach student AI use policies in higher education. *Husch Blackwell.* https://www.huschblackwell.com/newsandinsights/how-to-approach-student-ai-use-policies-in-higher-education

Instructional Design, Engagement, and Support (IDEAS). (2025, January 27). *AI's impact on higher education through the eyes of faculty: 4 key trends.* University of Massachusetts Amherst.

ISACA Now. (2024, July 24). AI ethics and why it matters. *ISACA Now Blog.* https://www.isaca.org/resources/news-and-trends/isaca-now-blog/2024/ai-ethics-and-why-it-matters

Jourdan, Z., Bilgin, O., & Campbell, M. (2024). Designing human-AI interactions to support human agency. *International Journal of Human-Computer Studies, 182*, 103171. https://doi.org/10.1016/j.ijhcs.2023.103171

Jumadinova, J. (n.d.). Cross-disciplinary approaches to artificial intelligence education in the liberal arts. *Allegheny College.* Retrieved April 25, 2025, from https://sites.allegheny.edu/jumadinova/research

Kim, J. S., Thu, K. Y., Ismayilzada, J., Park, J., Kim, E., Ahmad, H., An, N. M., Thorne, J., & Oh, A. (2025). *When Tom eats kimchi: Evaluating cultural bias of multimodal large language models in cultural mixture contexts.* arXiv. arXiv:2503.16826

Kotter, J. P., Akhtar, V., & Gupta, G. (2021). *Change: How organizations achieve hard-to-imagine results in uncertain and volatile times.* Wiley.

Le Dinh, T., Le, T. D., Uwizeyemungu, S., & Pelletier, C. (2025). Human-centered artificial intelligence in higher education: A framework for systematic literature reviews. *Information, 16*(3), 240. https://doi.org/10.3390/info16030240

Liaison International. (2024, December 20). *Demystifying AI: A guide to AI literacy in higher education enrollment management.* Liaison Resources Blog. https://www.liaisonedu.com/resources/blog/demystifying-ai-a-guide-to-ai-literacy-in-higher-education-enrollment-management/

Mollick, E., & Tambe, P. (2024, March 13). How AI and humans can work together in creative problem solving. *Harvard Business Review.*

National Institute of Standards and Technology (NIST). (2023). *Artificial Intelligence Risk Management Framework (AI RMF 1.0) (NIST AI 100-1).* U.S. Department of Commerce.

Organisation for Economic Co-operation and Development (OECD). (2023a). *Artificial Intelligence in Education: Key Insights from PISA and TALIS*. OECD Publishing.

Organisation for Economic Co-operation and Development (OECD). (2023b). *Artificial Intelligence in Science: Challenges, opportunities and the future of research*. OECD Publishing. https://doi.org/10.1787/a8d820bd-en

Organisation for Economic Co-operation and Development (OECD). (n.d.-a). *Artificial intelligence*. OECD Topics. Retrieved April 25, 2025, from https://www.oecd.org/en/topics/artificial-intelligence.html

Organisation for Economic Co-operation and Development (OECD). (n.d.-b). *OECD AI Principles overview*. OECD.AI Policy Observatory.

Pedreschi, D., et al. (2023). *Human-AI coevolution* [Preprint]. arXiv. https://doi.org/10.48550/arXiv.2306.13723

Ruffalo, N. L. (n.d.). Navigating the AI landscape in enrollment management: Governance and strategy. *RNL*. Retrieved April 25, 2025, from https://www.ruffalonl.com/ai-enrollment-resources

Sain, C. K., Mandloi, N. S., & Jadhav, P. M. (2025). Ethical issues in artificial intelligence. In *Artificial intelligence trends for data analytics using machine learning and deep learning approaches*. IGI Global.

Schroeder, R. (2025, February 5). The imperative for AI literacy in the new academic era. *Inside Higher Ed*. https://www.insidehighered.com/opinion/views/2025/02/05/imperative-ai-literacy-new-academic-era

Shneiderman, B. (2022). *Human-centered AI*. Oxford University Press.

State University of New York. (n.d.). Fact sheet: Artificial intelligence. *SUNY*. Retrieved April 25, 2025, from https://www.suny.edu/about/factsheets/artificial-intelligence

Stone, A. (2025, March 18). AI ethics in higher education: How schools are proceeding. *EdTech Magazine: Focus on Higher Education*. https://edtechmagazine.com/higher/article/2025/03/ai-ethics-higher-education-how-schools-are-proceeding-perfcon

Tao, Y., Viberg, O., Baker, R. S., & Kizilcec, R. F. (2024). Cultural bias and cultural alignment of large language models. *PNAS Nexus, 3*(9), pgae346. https://doi.org/10.1093/pnasnexus/pgae346

The World Bank. (2024). *Artificial intelligence for development: Emerging opportunities and challenges*. The World Bank. http://documents.worldbank.org/curated/en/099120224205026271/pdf/P1786161ad76ca0ae1ba3b1558ca4ff88ba.pdf

Tveit, K. C. (2025, February 20). A 5-step guide to creating a content governance strategy (with AI in mind). *Bigtincan*. https://www.bigtincan.com/resources/guide-to-creating-a-content-governance-strategy-with-ai/

U.S. Department of Education, Office of Educational Technology (OET). (2024, January). *Navigating artificial intelligence in postsecondary education: Building capacity for the road ahead*. U.S. Department of Education. https://portal.ct.gov/

das/-/media/das/ctedtech/publications/2025/2025-used-oet-archive/os-24-002833-ai-in-postsecondary-ed.pdf

UC Davis AI Council. (2025, February). *Campus-Based principles and proposed governance structures to support responsible use of artificial intelligence.* University of California, Davis.

UC Davis Leadership. (n.d.). *UC Davis Artificial Intelligence (AI) Council.* Retrieved April 25, 2025, from https://leadership.ucdavis.edu/about/councils-committees-boards/articifial-intelligence

UNESCO. (2021). *Recommendation on the ethics of artificial intelligence.* UNESCO Publishing. https://www.unesco.org/en/articles/recommendation-ethics-artificial-intelligence

UNESCO. (2023a). *Guidance for generative AI in education and research.* UNESCO. https://unesdoc.unesco.org/ark:/48223/pf0000386693

UNESCO. (2023b). *How generative AI is reshaping education in Asia-Pacific.* UNESCO.

Verhoeven, B., & Hor, T. (2025, February 12). A framework for human-centric AI-first teaching. *AACSB Insights.* https://www.aacsb.edu/insights/articles/2025/02/a-framework-for-human-centric-ai-first-teaching

Wargo, K., & Anderson, B. (2024, December 5). Striking a balance: Navigating the ethical dilemmas of AI in higher education. *EDUCAUSE Review.* https://er.educause.edu/articles/2024/12/striking-a-balance-navigating-the-ethical-dilemmas-of-ai-in-higher-education

Wu, D., Dempsey, P., Lin, C.-Y., Sharif, A., & Miller, L. (2024). Charting the course for AI governance in higher education: Insights from a system-wide task force. *Journal of Postsecondary Education and Disability, 37*(1), 7–24.

Wynants, S., Childers, G., De La Torre Roman, Y., Budar-Turner, D., & Vasquez, P. (2025, February). *ETHICAL principles AI framework for higher education.* California State University, Fullerton. https://fdc.fullerton.edu/_resources/pdfs/teaching/ethical-principles-ai-framework-for-higher-education-february-2025.pdf

Xie, B., Chen, T., Dameff, C., Dou, W., Fan, X., Fields, D. A., & Wang, M. (2024). Co-designing AI education curriculum with cross-disciplinary high school teachers. *Proceedings of the AAAI Conference on Artificial Intelligence, 38*(21), 23725–23733. https://doi.org/10.1609/aaai.v38i21.30360

17

Building an Enduring Culture of AI Literacy (Phase 4)

Introduction: Cultivating a Pervasive Culture of AI Literacy

The increasing integration of Artificial Intelligence (AI)—systems simulating human intelligence (Wang et al., 2025)—into society marks a significant transformation for higher education. Generative AI (GenAI), capable of producing novel content (Zheng et al., 2025), signals a move towards a Fifth Industrial Revolution (5IR) defined by enhanced human-machine and Human-AI Collaboration (HAIC), intended to augment human capabilities (Ziatdinov et al., 2024). While AI adoption can raise concerns about automation, this transition demands a proactive, human-centric approach to AI development and application, empowering individuals and enhancing their skills (Shneiderman, 2022). As detailed in Chap. 16, navigating this transition effectively requires HEIs to cultivate an institutional environment where AI literacy is fundamental for all members (Adobe, 2025; Lo, 2025), signifying a deep, systemic integration beyond surface-level training (Brady, 2024).

Building on the long-term governance structures from Chap. 16, this chapter focuses on embedding AI literacy into the institution's daily practices and organizational setting. Multifaceted AI literacy encompasses technical understanding, critical evaluation of AI tools and outputs, effective and ethical application, and equity awareness, including bias mitigation (Wang et al., 2025; Lo, 2025) (Fig. 17.1).

Fostering this culture requires institutions to critically evaluate and adapt comprehensive frameworks, such as EDUCAUSE's AI Literacy in Teaching and Learning (ALTL) (EDUCAUSE/Guidry, 2024) or UNESCO's AI Competency Frameworks (Miao

---

Supplementary Information The online version contains supplementary material available at https://doi.org/10.1007/978-3-032-05205-6_17.

D. Woodring, *Advancing Higher Education with Human-Centric AI Partnerships*,
https://doi.org/10.1007/978-3-032-05205-6_17

et al., 2024; Miao & Cukurova, 2024), to their specific context. The variety of frameworks highlights that AI literacy is an evolving concept. Institutions must therefore thoughtfully select and adapt a framework tailored to their unique needs, goals, objectives, desired user outcomes and deliverables (UODs), and sociocultural contexts (Gu & Ericson, 2025; Seung & Basham, 2024). Core principles for cultivating this institutional capacity include fostering collaboration, implementing safeguards against bias (Southgate, 2020), maintaining transparency, and promoting effective HAIC where AI augments human judgment (Kolbjørnsrud et al., 2023).

Critical evaluation and ethical application also demand perspectives from the humanities and social sciences (Michalak & Ellixson, 2025). The objective is to establish AI literacy as a core competency for all stakeholders (Lo, 2025), preparing them for an AI-infused future. Linking literacy initiatives to multilevel governance ensures strategic backing and resource allocation, facilitating long-term sustainability (DerSimonian & Montagnino, 2025).

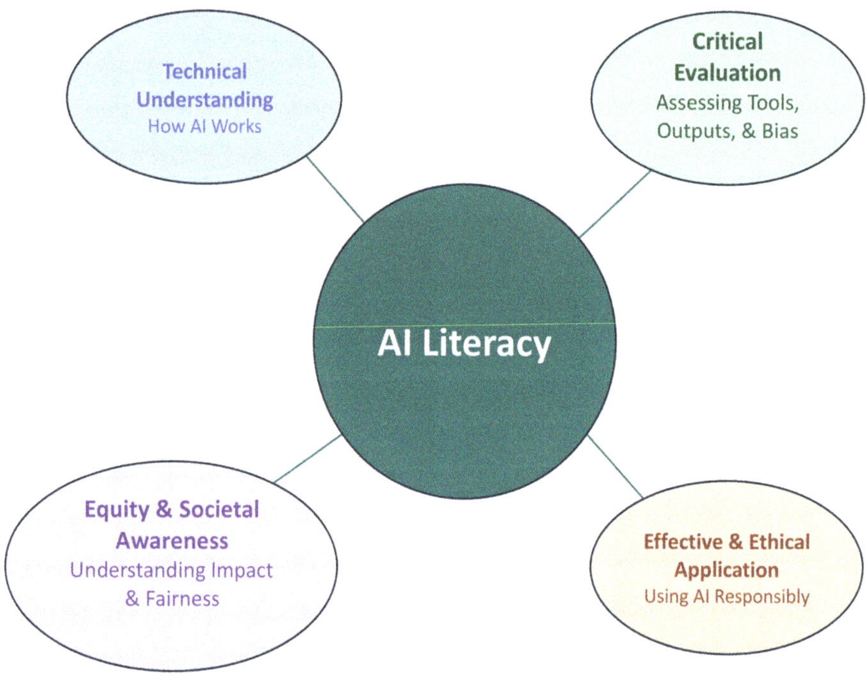

Fig. 17.1 Core components of AI literacy. (*Note*: A diagram with a central hub labeled "AI Literacy" connected to four segments: Technical Understanding, Critical Evaluation, Effective & Ethical Application, and Equity & Societal Awareness, outlining core components)

Workbook Prep 1. Envisioning an AI-Literate Institution

Reflection: What does a successful institutional environment fostering AI literacy look like in your specific context? What observable behaviors (among students, faculty, staff) would indicate this environment exists?

UX Workbook Link: See Chap. 17 UX Workbook, Activity 1.1 ("Defining Institutional AI Literacy Vision & Goals").

PALS Workbook Link & AI Partnership Guidance: Use Chap. 17 PALS Workbook prompts (Sec 1: Literacy Culture Vision). Input your institution's mission/priorities; ask your AI partner to brainstorm characteristics of the desired AI literacy environment that align with your values (referencing HCASR C1).

Example Prompt: "Our institutional priorities include [Value 1], [Value 2]. Referencing HCASR C1, brainstorm 3 characteristics of an institutional AI literacy environment aligned with these values (e.g., promoting AI for equity, fostering ethical AI research)."

Human Oversight Reminder: AI can suggest generic traits. Defining a meaningful vision requires human understanding of institutional identity, values, strategic aspirations, and stakeholder needs (HCASR C1 & 5).

Designing Institution-Wide AI Literacy Initiatives

An effective AI literacy environment requires a strategic, institution-wide approach, moving beyond isolated workshops. This involves assessing current needs and institutional readiness, selecting or adapting literacy frameworks, and integrating literacy goals into the broader institutional AI strategy. This positions AI literacy development as a significant institutional change initiative demanding thoughtful management (Brady, 2024; DerSimonian & Montagnino, 2025).

Needs Assessment for Institutional Readiness

Understanding the existing landscape is crucial before launching initiatives. This involves assessing current AI literacy levels (knowledge, skills, attitudes) and institutional readiness across diverse stakeholder groups (Ellucian, 2025; Campus Technology Contributor, 2025) using methods like surveys, focus groups, interviews, and analysis of existing feedback. Identifying specific gaps—such as in technical understanding, ethical awareness, practical skills (Adobe, 2025), or institutional factors like resistance to change—allows for targeted program design. This aligns with design thinking's emphasis on user context. Proactively addressing institutional factors through change management strategies is vital for buy-in and successful implementation (Brady, 2024).

Adopting and Adapting AI Literacy Frameworks

A clear institutional understanding of core AI competencies, guided by established frameworks, facilitates comprehensive AI literacy (Seung & Basham, 2024). Frameworks like EDUCAUSE's ALTL or UNESCO's AI Competency Frameworks offer valuable starting points by typically covering AI principles, critical evaluation (bias, impact), effective and ethical application, and societal implications. Rather than providing extensive summaries here, the focus is on the institutional process: HEIs must select, adapt, and apply these frameworks contextually. This involves choosing or customizing a framework that aligns with the institution's specific context, values, strategic goals, and stakeholder needs (EDUCAUSE/Guidry, 2024). This selection process acknowledges AI literacy as an evolving field (Gu & Ericson, 2025; Seung & Basham, 2024) and ensures the chosen framework is adaptable. The selected framework then provides a common language and structure for developing curriculum, training, and assessment across the institution (Seung & Basham, 2024) (Table 17.1).

Strategic Planning for Literacy Integration

Building institutional AI literacy capacity requires integration into the overarching strategic AI plan (Brady, 2024), connecting literacy initiatives with governance (Chap. 16) and sustainability roadmaps (DerSimonian & Montagnino, 2025). Strategic planning involves setting clear institutional AI competency goals, allocating resources, defining timelines, and establishing key performance indicators (KPIs) (Campus Technology Contributor, 2025). These metrics should link to the institution's AI Maturity Model (UOD 8, Chap. 16) to track literacy growth. Frameworks for HEIs (Jisc, 2024) or other sectors (Veritis, 2024; Willems, 2025) can inform this. Placing literacy planning under the AI Governance Council ensures alignment and resource

Table 17.1 Comparison of selected AI literacy framework dimensions

Feature/ dimension	EDUCAUSE ALTL framework (EDUCAUSE/Guidry, 2024)	UNESCO AI competency frameworks (students & teachers) (Miao et al., 2024; Miao & Cukurova, 2024)
Primary focus	Teaching and Learning contexts within Higher Education.	Broad educational contexts (potentially K–12 through Higher Ed/Lifelong); separate frameworks for students and teachers.
Core components/ dimensions	"Often conceptualized around understanding AI fundamentals, critical evaluation, effective/ethical application, managing AI tools, awareness of societal impact. (Specifics vary as framework evolves)."	"Students: Understanding AI, Using AI Effectively & Responsibly, AI Ethics & Human Agency, Creating with AI. Teachers: Understanding AI & Education, Curriculum & Assessment, Pedagogy, Administration, Professional Learning."
Emphasis on ethics/equity	Strong emphasis on ethical usage, bias awareness, and responsible application.	"Central focus on human rights, ethics, inclusion, diversity, gender equality, human agency throughout competencies."
Technical depth	Generally focuses on conceptual understanding and practical application over deep technical expertise.	Varies by competency; includes understanding basic principles but also potentially advanced application/creation depending on level/role.
Pedagogical application	Explicit focus on integrating AI into teaching practices and curriculum design.	"Teacher framework explicitly addresses AI-enhanced pedagogy, assessment strategies, and supporting diverse learners. Student framework focuses on using AI for learning."
Intended competency level	"Applicable across various levels (foundational to advanced) for faculty, staff, students within HEI roles."	Often structured with proficiency levels; aims to guide development from basic understanding to more sophisticated application/ creation/pedagogical integration.

Note. This table provides a high-level comparison based on available summaries and framework documents. Institutions should consult the primary sources and conduct their own analysis to select or adapt the framework best suited to their specific context, mission, and goals

advocacy (DerSimonian & Montagnino, 2025). Utilizing frameworks like the synergy between design thinking and change management, and involving diverse stakeholders, fosters buy-in and ensures plans are feasible and relevant (DerSimonian & Montagnino, 2025).

Workbook Prep 2. Framing the Institutional Literacy Strategy

Reflection: Based on your institutional AI vision and governance plan (Chap. 16), what are the top 1–2 strategic priorities for building AI literacy capacity in the next 1–3 years? Justify your choices.

UX Workbook Link: See Chap. 17 UX Workbook, Activity 1.2 ("Selecting/Customizing AI Literacy Competency Frameworks") & 1.3 ("Outlining Long-Term AI Literacy Integration Plan").

PALS Workbook Link & AI Partnership Guidance: Use Chap. 17 PALS Workbook prompts (Sec 1: Literacy Strategy Planning). Input your selected framework/priorities; ask your AI partner for measurable objectives for the Literacy Integration Plan that align the framework and goals (referencing HCASR C1 & C5).

Example Prompt: "Our selected framework emphasizes [Dimension 1, Dimension 2]. Our top priority is [X]. Referencing HCASR C1 & C5, draft 2 SMARTIE objectives for our AI Literacy Integration Plan connecting the framework and priority for the next 2 years."

Human Oversight Reminder: AI can assist in formulating objectives. Ensuring they are strategic, measurable, achievable, relevant, time-bound, inclusive, and equitable (SMARTIE) requires human planning expertise, leadership validation, and stakeholder consultation (HCASR C1 & 5).

Implementing Foundational AI Literacy Training

Establishing baseline AI literacy across the institution is vital for a shared understanding and responsible AI use. This involves integrating foundational training into onboarding and early professional development, establishing common vocabulary, and setting initial expectations for responsible AI interaction (Ellucian, 2024).

Onboarding Programs for Students and Employees

Integrating AI literacy into standard onboarding provides foundational knowledge from the start (Ellucian, 2024). Brief modules should introduce core AI concepts, common campus AI applications, fundamental ethical principles (bias, privacy, academic integrity) (Wang et al., 2025), institutional policies, and the HAIC concept. Training should emphasize critical engagement and responsible use (DebiasMe Contributors, 2025; University of Lincoln, 2025), aiming for baseline awareness of responsible AI interaction within the institutional context (Wang et al., 2025).

Foundational Faculty Development on AI Ethics & Pedagogy

Faculty require specific foundational training on AI's implications for teaching, research, and professional responsibilities (Mercer University Center for Teaching and Learning, n.d.). Initial workshops should address pedagogical benefits and ethical pitfalls (Wang et al., 2025). Guidance on critically evaluating AI tools for pedagogy (EDUCAUSE/Guidry, 2024), understanding institutional AI academic integrity policies (Gander & Harris, 2024), and navigating ethics in AI-assisted research are essential (Gander & Harris, 2024). This equips faculty to make informed decisions about AI integration (Danault, 2025; Gander & Harris, 2024).

Foundational Staff Training on Policy and Responsible Use

Staff in various units need role-tailored foundational training, especially if handling sensitive data or using AI in operational workflows (Microsoft, 2025). Training should prioritize understanding institutional policies on data security, privacy, and acceptable AI use (Ellucian, 2024). Awareness of compliance basics (Generation AI, 2025; Element451, 2025) and recognizing operational risks, like bias in AI-driven systems (Southgate, 2020), are critical. Guidelines for responsible use of approved AI tools to enhance efficiency while upholding ethics and security are central.

Emphasizing Human-AI Collaboration (HAIC) from the Start

Across all foundational training, the HAIC principle is essential (University of Lincoln, 2025). AI must be framed as a tool requiring human guidance, critical evaluation, and ethical oversight, not an autonomous replacement for human judgment (Kolbjørnsrud et al., 2023). Training should highlight AI's strengths (Zheng et al., 2025) and weaknesses (EDUCAUSE/Guidry, 2024; Rulandari & Silalahi, 2025*). Introducing basic HAIC strategies, like effective prompting, iterative refinement, and cross-checking AI-generated information (EDUCAUSE/Guidry, 2024), provides practical skills. Promoting metacognitive awareness—encouraging reflection on AI interaction and output quality (DebiasMe Contributors, 2025)—reinforces human agency (Rulandari & Silalahi, 2025*) and aligns with change management by building competence (Kotter et al., 2021). The HAICS Guide provides a detailed framework for such collaboration.

Workbook Prep 3. Designing Foundational Training

Reflection: What single most important AI concept, skill, or ethical principle should every member of your institutional community understand after completing foundational training? Why is this element crucial for your specific context?

UX Workbook Link: See Chap. 17 UX Workbook, Section 2 ("Designing Foundational Training Initiatives").

PALS Workbook Link & AI Partnership Guidance: Use Chap. 17 PALS Workbook prompts (Sec 2: Foundational Training Design). Select a target audience (students, faculty, or staff) and input a core learning objective. Ask your AI partner to help outline key points and potential activities for a brief foundational training module focused on that objective (referencing HCASR C5).

Example Prompt: "Objective (Undergraduate Module): Understand the importance of critically evaluating AI-generated information. Referencing HCASR C5, outline 3 key points and 1–2 brief activities for a 30-minute online module addressing this."

Human Oversight Reminder: AI can suggest standard outlines. Designing effective training requires human expertise in pedagogy, instructional design, and understanding the specific audience's prior knowledge and needs. It also necessitates ensuring alignment with institutional policies and ethical guidelines (HCASR C4 & 5).

Embedding Applied AI Literacy Across the Institution

Lasting institutional capacity requires systematically embedding applied AI literacy skills and critical perspectives into core academic and professional activities, moving beyond awareness to active, contextualized engagement (Lo, 2025). This involves curricular integration, role-specific training, and robust competency assessment (EDUCAUSE/Guidry, 2024), all aligned with the institution's AI literacy framework (UOD 1).

Integrating Literacy into Curriculum

Widespread student AI literacy necessitates direct curricular integration (Michalak & Ellixson, 2025). Strategies include dedicated modules in general education or first-year programs (Yoon, 2024; Hazari, 2024). More impactfully, institutions should support departments in integrating discipline-specific AI literacy into existing courses (EDUCAUSE/Guidry, 2024; Salhab, 2024). This contextualized approach enhances relevance (Michalak & Ellixson, 2025). Examples include assignments requiring ethical AI use for discipline-specific research or creative work; activities focused on critically evaluating AI applications relevant to their field (Lo, 2025; Tzirides et al., 2024); or case studies on AI's ethical implications in future professions (Hazari, 2024). Effective integration requires support from teaching centers, instructional designers, and libraries (Mercer University Center for Teaching and Learning, n.d.).

Developing Role-Specific Applied AI Literacy Modules

Different HEI roles need distinct applied AI skills, requiring targeted professional development (Ellucian, 2024). These modules should build on foundational training and align with the institutional AI literacy framework.

- **For Faculty:** Training on advanced pedagogical AI uses (e.g., AI-integrated assignments fostering critical thinking, AI for personalized feedback) (Mercer University Center for Teaching and Learning, n.d.); leveraging AI for research; strategies for AI-resistant assessments or ethically evaluating AI in student work (Rulandari & Silalahi, 2025*).
- **For Staff:** Workshops on using approved AI for functions like student advising (Ellucian, 2024); training on responsible administrative task automation (Microsoft, 2025); skills for managing AI vendors (Ellucian, 2024).
- **For Administrators/Leadership:** Seminars on AI for strategic decision-making (Brady, 2024); managing large-scale AI deployment risks; overseeing AI governance (DerSimonian & Montagnino, 2025).

Assessing AI Literacy Competencies

Robust assessment is essential for understanding initiative effectiveness and tracking progress (EDUCAUSE/Guidry, 2024). Given emerging universal assessment instruments (Gander & Harris, 2024; Gu & Ericson, 2025) and AI literacy's multifaceted nature (Wang et al., 2025), a mix of methods is recommended.

- **Surveys:** Pre/post-training self-assessments (EDUCAUSE/Guidry, 2024); adapted validated scales like MAILS (Carolus et al., 2023).
- **Knowledge Tests:** Assessing understanding of AI concepts, ethics, or policies (EDUCAUSE/Guidry, 2024); objective tests like AICOS or GLAT (Markus et al., n.d.; Jin et al., 2024) could be adapted.
- **Performance-Based Tasks:** Applying AI skills, evaluating outputs, analyzing cases, or creating AI-informed artifacts (EDUCAUSE/Guidry, 2024); task-based assessments like the "A-Factor" battery (Li et al., n.d.) offer models. For instance, assessing competencies through such tasks can be enhanced by applying HAICS principles like Ethical & Effective Goal Setting (EHCS from the HAICS Guide) to ensure tasks are well-defined and ethically sound.
- **Portfolio Reviews:** Analyzing work involving AI use against rubrics (EDUCAUSE/Guidry, 2024).
- **Rubrics:** Developing specific rubrics for AI literacy in projects (Zhou et al., 2025*).
- **Qualitative Methods:** Observations, interviews, or focus groups for deeper insights (EDUCAUSE/Guidry, 2024).

Assessment results should inform program refinement and individual learning pathways (Ellucian, 2024), contributing to tracking institutional progress via the AI Maturity Model (Jisc, 2024).

Workbook Prep 4. Planning Applied Literacy Integration

Reflection: Considering your specific role within the institution, what is one practical, applied AI skill that, if developed, would significantly enhance your effectiveness or allow you to better serve the institutional mission?

UX Workbook Link: See Chap. 17 UX Workbook, Section 3 ("Embedding Applied Literacy…").

PALS Workbook Link & AI Partnership Guidance: Use Chap. 17 PALS Workbook prompts (Sec 3: Applied Literacy Integration). Input your role and the desired applied skill. Ask your AI partner for learning activities or components for a potential training module focused on that skill (referencing HCASR C4 & C5).

Example Prompt: "Role: Faculty. Skill: Designing course assignments that ethically incorporate student use of GenAI tools. Referencing HCASR C4 & C5, brainstorm 2 workshop activities (e.g., analyze sample AI-integrated assignments, draft adaptable syllabus policies)."

Human Oversight Reminder: AI can generate activity ideas. Designing effective applied skill development requires deep pedagogical or professional expertise relevant to the role, understanding of the specific tools and contexts, nuanced ethical consideration, and skilled facilitation, all necessitating human design and oversight (HCASR C4 & 5).

Fostering Communities of Practice and Continuous Learning

A sustainable AI literacy environment thrives on ongoing learning, collaboration, and knowledge sharing. Accessible resources and vibrant communities of practice are key. This involves creating central hubs, supporting peer-led groups, empowering champions, and leveraging collaborative events.

Establishing AI Learning Hubs and Resources

A central, accessible repository (website or LMS section) is crucial for self-directed learning (Truckee Meadows Community College, AI & Emerging Tech Committee, n.d.). Hubs should provide curated tutorials, institutional policies, ethical guidelines (Harvard University, Office of Academic Integrity and Student Conduct, n.d.), best practices, training links, FAQs, and support contacts (Saint Mary's University, Studio for Teaching and Learning, n.d.; Truckee Meadows Community College, AI & Emerging Tech Committee, n.d.). Maintenance requires keeping content organized, user-friendly, accessible, and updated (Truckee Meadows Community College, AI & Emerging Tech Committee, n.d.). Consider incorporating "smart spaces" or "sandboxes" for experimentation.

Cultivating Cross-Disciplinary Communities of Practice (CoPs)

CoPs offer spaces for cross-disciplinary connection, strategy sharing, collaborative problem-solving, and collective expertise building (ASU Enterprise Technology, 2024). Institutions can support CoPs focused on themes like AI Ethics, Pedagogical GenAI Uses, or Discipline-Specific Applications by providing communication platforms, promotion, meeting spaces, resources, and potential small grants (ASU Enterprise Technology, 2024). CoPs foster organic, peer-driven learning (Ellucian, 2024). The design and facilitation of these CoPs can benefit from HAICS principles, particularly HCASR Component 4 (Human-Centric Implementation & Stakeholder Engagement), by ensuring participatory design and co-regulated activities.

Promoting Peer Mentoring and AI Champions

Empowering "AI Champions"—early adopters and enthusiastic AI users—within departments provides localized support (Chap. 7). Champions can mentor, share best practices, and provide feedback (University of Maryland, Baltimore, LEAPS Initiative, 2024). Institutions can recognize champions (UOD 10) and provide resources and training. Encouraging broader peer mentoring strengthens the learning network (Kolbjørnsrud et al., 2023). Change Management principles highlight such internal champions for driving adoption (Kotter et al., 2021).

Leveraging Design Thinking Events for Collaborative Innovation

Events like AI-focused hackathons or design sprints can stimulate collaboration and generate innovative solutions related to AI literacy, ethics, or application (Cornell University, 2025*; Kelly, 2025). Frameworks like the synergy between design thinking and change management can guide participants through empathizing, defining, ideating, prototyping, and testing. These events build community, foster experimentation, enhance AI literacy, and make responsible AI innovation visible (Cornell University, 2025*). Institutions should encourage exploring various innovation frameworks aligned with their unique contexts.

Workbook Prep 5. Planning for Community Building

Reflection: Which community-building initiative (e.g., establishing the AI Learning Hub, launching a specific CoP, implementing the AI Champion program, hosting a design thinking event) do you believe would have the most significant immediate impact on fostering AI literacy culture within your specific unit, department, or the institution as a whole? Why?

UX Workbook Link: See Chap. 17 UX Workbook, Section 4 ("Fostering Learning Communities…").

PALS Workbook Link & AI Partnership Guidance: Use Chap. 17 PALS Workbook prompts (Sec 4: Community Building). Select one community initiative. Ask your AI partner for potential initial activities, discussion topics, or resource categories relevant to launching that initiative (referencing HCASR C4).

Example Prompt: "For launching a new Community of Practice focused on 'AI in the Humanities,' referencing HCASR C4, brainstorm 4 initial activities or discussion topics that would likely engage faculty and address relevant ethical and practical issues in their disciplines."

Human Oversight Reminder: AI can suggest generic activities or topics. Designing successful community initiatives requires human understanding of the target audience's specific interests and concerns, effective facilitation skills, strategies for creating inclusive and psychologically safe environments, and alignment with institutional support structures (HCASR C4 & 5).

Monitoring, Evaluating, and Sustaining the AI Literacy Culture

Building institutional AI literacy capacity is an ongoing process needing continuous monitoring, evaluation, and adaptation (Brady, 2024). This involves regular checkpoints, systematic feedback gathering, linking literacy to institutional assessment and maturity models, and securing long-term commitment (Jisc, 2024).

Checkpoints for Literacy Program Effectiveness

Regular checkpoints are necessary to assess AI literacy initiative progress, reach, and effectiveness (UOD 1) (Ellucian, 2024). These reviews should evaluate participation rates, stakeholder awareness of resources, impact on skills and practices (UOD 4), and alignment with strategic goals (Activity 1.1). Data sources include training records, Hub analytics, workshop feedback, CoP reports, and AI Champion input (Ellucian, 2024). Findings should inform program adjustments.

Gathering Feedback for Continuous Improvement

Ongoing, actionable feedback is crucial for refining AI literacy programs (Ellucian, 2024). Beyond formal checkpoints, institutions should use accessible channels like embedded surveys post-training (Northern Michigan University AI Literacy Initiative, n.d.), dedicated Hub feedback forms, CoP discussion forums, or specific questions in broader institutional surveys. Systematically synthesizing this feedback provides insights into user experiences, unmet needs, and successes (Ellucian, 2024). Closing the feedback loop by communicating how input informed changes maintain engagement.

Linking Literacy Culture to the Institutional AI Maturity Model

Progress in AI literacy culture should be tracked within the institution's AI Maturity Model (UOD 8, Chap. 16). This involves incorporating AI literacy metrics into the model's dimensions, using HEI frameworks (Jisc, 2024) or adapting others (Willems, 2025). Metrics could include training completion percentages, CoP participation, self-reported competency gains (UOD 1), AI integration in syllabi, and qualitative indicators of ethical awareness. Regularly assessing these metrics tracks institutional progress and demonstrates literacy initiative impact on AI readiness (Jisc, 2024).

Ensuring Long-Term Sustainability

Sustaining a vibrant AI literacy culture requires embedding efforts into ongoing operations and securing long-term commitment (Gu & Ericson, 2025). Key factors include stable funding in operational budgets; integrating AI literacy support into relevant job descriptions (CTL, IT, Libraries); establishing regular review cycles for training content, Hub materials, and policies; and explicitly including AI literacy goals in institutional strategic planning. Embedding AI literacy into routines, values, and priorities is key to its endurance (Ellucian, 2024). Navigating bureaucracy and budget cycles needs proactive planning from AI leadership (Willems, 2025).

Workbook Prep 6. Evaluating Literacy Initiatives

Reflection: Beyond tracking participation or completion rates, how could your institution effectively measure whether AI literacy initiatives are genuinely changing behaviors and fostering a more responsible and critical institutional culture regarding AI use? What qualitative indicators might be particularly valuable?

UX Workbook Link: See Chap. 17 UX Workbook, Section 5 ("Evaluating Literacy Initiatives…").

PALS Workbook Link & AI Partnership Guidance: Use Chap. 17 PALS Workbook prompts (Sec 5: Literacy Evaluation). Input a specific literacy initiative (e.g., Faculty CoP). Ask your AI partner for potential qualitative methods or indicators suitable for evaluating the cultural impact of that initiative (referencing HCASR C5).

Example Prompt: "For evaluating the cultural impact of our 'Faculty CoP on Ethical AI Pedagogy,' referencing HCASR C5, brainstorm 3 qualitative methods or indicators beyond attendance (e.g., analysis of shared resources/discussions, tracking changes in reported teaching practices)."

Human Oversight Reminder: AI can suggest generic evaluation methods. Designing meaningful evaluations for cultural change requires human expertise in qualitative research methodologies, assessment design appropriate for the context, deep understanding of the program's goals, and the ability to interpret nuanced qualitative data effectively (HCASR C5).

The Strategic Value of an AI-Literate Institution

Institutionalizing an AI literacy culture offers significant strategic advantages, reinforcing the HEI's mission and enhancing its resilience, relevance, and competitiveness.

Enhancing Workforce Readiness and Institutional Competitiveness

An AI-literate campus directly enhances workforce readiness. Graduates with disciplinary knowledge plus practical AI skills, critical evaluation abilities, ethical reasoning, and HAIC competencies are better prepared for modern workplace demands (White House, 2025; Gander & Harris, 2024; Adobe, 2025). This strengthens the institution's value to students and employers (AI Accelerator Institute, 2025). Internally, AI-literate faculty and staff boost institutional agility and effectiveness (The EvoLLLution Contributor, 2025), enabling responsible tool adoption, pedagogical innovation (Danault, 2025), and improved operational efficiency. This collective capacity enhances institutional competitiveness (AI Accelerator Institute, 2025).

Strengthening Academic Integrity, Equity, and Ethical Climate

Widespread AI literacy strengthens academic integrity. Shared understanding of ethical AI use, institutional policies, and AI citation practices reduces unintentional misconduct (Ellucian, 2024). This understanding, fostered by clear communication and education (Harvard University, Office of Academic Integrity and Student Conduct, n.d.), normalizes integrity. Critical AI literacy also promotes equity by educating the community to recognize and mitigate biases in AI systems, reducing the risk of perpetuating disparities (Southgate, 2020). This focus on fairness strengthens the ethical climate and supports responsible AI adoption (Campus Technology Contributor, 2025).

Boosting Institutional Innovation, Reputation, and Adaptability

An AI-literate environment fosters responsible innovation (Danault, 2025; The EvoLLLution Contributor, 2025). Confident, competent, and ethically grounded stakeholders are more likely to experiment with novel AI applications in teaching, research, and administration (Digital Education Council, 2025). This grassroots innovation enhances institutional dynamism. Institutions known for thoughtful, ethical AI engagement enhance their reputation as leaders, attracting talent and support, bolstering long-term adaptability (Danault, 2025).

Workbook Prep 7. Communicating Literacy Culture Value

Reflection: When communicating the importance of AI literacy initiatives to stakeholders who might be resistant or skeptical, which strategic value aspect (enhancing workforce readiness, strengthening integrity/equity, or boosting innovation/reputation) do you believe would be the most persuasive argument within your specific institutional context? Why?

UX Workbook Link: See Chap. 17 UX Workbook, Section 6 ("Communicating Value…").

PALS Workbook Link & AI Partnership Guidance: Use Chap. 17 PALS Workbook prompts (Sec 6: Communicating Value). Input a specific stakeholder concern. Ask your AI partner for talking points that frame the value of AI literacy in a way that addresses that concern, drawing on strategic value arguments and change management principles (referencing HCASR C5, Kotter et al., 2021).

Example Prompt: "Concern: Faculty worry that focusing on AI literacy detracts from core disciplinary learning. Referencing Section 6 arguments and HCASR C5/Kotter principles, draft 2 talking points emphasizing how AI literacy actually enhances disciplinary learning and prepares students for future disciplinary work."

Human Oversight Reminder: AI can help draft persuasive points. Effectively addressing stakeholder concerns requires human skills like empathy, understanding the root causes of resistance, tailoring messages appropriately, building trust through consistent action and dialogue, and strategic communication planning (HCASR C4 & 5).

Outcomes & Deliverables (UODs) for Building an AI Literacy Culture

The activities outlined in Chap. 17 and operationalized through its companion UX Workbook are designed to produce a suite of key User Outcomes and Deliverables (UODs). These UODs provide the concrete resources, plans, and strategies necessary for systematically cultivating and sustaining an institutional culture of AI literacy:

- **UOD 1: Institutional AI Literacy Integration Plan (V1):** The overarching strategic document outlining the institutional vision, goals, chosen competency framework, integration strategies across different areas (training, curriculum, community), implementation timeline, evaluation metrics, and sustainability plan for AI literacy.
- **UOD 2: Foundational AI Training Module Outlines (V1):** Design outlines for core introductory training content intended for onboarding new students, faculty, and staff, covering AI basics, ethics, institutional policies, and HAIC principles.
- **UOD 3: Applied AI Literacy Program Designs (V1):** Detailed plans for role-specific training pathways (e.g., for faculty researchers, student advisors, IT staff) and strategies for integrating applied AI literacy competencies into the academic curriculum across various disciplines.
- **UOD 4: AI Literacy Assessment Toolkit (V1):** A collection of tools, methods, and guidelines for assessing AI literacy competencies and evaluating the effectiveness of literacy initiatives (e.g., survey instruments, rubric templates, performance task examples).
- **UOD 5: AI Learning Hub Plan & Resource Outline (V1):** A blueprint detailing the structure, content categories, curation process, and maintenance plan for a centralized institutional AI resource hub.
- **UOD 6: Community of Practice (CoP) Charter Templates & Support Plan (V1):** Standardized templates for establishing AI-focused CoPs and an outline of institutional mechanisms for supporting these peer-led learning communities.
- **UOD 7: AI Champion Program Description & Support Plan (V1):** A plan defining the role, selection process, support structures, and recognition for departmental or unit-based AI Champions who serve as local resources and advocates.
- **UOD 8: AI Literacy Evaluation Plan & Checkpoint Procedures (V1):** A formal plan outlining the procedures, metrics, and schedule for regular

institutional checkpoints to monitor the progress, impact, and effectiveness of AI literacy initiatives.

- **UOD 9: Communication Plan for AI Literacy Initiatives (V1):** A strategic plan detailing key messages, target audiences, communication channels, and a cadence for promoting AI literacy programs, sharing resources, and highlighting successes.

These deliverables, when developed collaboratively and implemented effectively, provide the practical tools and strategic direction necessary for cultivating and sustaining a vibrant, ethical, and impactful AI literacy culture throughout the institution.

Conclusion: Sustaining an Ethical and AI-Literate Future

Building a pervasive AI literacy culture culminates the strategic efforts in this book, focusing on the human dimension of AI integration. This requires ongoing institutional commitment for navigating an AI-pervaded future responsibly (Lo, 2025). Cultivating this culture needs a holistic approach, guided by adapted frameworks (EDUCAUSE/Guidry, 2024), that systematically develops technical understanding, practical application skills, critical evaluation, and robust ethical reasoning across the university community (Wang et al., 2025).

Success depends on coordinated strategies embedding AI literacy into curriculum (Michalak & Ellixson, 2025), offering continuous professional development (Ellucian, 2024), and fostering collaborative learning environments like CoPs and resource hubs (Truckee Meadows Community College, AI & Emerging Tech Committee, n.d.). This demands human-centered design and thoughtful change management (Brady, 2024). By prioritizing and committing to long-term critical and ethical AI literacy, HEIs fulfill their mission: preparing students for the future (Gander & Harris, 2024), advancing knowledge responsibly, ensuring equity (Southgate, 2020), and strengthening institutional capacity for adaptation and leadership in the age of AI (Danault, 2025).

Looking Ahead: Beyond the Book

This book presented an adaptable four-phase plan for HEIs' ethical and effective GenAI integration, emphasizing human-centric strategies, multilevel stakeholder alignment, and responsible Human-AI Collaboration (Fig. 17.2).

AI integration is ongoing, as technologies and implications evolve rapidly (Lo, 2025). The provided frameworks (HCASR, RSA, EHCS, EAO, AHIPs, TLARA structure), strategies, and workbooks offer a robust foundation, but institutions must remain vigilant, adaptive, and committed to continuous learning (Lo, 2025). The "living documents"—plans, policies, guidelines, resources—require regular review and updates based on emerging technologies, ethical standards, regulations, and community insights (Ellucian, 2024).

The enduring challenge is sustaining momentum, nurturing responsible innovation, and continually interrogating how AI can best serve the institution's core mission—advancing knowledge and empowering learners—equitably, ethically, and human-centrically. Success lies not merely in initial AI adoption, but in the institution's ongoing capacity to thoughtfully shape its future within the complex academic and societal context.

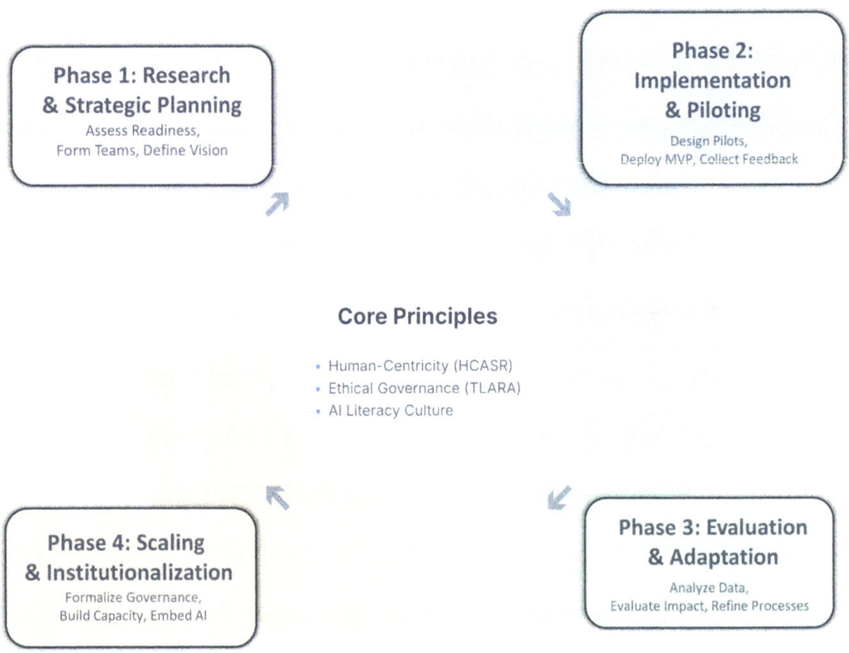

Fig. 17.2 The institutional AI integration lifecycle: a continuous journey

Long-Term Scaling, Sustainability & AI Governance: Sustaining a TLARA-Informed AI Literate Culture

Chapter 17, the culmination of your journey through this book, has focused on the vital, ongoing mission of cultivating a pervasive culture of AI literacy for all stakeholders within your institution. This is not a one-time initiative but a sustained commitment to ensuring that students, faculty, staff, and administrators possess the multifaceted competencies—technical understanding, critical evaluation, ethical application, and equity awareness—to navigate the evolving AI context responsibly across all TLARA (Teaching, Learning, Assessment, Research, Administration) domains (Lo, 2025; Wang et al., 2025). Embedding AI literacy deep within the institutional sociocultural context, across all levels and spaces, is fundamental for the ethical, effective, and sustainable use of AI now and into the future. Both the Chap. 17 UX and PALS Workbooks offer strategies for ethically partnering with AI to design and implement these crucial cultural initiatives.

As AI continues its rapid evolution, the true measure of successful integration in HEIs will be the adaptability and informed engagement of their communities (IDEAS, 2025). The PALS Workbook for Chap. 17, through "Long-Term Scaling, Sustainability & AI Governance: Iteratively Developing Your Custom TLARA-Informed AI Guidelines," provides a final customizable prompt to assist this endeavor. Using your GenAI expert advisor and focusing on "Topics for PALS TLARA Focus"—such as "Embedding AI Literacy Across TLARA Domains" and "Fostering Continuous Learning for a Thriving AI Literacy Culture"—you will refine your TLARA guidelines to ensure they actively support and are, in turn, supported by this vibrant literacy culture. This ongoing process ensures your "Living Document" and governance structures remain dynamic, empowering your institution to lead with ethical foresight in the age of AI (DerSimonian & Montagnino, 2025).

References

Adobe. (2025, March 7). Connecting AI literacy, learning outcomes, and career-readiness. *Adobe for Education Blog*. https://www.adobeforeducation.com/post/connecting-ai-literacy-learning-outcomes-and-career-readiness

AI Accelerator Institute. (2025, April 2). AI literacy: Essential for today's workforce and businesses. *AI Accelerator Institute Blog*. https://www.aiacceleratorinstitute.com/ai-literacy-essential-for-todays-workforce-and-businesses/

ASU Enterprise Technology. (2024, August 9). *LX AY 2023* [Annual Report]. Issuu. https://issuu.com/asu_uto/docs/et-lx-ay23

Brady, C. (2024, December 12). A strategic framework for action for leading AI integration in higher education. *Dr. Claire Brady Blog*. https://www.drclairebrady.com/post/a-strategic-framework-for-action-for-leading-ai-integration-in-higher-education

Campus Technology Contributor (DerSimonian & Montagnino). (2025, March 26). Crafting thoughtful AI policy in higher education: A guide for institutional leaders. *Campus Technology*. https://campustechnology.com/articles/2025/03/26/crafting-thoughtful-ai-policy-in-higher-education-a-guide-for-institutional-leaders.aspx

Carolus, A., Koch, M. J., Straka, S., Latoschik, M. E., & Wienrich, C. (2023). MAILS-Meta AI literacy scale: Development and testing of an AI literacy questionnaire based on well-founded competency models and psychological change- and meta-competencies. *Computers in Human Behavior: Artificial Humans, 1*(2), 100014. https://doi.org/10.1016/j.chbah.2023.100014

Cornell University. (2025). Cornell AI initiative. Retrieved May 11, 2025, from https://ai.cornell.edu/

Danault, R. R. (2025, March 23). *Adapting to AI: The evolving role of faculty in higher education*. University of New Hampshire Scholars Repository. https://scholars.unh.edu/faculty_pubs/2156/

DebiasMe Contributors (Sidky, A. et al.). (2025, April 23). *DeBiasMe: De-biasing Human-AI Interactions with Metacognitive AIED (AI in Education) Interventions* [Preprint]. arXiv. https://arxiv.org/html/2504.16770

DerSimonian, R., & Montagnino, C. (2025, March 26). Crafting thoughtful AI policy in higher education: A guide for institutional leaders. Campus Technology. https://campustechnology.com/articles/2025/03/26/crafting-thoughtful-ai-policy-in-higher-education-a-guide-for-institutional-leaders.aspx

Digital Education Council. (2025, March 4). *Breaking Faculty Barriers to AI Literacy: Adoption, Incentives, and Scaling Strategies*. Digital Education Council Blog. https://www.digitaleducationcouncil.com/post/breaking-faculty-barriers-to-ai-literacy-adoption-incentives-and-scaling-strategies

EDUCAUSE/Guidry, K. R. (2024). *AI literacy in teaching and learning: A durable framework for higher education*. EDUCAUSE.

Element451. (2025, February 28). AI, student data, and FERPA compliance: Why Element451 is the trusted choice for higher education. *Element451 Blog*. https://element451.com/blog/ai-student-data-and-ferpa-compliance

Ellucian. (2024, December 16). AI in higher education: Implementation strategies for institutions and tech partners. *Higher Ed Dive*. https://www.highereddive.com/spons/ai-in-higher-education-implementation-strategies-for-institutions-and-tech/735434/

Ellucian. (2025, date unspecified). Integrating AI skills prepares graduates for workforce needs. *Ellucian Blog*. https://www.ellucian.com/blog/empower-graduates-workforce-ai-literacy-and-ethics

Gander, T., & Harris, G. (2024). Understanding AI literacy for higher education students: Implications for assessment. In *Proceedings of the New Zealand AI in Higher Education Symposium 2024*. ResearchGate. https://www.researchgate.net/publication/389693740_Understanding_AI_Literacy_for_Higher_Education_Students_Implications_for_Assessment

Generation AI. (2025, March 11). FERPA & AI: What Higher Ed needs to know. *Generation AI Podcast/Blog*. https://generationaishow.com/episodes/making-sense-of-ferpa-compliance-in-the-age-of-ai

Gu, X., & Ericson, B. J. (2025, March 28). *AI Literacy in K-12 and higher education in the wake of generative AI: An integrative review* [Preprint v3]. arXiv. https://arxiv.org/html/2503.00079v3

Harvard University, Office of Academic Integrity and Student Conduct. (n.d.). *Academic integrity and teaching with(out) AI*. Retrieved May 11, 2025, from https://oaisc.fas.harvard.edu/academic-integrity-and-teaching-without-ai/

Hazari, S. (2024). Justification and roadmap for artificial intelligence (AI) Literacy courses in higher education. *Journal of Educational Research and Practice, 14*(1), 106–118. https://doi.org/10.5590/JERAP.2024.14.1.07

Instructional Design, Engagement, and Support (IDEAS). (2025, January 27). *AI's impact on higher education through the eyes of faculty: 4 key trends*. University of Massachusetts Amherst News. https://www.umass.edu/ideas/news/ais-impact-higher-education-through-eyes-faculty-4-key-trends

Jin, Y., Martinez-Maldonado, R., Gašević, D., & Yan, L. (2024). *GLAT: The generative AI literacy assessment test*. arXiv. https://doi.org/10.48550/arXiv.2411.00283

Jisc. (2024, June 11). New toolkit for colleges and universities sets pathway for effective adoption of AI. *Jisc News*. https://www.jisc.ac.uk/news/all/new-toolkit-for-colleges-and-universities-sets-pathway-for-effective-adoption-of-ai

Kelly, R. (2025, March 10). Cal Poly Pomona launches AI and Innovation Center. *Campus Technology*. https://campustechnology.com/Articles/2025/03/10/Cal-Poly-Pomona-Launches-AI-and-Innovation-Center.aspx

Kolbjørnsrud, V., Thomas, R. J., & Amico, R. (2023). Designing the intelligent organization: Six principles for human-AI collaboration. *California Management Review, 66*(2), 44–64. https://doi.org/10.1177/00081256231211020

Kotter, J. P., Akhtar, V., & Gupta, G. (2021). *Change: How organizations achieve hard-to-imagine results in uncertain and volatile times*. Wiley.

Li, N., Deng, W., & Chen, J. (n.d.). *From G-Factor to A-Factor: Establishing a psychometric framework for AI literacy* [Preprint]. ResearchGate.

Lo, L. S. (2025, Spring). AI literacy: A guide for academic libraries. *College & Research Libraries News, 86*(3), 120. https://doi.org/10.5860/crln.86.3.120

Markus, A., Carolus, A., & Wienrich, C. (n.d.). *Objective measurement of AI literacy: Development and validation of the AI competency objective scale (AICOS)* [Preprint]. OSF. https://osf.io/ehk8u/?view_only=df9a6ea06d1446659437a73946f68e5c

Mercer University Center for Teaching and Learning. (n.d.). *Center for teaching and learning*. Retrieved May 11, 2025, from https://ctl.mercer.edu/

Miao, F., & Cukurova, M. (2024). *AI competency framework for teachers*. UNESCO. https://unesdoc.unesco.org/ark:/48223/pf0000390111

Miao, F., Shiohira, K., & Lao, N. (2024). *AI competency framework for students*. UNESCO. https://unesdoc.unesco.org/ark:/48223/pf0000390100

Michalak, R., & Ellixson, D. (2025). Fostering ethical AI integration in first-year writing: A case study on human-tool collaboration in artificial intelligence literacy. *Journal of Library Administration, 65*(3), 361–377. https://doi.org/10.1080/01930826.2025.2468136
Microsoft. (2025, March 31). Personalizing education and enhancing administration with AI. *Microsoft in Business Blogs*. https://www.microsoft.com/en-us/industry/microsoft-in-business/future-of-work/2025/03/31/personalizing-education-and-enhancing-administration-with-ai/
Northern Michigan University AI Literacy Initiative. (n.d.). *General - getting started with ChatGPT*. Retrieved May 11, 2025, from https://nmu.edu/ai-literacy-initiative/getting-started-chatgpt
Rulandari, N., & Silalahi, A. D. K. (2025). Achieving effectiveness of public service in AI-enabled service from public value theory: does human–AI collaboration matters? *Transforming Government: People, Process and Policy, 19*(2), 428–452. https://doi.org/10.1108/TG-01-2025-0004
Saint Mary's University, Studio for Teaching and Learning. (n.d.). *Artificial intelligence in teaching and learning*. Retrieved May 11, 2025, from https://studio.smu.ca/ai-in-teaching-and-learning
Salhab, R. (2024). AI literacy across curriculum design: Investigating college instructors' perspectives. *Online Learning, 28*(2), 22–47. https://doi.org/10.24059/olj.v28i2.4426
Seung, Y., & Basham, J. (2024, October 21). Conceptualizing AI literacy: A critical skill for the 21st century. *CIDDL Blog*. https://ciddl.org/conceptualizing-ai-literacy-a-critical-skill-for-the-21st-century/
Shneiderman, B. (2022). *Human-centered AI*. Oxford University Press.
Southgate, E. (2020, July). *Artificial intelligence, ethics, equity and higher education: A 'beginning-of-the-discussion' paper*. National Centre for Student Equity in Higher Education. https://www.ncsehe.edu.au/app/uploads/2020/07/Southgate_AI-Equity-Higher-Education_FINAL.pdf
The EvoLLLution Contributor (Swartsel, A.). (2025, April 23). Shaping workforce readiness with AI and agility. *The EvoLLLution*. https://evolllution.com/shaping-workforce-readiness-with-ai-and-agility
Truckee Meadows Community College, AI & Emerging Tech Committee. (n.d.). Faculty how-to. *AI Learning Hub*. Retrieved May 11, 2025, from https://www.tmcc.edu/committees/ai-emerging-tech-committee/ai-learning-hub/faculty-how-to
Tzirides, A. O., Zapata, G., Kastania, N. P., et al. (2024). Combining human and artificial intelligence for enhanced AI literacy in higher education. *Computers and Education Open, 6*, 100184. https://doi.org/10.1016/j.caeo.2024.100184
University of Lincoln. (2025, April). *AI teaching, learning, and assessment policy* (Version 1.0). https://legal-compliance.lincoln.ac.uk/files/2025/04/AI-Teaching-Learning-and-Assessment-Policy-April-2025.pdf

University of Maryland, Baltimore, Leaders in Education: Academy of Presidential Scholars (LEAPS). (2024?). *2024–25 instructional innovation seed grants*. Retrieved May 11, 2025, from https://www.umaryland.edu/leaps/leaps-symposium/2024-leaps-symposium/2024-25-instructional-innovation-seed-grants/

Veritis. (2024). AI Maturity Model – A CEO's guide to scaling AI for success. *Veritis Blog*. https://www.veritis.com/blog/ai-maturity-model-a-ceos-guide-to-scaling-ai-for-success/

Wang, K., Cui, W., & Yuan, X. (2025). Artificial intelligence in higher education: The impact of need satisfaction on artificial intelligence literacy mediated by self-regulated learning strategies. *Behavioral Sciences, 15*(2), 165. https://doi.org/10.3390/bs15020165

White House. (2025, April 23). *Advancing Artificial Intelligence Education for American Youth* [Presidential Document]. https://www.whitehouse.gov/presidential-actions/2025/04/advancing-artificial-intelligence-education-for-american-youth/

Willems, W. J. M. (2025). *From principles to practice: A tailored maturity model for responsible AI in public sector organisations* [Master's thesis, University of Twente]. University of Twente Student Theses. https://purl.utwente.nl/essays/106082

Yoon, J. (2024, August). Artificial intelligence humanities as a subject within general education. *The Korean Association of General Education, 18*(4), 11–22. https://doi.org/10.46392/kjge.2024.18.4.11

Zheng, J., Hong, Y., & Richter, A. (2025, January 7). Artificial trailblazing - How human-AI collaboration transforms organizational innovation practices. In *Proceedings of the 58th Hawaii International Conference on System Sciences (HICSS)*. ScholarSpace. https://hdl.handle.net/10125/108874

Zhou, X., Li, Y., Chai, C. S., & Chiu, T. K. F. (2025). Defining, enhancing, and assessing artificial intelligence literacy and competency in K-12 education from a systematic review. *Interactive Learning Environments*. Advance online publication. https://doi.org/10.1080/10494820.2025.2487538

Ziatdinov, R., Atteraya, M. S., & Nabiyev, R. (2024, February 2). The fifth industrial revolution as a transformative step towards Society 5.0. *Societies, 14*(2), 19. https://doi.org/10.3390/soc14020019

Appendices

Appendix A. Strategic Approaches to Artificial Intelligence Systems in Higher Education: Build, Buy, or Both?

Introduction

As Higher Education Institutions (HEIs) explore the integration of Artificial Intelligence (AI), particularly Generative AI (GenAI), a fundamental strategic decision emerges regarding how to acquire these capabilities. Institutions must determine whether to primarily develop their own proprietary AI systems ("Build"), integrate existing third-party solutions offered by vendors ("Buy"), or pursue a "Hybrid" approach that combines elements of both strategies (Brighteye Ventures, 2024; Cameron, 2024; Grajek et al., 2024). This choice carries significant implications for strategic control, cost structures, customization potential, technical requirements, data governance, and alignment with institutional mission and risk tolerance (Culotta & Mattei, 2023; Inclusion Cloud, 2024; Nguyen, 2024). Making an informed decision requires careful analysis of the institution's specific context and objectives.

D. Woodring, *Advancing Higher Education with Human-Centric AI Partnerships*,
https://doi.org/10.1007/978-3-032-05205-6

Defining the Approaches

Build (Proprietary Development): This approach involves the institution investing in the internal development of custom AI systems. This could range from building specific applications using foundational models to potentially developing unique models, often requiring significant internal expertise and infrastructure (O'Connell, 2024; Paykamian, 2023).

Buy (Third-Party Integration): This strategy focuses on procuring and integrating AI solutions developed by external vendors. Examples include subscribing to enterprise versions of large language models (LLMs) like ChatGPT Edu or Google Cloud AI platforms, or implementing specialized AI tools for specific functions like student support or administrative tasks (Coffey, 2024a; Rispens, 2024; Rutgers Business School, 2024).

Both (Hybrid Approach): This involves strategically combining elements of building and buying. An institution might use a vendor's platform (e.g., Microsoft Azure OpenAI service) as a foundation but build custom applications or integrations on top of it, or it might buy solutions for some needs while building custom tools for others where control or specificity is critical (Attridge, 2024; Coffey, 2024b; Rutgers Business School, 2024).

Core Considerations for Each Approach

The selection of an approach involves weighing several critical factors. Table A1 summarizes key considerations:

Strategic Control: Building offers the highest control over features, development roadmaps, and updates. Buying entails dependency on vendor roadmaps. Hybrid approaches provide moderate control, focused on integration points and potentially customized layers (Main Text Table 6.2).

Data Privacy and Security: Proprietary systems generally offer greater control over data, potentially simplifying compliance with regulations like FERPA, as data often remains in-house (O'Connell, 2024; Paykamian, 2023). Third-party solutions necessitate rigorous vendor vetting regarding data handling, security protocols, and compliance; data may reside outside institutional control (Coffey, 2024a; Rispens, 2024). Hybrid models require careful management of data flows between internal and external components (Attridge, 2024).

Customization: Building allows for maximum tailoring to specific institutional needs, unique workflows, or specialized data sources. Buying typically offers limited customization beyond configuration options. Hybrid approaches

Table A1 Comparison of considerations for proprietary vs. third-party AI solutions in higher education

Consideration	Proprietary ("Build") approach	Third-party ("Buy") approach	Hybrid approach
Strategic control	High: Full control over features, development roadmap, updates.	Low: Dependent on vendor roadmap, features, updates.	Moderate: Control over integration, potential customization.
Data privacy/ security	High: Data typically remains in-house (e.g., FERPA compliance easier).	Moderate/low: Requires rigorous vetting; data may leave institution; reliance on vendor security.	Varies: Depends on specific mix; careful vetting still needed.
Customization	High: Tailored specifically to institutional needs, workflows, data.	Low: Generally standardized; limited customization options.	Moderate: Can integrate best-of-breed tools; customization focus on integration.
Initial cost	High: Significant investment in infrastructure, personnel, development.	Low/moderate: Typically subscription-based; lower upfront cost.	Varies: Can be lower than full build, higher than pure buy.
Ongoing cost	High: Maintenance, updates, personnel, infrastructure upkeep.	Moderate/high: Subscription fees, potential integration costs.	Moderate/high: Mix of internal upkeep and vendor fees.
Time to deployment	Slow: Requires significant development time.	Fast: Relatively quick to implement off-the-shelf solutions.	Moderate: Faster than full build, slower than pure buy.
Expertise required	High: Needs specialized AI/ML, data science, IT infrastructure staff.	Low: Primarily requires vendor management, integration skills.	Moderate: Requires both internal integration expertise and vendor management.
Scalability	Moderate: Scaling depends on internal infrastructure/ resources.	High: Vendor typically handles scaling infrastructure.	Moderate/high: Depends on vendor scalability and internal integration capacity.
Vendor dependency	Low: Independent of external vendors for core functionality.	High: Reliant on vendor viability, support, pricing changes.	Moderate: Dependent on multiple vendors; less reliant on a single one.

(*continued*)

Table A1 (continued)

Consideration	Proprietary ("Build") approach	Third-party ("Buy") approach	Hybrid approach
Access equity	Potentially high: Can be designed for broad institutional access.	Potential risk: Tiered pricing or licensing may limit access (McIntosh, 2024).	Varies: Depends on specific licensing/ integration choices.
Innovation speed	Moderate/slow: Innovation pace set by internal capacity.	Potentially fast: Leverages vendor's R&D and market innovations.	Moderate: Can leverage vendor innovation but integration takes time.

Note. This table summarizes key factors influencing the strategic decision between developing proprietary AI systems ("Build"), integrating existing third-party solutions ("Buy"), or using a hybrid approach in higher education. The optimal choice depends on institutional context, resources, risk tolerance, strategic goals, and specific use cases (e.g., Attridge, 2024; Coffey, 2024a, 2024b; Grajek et al., 2024; O'Connell, 2024; Paykamian, 2023)

allow integration of potentially best-of-breed tools, with customization focused on the integration layer or specific in-house components (Main Text Table 6.2).

Cost (Initial and Ongoing): Building requires significant upfront investment in personnel (AI/ML experts, data scientists, infrastructure staff), development resources, and computing infrastructure, along with high ongoing maintenance costs (Attridge, 2024; Coffey, 2024b; O'Connell, 2024). Buying usually involves lower initial costs (often subscription-based) but potentially high recurring fees and integration expenses (Paykamian, 2023). Hybrid costs vary, potentially offering a balance but involving both internal upkeep and vendor fees (Rutgers Business School, 2024).

Time to Deployment: Building is typically a slow process requiring considerable development time. Buying allows for relatively rapid deployment of off-the-shelf solutions. Hybrid deployment speed is moderate, faster than a full build but slower than a pure buy approach (Main Text Table 6.2).

Expertise Required: Building demands high levels of specialized internal technical expertise. Buying primarily requires vendor management and integration skills. Hybrid models necessitate a combination of internal integration expertise and effective vendor management (O'Connell, 2024; Paykamian, 2023).

Scalability: Scalability in the build approach depends on internal infrastructure capacity. Vendors typically handle infrastructure scaling in the buy approach. Hybrid scalability depends on both vendor capabilities and internal integration capacity (Main Text Table 6.2).

Vendor Dependency: Building offers low dependency. Buying creates high reliance on a vendor's viability, support model, pricing, and strategic direction. Hybrid approaches involve dependency on multiple vendors but may reduce reliance on any single provider (Main Text Table 6.2).

Access Equity: Proprietary systems can potentially be designed for broad, equitable institutional access. Third-party solutions may involve tiered pricing or licensing models that could create access barriers for some users (McIntosh, 2024). Hybrid access equity depends on the specific licensing and integration choices made.

Innovation Speed: Building sets the innovation pace based on internal capacity. Buying potentially allows leveraging rapid vendor R&D and market innovations. Hybrid approaches can leverage vendor innovation, but integration timelines affect the overall pace (Main Text Table 6.2).

Illustrative Examples from Higher Education

Recent initiatives highlight these different strategies. Some universities, like the University of Michigan with its U-M GPT suite and UC San Diego with TritonGPT, have invested in building custom, on-premises platforms primarily to ensure data privacy, security, and greater control over the AI environment (Attridge, 2024; Coffey, 2024b; O'Connell, 2024; Paykamian, 2023). Others, such as Arizona State University through its collaboration with OpenAI for ChatGPT Enterprise access, or Rutgers Business School partnering with Google Cloud AI, represent a "Buy" strategy, leveraging advanced vendor models and platforms to provide broad access quickly (Coffey, 2024a; Fore, 2024; Rispens, 2024; Rutgers Business School, 2024). Hybrid approaches are also emerging, where institutions might use platforms like Microsoft Azure OpenAI service as a foundation upon which they build tailored applications or integrations, aiming for a balance of vendor capabilities and institutional control (Attridge, 2024; Coffey, 2024b).

Factors Influencing the Decision

The optimal strategy—Build, Buy, or Both—is highly dependent on the specific institutional context. Key factors demanding careful analysis include available financial resources and personnel expertise, the institution's tolerance for risks associated with data control and vendor dependency, the degree of customization required for planned AI applications, critical data governance and privacy requirements (including regulatory compliance), and the alignment of the approach with overarching institutional strategic goals (Katsamakas et al., 2024; Paykamian, 2023).

Conclusion

Choosing whether to build proprietary AI systems, integrate third-party solutions, or adopt a hybrid model is a significant strategic decision for HEIs. Each path presents distinct advantages and disadvantages concerning control, cost, customization, speed, and risk. A thorough analysis of institutional needs, resources, strategic priorities, and ethical commitments is necessary to select the approach, or combination of approaches, that best positions the institution to leverage AI effectively and responsibly. This decision, considered early in the planning process, provides crucial framing for subsequent resource allocation, technical planning, and governance development.

Appendix B. Glossary

4IR (Fourth Industrial Revolution): The period (roughly 2000–2020) characterized by hyper-connectivity, the widespread adoption of artificial intelligence (AI), the Internet of Things (IoT), and big data, significantly reshaping society and industries.

5IR (Fifth Industrial Revolution): The current and emerging era defined by deeper human-AI collaboration, synergy, and co-evolution, moving beyond the 4IR's focus on connectivity and automation to redefine intelligence, work, and education.

Accessibility Testing: Evaluating an AI tool or system to ensure it is usable by people with diverse abilities, often measured against standards like the Web Content Accessibility Guidelines (WCAG).

Adaptation (AI Context): The ongoing process of modifying AI tools, processes, policies, and competencies in response to feedback, performance data, technological changes, and evolving institutional needs.

AGI (Artificial General Intelligence): A form of artificial intelligence that possesses the ability to understand, learn, and apply knowledge across a wide range of tasks at a level comparable to human intelligence, if not surpassing it.

Agile Methodology: An iterative approach to project management and software development that emphasizes flexibility, collaboration, customer feedback, and rapid, incremental delivery of functional components (Relevant to iterative piloting).

AHIPs (Authentic High-Impact Practices): A pedagogical and andragogical framework that integrates established High-Impact Practices (HIPs) with principles of authentic teaching, learning, and assessment to create meaningful, complex, real-world educational experiences designed to foster higher-order thinking and skills not easily replicated by GenAI.

AI (Artificial Intelligence): Technology enabling computers and machines to simulate human intelligence and problem-solving capabilities; the theory and development of computer systems able to perform tasks that normally require human intelligence, such as visual perception, speech recognition, decision-making, and translation between languages. In the context of the source materials, it often refers to the simulation and replication of human intelligence by machines. *(Merged definition)*

AI Ambassadors: Individuals from various units lacking formal authority but possessing interest, early adoption experience, peer credibility, and enthusiasm for AI, serving as bridges for Smart Teams. (Compare with AI Champion)

AI Champion: Individuals within departments or units who are early adopters and advocates for AI, serving as informal mentors and local resources to support peer learning and promote AI literacy. Used in the context of scaling AI initiatives. (Compare with AI Ambassadors)

AI Competency Frameworks (e.g., EDUCAUSE ALTL, UNESCO): Structured outlines of the knowledge, skills, and attitudes that constitute AI literacy, often tiered by proficiency levels, used to guide curriculum development, training, and assessment.

AI Council (Permanent): The main chapter and workbooks emphasize transitioning to permanent governance. The formalized, cross-functional, and empowered institutional body responsible for long-term strategic direction, ethical oversight, policy development and approval, and compliance monitoring related to AI. (See also AI Ethics Board / Governance Council)

AI Ethics and Sustainability Strategy: A key deliverable related to scaling AI.

AI Ethics Board/Governance Council: An institutional body responsible for overseeing the ethical development, deployment, and use of Artificial Intelligence, including reviewing AI pilot protocols and risk registers. (See also AI Council (Permanent))

AI Ethics TLARA Structure/Scaffold: The adaptable organizational structure (Teaching, Learning, Assessment, Research, Administration) used by

HEIs to collaboratively develop, embed, and refine their own responsible, equitable, and human-centric AI governance principles and policies. This is presented as a scaffold rather than a fixed framework.

AI Governance Framework: The comprehensive system of principles, policies, structures (e.g., AI Council, Smart Teams), processes, and roles established to guide the ethical, effective, and responsible development, deployment, and use of AI within an institution.

AI Impact Assessment: An evaluation to determine the effects of an AI pilot or system across various dimensions, including pedagogical effectiveness, ethical compliance (fairness, transparency, privacy, bias mitigation), accessibility, impact on roles/skills, and alignment with institutional values.

AI-Integrated Curriculum & Policy Playbook: A key deliverable related to scaling AI.

AI Learning Hub: A centralized institutional repository (often web-based) providing curated resources, tutorials, policies, ethical guidelines, best practices, and support for AI literacy development.

AI Literacy: The set of competencies that enables individuals to critically understand, use, evaluate, and create with AI technologies in an ethical and effective manner. A multifaceted competency encompassing technical understanding of AI, critical evaluation of AI tools and outputs, ethical application, effective Human-AI Collaboration (HAIC), and awareness of equity considerations and potential biases. It is crucial for assessment and auditing. *(Merged definition)*

AI Literacy Audits: The specific concept of auditing AI Literacy. Various types of assessments are mentioned, including objective, performance-based, and self-report.

AI Literacy Culture: An institutional environment where AI literacy is pervasively understood, valued, practiced, and continuously developed by all stakeholders, embedded in daily operations, decision-making, and the institutional ethos.

AI Long-Term Roadmap: A key deliverable reflecting institutionalized planning for AI.

AI Maturity Model: A framework used to assess an institution's current capabilities and level of sophistication regarding the adoption, implementation, and management of AI, often depicting stages of progression. Frameworks used to assess an HEI's progress and capabilities across dimensions like AI strategy, data governance, technological infrastructure, workforce skills, and ethical considerations, guiding targeted improvements and adaptation efforts. An institutional version is a key deliverable. *(Merged definition)*

AI Pilot Strategy Document: A comprehensive blueprint for an AI pilot project, detailing its objectives, scope, target users, AI tool(s), timeline, activities, success metrics, data collection plan, risk assessment, ethical protocols, responsible team, and strategic alignment.

AI Readiness Evaluation: The systematic assessment of an institution's capacity across key dimensions (e.g., data, technology, culture, skills, governance) to effectively and responsibly integrate AI.

AI Scalability and Compliance Report: A key deliverable for monitoring scaled AI initiatives.

AI Sourcing Strategy: The strategic decision-making process regarding whether an institution will primarily build proprietary AI systems ("build"), integrate third-party solutions ("buy"), leverage open-source models ("adapt"), or use a hybrid approach, considering factors like cost, control, expertise, and risk.

AI Sustainability Roadmap (Long-Term): A key deliverable; a strategic plan outlining phases, milestones, resources, and evaluation points for ensuring the enduring and ethical integration of AI over an extended period (e.g., 5–10 years).

AI Vision Statement: A clear, concise, and aspirational declaration that articulates the desired long-term future state for ethical Human-AI collaboration and synergy within a Higher Education Institution, aligned with its mission and values.

AI-Centric Collaboration (Mode of HAIC): A mode of human-AI collaboration where AI systems take the primary role in performing tasks or making decisions, with minimal human input or oversight. Generally suitable for routine or highly data-intensive tasks where human nuance is less critical.

Applied AI Literacy: The ability to use AI tools and AI-related knowledge effectively and ethically within specific disciplinary, professional, or contextual tasks, moving beyond foundational awareness.

ASI (Artificial Superintelligence): A hypothetical form of artificial intelligence that would possess intelligence far surpassing that of the brightest and most gifted human minds across virtually all fields of endeavor.

Authentic Assessment: Evaluation methods that require students to apply knowledge and skills to real-world tasks or complex problems, demonstrating their learning in a contextually meaningful way, as opposed to rote memorization or standardized testing.

AutoGPT: An example of an autonomous AI agent that can take a high-level goal and independently break it down into sub-tasks, execute those tasks using various tools (like web browse or file manipulation), and iterate until the goal is achieved.

Baseline Institutional Readiness Metrics: Quantitative and qualitative data captured before an AI pilot begins, documenting the pre-pilot state of relevant Key Performance Indicators (KPIs) and other measures (e.g., user satisfaction, AI literacy levels) for later comparison to assess impact.

Bias (in AI / AI): Systematic prejudice or favoritism in AI algorithms or the data they are trained on, leading to unfair, inaccurate, or discriminatory outcomes against certain individuals or groups. Systematic errors or skewed results in AI outputs, often reflecting biases present in the training data or algorithm design. *(Merged definition)*

Bias Creep: Mentioned in Chap. 13. Ensure this concept is clear. *(Definition requires further clarification from source or context)*

Capacity Building (AI Context): Systematically developing institutional and stakeholder ability to use and manage AI effectively and ethically at scale.

Change Management (CM): Frameworks for managing the organizational and human side of change, such as Kotter's 8-step process. Referenced in relation to the synergy with Design Thinking. A structured approach to transitioning individuals, teams, and organizations from a current state to a desired future state, often involving principles to manage resistance and foster adoption, such as those outlined by Kotter. *(Merged definition)*

Chatbot: A computer program designed to simulate human conversation through voice or text commands, often used for customer service, information retrieval, or task assistance.

ChatGPT: A specific generative AI language model developed by OpenAI, capable of understanding and generating human-like text based on the prompts it receives.

Chief AI Officer (CAIO): A specific leadership role mentioned for overseeing AI strategy.

Claude (AI Model): A family of large language models developed by Anthropic, designed with a focus on safety and helpfulness.

Co-evolution (Human-AI): A dynamic and reciprocal process where humans and AI systems continuously influence and shape each other's development, capabilities, and behaviors over time.

Co-Regulated Learning (CRL): A learning process where regulation (monitoring, strategizing, evaluating) is shared between individuals, such as a student and a teacher, a student and a peer, or a human and an AI, involving mutual support and adjustment.

Collective Intelligence (CI): Shared or group intelligence that emerges from the collaboration, collective efforts, and competition of many individuals (and potentially AI agents).

Communities of Practice (CoPs) (AI Context): Groups fostering shared learning and best practices for AI. Groups of stakeholders who share an interest in AI and collaborate to learn, share best practices, troubleshoot challenges, and build collective expertise related to AI literacy and application. *(Merged definition)*

Competency Mapping (AI Context): The process of identifying key skills and competencies required for specific roles or tasks (especially related to AI) and comparing them against current capabilities to identify gaps. Defining and linking AI-related skills and knowledge to specific roles within the institution. *(Merged definition)*

Constructive Alignment: An educational design principle where teaching methods and assessment tasks are deliberately aligned with the intended learning outcomes.

Continuous Evaluation Framework: A central concept in Chap. 13. *(Definition requires further clarification from source or context)*

Continuous Improvement Cycle (AI Context): An ongoing, iterative process of planning, implementing, monitoring, evaluating, and refining AI initiatives and related practices to enhance effectiveness, ethics, and alignment with goals. (Relates to Design Thinking cycles and PDCA mentioned in HAICS Guide).

Continuous Quality Improvement Scorecard: Mentioned as a tool for tracking metrics.

Cultural Adaptation (AI Context): The process by which an institution's shared values, beliefs, norms, and practices evolve to effectively and ethically integrate AI into its daily operations and strategic vision.

Cultural Prompting: Explicitly defining the desired cultural context within an AI prompt to improve the relevance and appropriateness of the response for specific sociocultural settings.

Customized/Domain-Specific Ethical Frameworks (AI): AI ethical guidelines and policies developed by an HEI, often using a structure like TLARA, tailored to its specific mission, values, stakeholder needs, sociocultural context, and the unique challenges within its core operational domains.

Daily Feedback Logs: A mechanism for pilot participants to provide regular, often brief, contextualized feedback on their ongoing experiences with an AI tool or process, capturing real-time insights, challenges, and suggestions. A specific mechanism mentioned for continuously collecting qualitative feedback from pilot participants regarding their daily experiences, challenges, and suggestions related to the AI tool or process. Its practical application in data collection is central. *(Merged definition)*

Data Governance: The overall management of the availability, usability, integrity, and security of data used in an organization, crucial for AI systems.

Data Integration (Mixed Methods): The process of combining quantitative data (e.g., KPIs, usage statistics) and qualitative data (e.g., feedback from Daily Logs, interview themes) to create a more comprehensive and nuanced understanding of pilot outcomes. This can involve triangulation, using qualitative insights to explain quantitative trends, or developing a richer narrative.

Deliverables (Chap. 7 Specific): Process Deliverables: Smart Team Charters, Stakeholder Maps & Analyses, Stakeholder Engagement Plan Outlines. Initial Work Product Deliverables: Ethical AI Policy Scopes (V1), AI Risk Management Guide Scopes (V1), Stakeholder Use Case Inventories (V1).

Design Thinking (DT): A human-centered approach to innovation and problem-solving that integrates the needs of people, the possibilities of technology, and the requirements for business success. It emphasizes empathy, ideation, prototyping, and iteration. A human-centered approach to innovation that uses empathy, ideation, and experimentation to solve complex problems. Referenced in relation to the synergy with Change Management. *(Merged definition)*

Dual System (Leadership & Management): In the context of change (like AI pilots), the need to balance visionary leadership (inspiring, supporting, championing) with operational management (coordinating, monitoring, problem-solving) for successful implementation (Kotter et al., 2021).

Dynamic Stakeholder Coordination/Engagement: Sustained, responsive, and collaborative interaction with diverse stakeholders to gather input, share information, and co-create solutions for the ongoing adaptation and refinement of AI integration.

EAO (Enhancing Academic Outcomes): A framework focused on strategically leveraging GenAI technologies to directly improve measurable student success metrics, such as grades, retention, assessment scores, and competency mastery, through responsible and regulated integration.

EHCS (Enhancing Human-Centric Skills): A pedagogical framework for using GenAI to actively cultivate essential human competencies (e.g., critical thinking, creativity, ethical reasoning) by designing learning experiences that require active human engagement and critical evaluation of AI contributions.

Emergence (in systems): The arising of novel and coherent structures, patterns, and properties during the process of self-organization in complex systems. These properties are not present in the individual components but arise from their interactions.

Empathy Mapping: A collaborative UX tool used to gain deeper insight into stakeholder perspectives by visualizing what they say, think, feel, and do regarding a specific topic (like AI).

Epistemology: A branch of philosophy concerned with the theory of knowledge, exploring the nature of knowledge, justification, and the rationality of belief.

Equity and Accessibility Review: A specific component of the readiness assessment focused on identifying potential barriers and disparate impacts of AI integration on diverse stakeholder groups.

Ethical Foresight (AI Context): The proactive process of anticipating future ethical dilemmas, societal impacts, and governance needs arising from next-generation AI systems and evolving technological capabilities.

Ethical Governance (AI): The system of rules, practices, and processes an institution puts in place to ensure that AI is used responsibly, ethically, and in alignment with its values and applicable regulations.

Explainable AI (XAI): Artificial intelligence systems designed so that their actions, decisions, and predictions can be understood by humans, promoting transparency, trust, and accountability.

Feedback Loops (in Human-AI systems / AI Adaptation): Iterative processes where the output or action of one agent (human or AI) influences the subsequent input or action of the other, leading to mutual learning, adaptation, and refinement of the collaborative process or system. Planned mechanisms for systematically gathering, analyzing, and responding to user input and experiences during an AI pilot to guide iterative improvements and address concerns. Established channels and processes for continuously collecting, analyzing, and acting upon feedback from users and other stakeholders regarding AI tools, policies, and support, to drive refinement. *(Merged definition)*

Foundational AI Literacy Training: Introductory programs designed to provide all stakeholders (students, faculty, staff) with a baseline understanding of AI concepts, ethical principles, institutional policies, and HAIC.

Friction Point / Emerging Issues Log: A specific deliverable from the UX workbook.

Gap Analysis (Needs vs. Readiness): A process of comparing the identified needs of stakeholders and the institution regarding AI with the current state of institutional readiness, to pinpoint critical areas requiring attention.

Gemini (AI Model): A family of multimodal large language models developed by Google DeepMind.

GenAI (Generative Artificial Intelligence): A type of artificial intelligence capable of generating novel content, including text, images, audio, video, and code, based on patterns and structures learned from existing data. Generative AI (GenAI) is a subset of AI capable of creating new content. *(Merged definition)*

Guided Prompt: A structured prompt provided in the PALS workbook (Step 3) designed to retrieve foundational external research on chapter topics.

HAIC (Human-AI Collaboration): Dynamic partnerships where humans and AI systems combine complementary skills, roles, and responsibilities to achieve shared goals, leveraging the unique strengths of each. The principle and practice of humans and AI systems working together synergistically, leveraging the strengths of each to achieve goals, with an emphasis on human agency and oversight. Its role in training for literacy is key in Ch17. *(Definition enhanced with Ch17 context)*

HAIC TLA & Research Playbook: A key deliverable; a consolidated institutional resource documenting best practices, guidelines, and ethical considerations for Human-AI Collaboration across Teaching, Learning, Assessment, and Research.

HAICS Guide (Human-GenAI Collaboration & Synergy Guide): The companion guide detailing core frameworks (HCASR, HCAS, RSA, etc.) and principles used throughout the textbook. Contains an extensive glossary.

HCAS (Human-Centric AI Synergy): A foundational conceptual framework that translates AI design principles into practical, use-centric applications for higher education, emphasizing the leveraging of complementary human and AI strengths while being guided by human-centric values. A core principle focused on using AI to empower humans and augment their capabilities, leveraging mutual strengths while mitigating weaknesses. An approach to AI development and use that prioritizes empowering humans rather than replacing them, augmenting human intelligence, and leveraging the complementary strengths of humans and AI. (See HAICS Guide). *(Merged definition)*

HCASR (Human-Centric AI Synergy & Regulation): An overarching process or framework that combines Human-Centric AI Synergy (HCAS) with principles of regulation (like those in the RSA framework) to guide the ethical and effective integration of GenAI. A five-component, nonlinear framework (C1–C5) for guiding ethical and effective human-GenAI collaboration and synergy. A customizable five-component framework for ethically and effectively using and interacting with GenAI, optimizing human-GenAI collaboration. Components: C1: Ethical & Effective Goal Setting; C2: Ethical & Effective Prompt Engineering; C3: Rapid Research & Review (RRR); C4: Human-Centric Implementation & Stakeholder Engagement; C5:

Continuous Improvement & Advanced Human–AI Synergy. (See HAICS Guide). *(Merged definition including components)*

HCD (Human-Centered Design): An approach that prioritizes human needs, capabilities, and behaviors in the design process. An approach that prioritizes human needs in the design process. (Referenced in PALS Prompt context). *(Merged definition)*

HEI (Higher Education Institution): Universities, colleges, and other post-secondary educational institutions.

Heutagogy: A self-determined learning approach where learners are the primary agents in their own learning path, including the design of the learning questions, resources, processes, and assessment.

HHAIR (Hybrid Human-AI Regulation): Models of regulation in learning or task completion where both human judgment and AI systems contribute to monitoring, evaluating, and adjusting the process, ensuring human oversight guides AI contributions.

High-Impact Practices (HIPs): Educational practices recognized by research to increase rates of student retention and student engagement, such as first-year seminars, learning communities, undergraduate research, internships, and capstone projects.

Human-AI Co-Creation: A collaborative process where humans and AI agents work together to produce novel outputs or solutions that extend beyond what either could create independently.

Human-AI Partnership: A collaborative relationship where humans and AI systems work interdependently, sharing tasks and responsibilities to achieve common objectives, leveraging each other's unique capabilities.

Human-AI Synergy: The combined effect of human and AI collaboration that produces an outcome greater than the sum of their individual capabilities, leading to enhanced performance, creativity, or problem-solving.

Human-Centric AI (H-CAI): An approach to the design, development, and deployment of AI systems that prioritizes human well-being, values, agency, and oversight, ensuring that AI augments human capabilities and serves human goals ethically and effectively.

Human-Centric AI Collaboration (Mode of HAIC): A mode of human-AI collaboration where human users retain primary control and decision-making authority, with AI serving in a supportive, assistive, or advisory capacity.

Hybrid Human-AI Intelligence: The synergistic integration of human cognitive abilities (such as reasoning, creativity, ethical judgment) with the computational power and data processing capabilities of artificial intelligence systems to achieve enhanced collective intelligence and performance.

IA (Intelligence Augmentation): The use of technology, particularly AI, to enhance and extend human intelligence and cognitive capabilities, rather than aiming to replace human intelligence entirely.

Implementation Roadmap (for AI): A detailed, phased plan outlining the key initiatives, activities, milestones, timelines, resource allocations, and responsibilities for operationalizing the institutional AI strategy.

Influence/Interest Grid (Power/Interest Grid): A tool for categorizing stakeholders based on their level of influence (power) and interest in an initiative to guide engagement strategies.

Initial Pilot Performance Evaluation: The preliminary assessment of an AI pilot's success conducted immediately after the pilot's conclusion, based on the collected and synthesized data, measured against predefined pilot-specific goals, KPIs, and benchmarks.

Institutional AI Literacy Integration Plan: A strategic document outlining an HEI's vision, goals, competency framework, implementation strategies (training, curriculum, community), evaluation methods, and sustainability plan for fostering AI literacy.

Institutional AI Maturity Model: A key deliverable; a customized framework with defined levels and criteria for assessing an HEI's progress over time in AI literacy, ethical practice, integration, and governance.

Institutionalization (AI Context): Defined in the main chapter as embedding AI as a fundamental component of standard operating procedures, culture, policies, and infrastructure.

Integrated and Sustained Governance Structures (AI): Permanent, embedded AI oversight bodies (e.g., AI Councils, ethics subcommittees) with clear mandates, authority, and resources, designed for long-term operation and adaptation.

Interdisciplinarity: An approach to knowledge and learning that involves integrating information, data, techniques, tools, perspectives, concepts, and/or theories from two or more disciplines or bodies of specialized knowledge to advance fundamental understanding or to solve problems whose solutions are beyond the scope of a single discipline.

IoT (Internet of Things): A network of interconnected physical devices, vehicles, home appliances, and other items embedded with electronics, software, sensors, actuators, and connectivity which enables these objects to connect and exchange data.

Iteration (Iterative Piloting Process/Cycle): A cyclical process of designing, deploying an MVP, testing, collecting user data/feedback, evaluating, refining the AI tool/process, and then redeploying. Central to the pilot phase.

Iterative Policy Development: An approach to creating and updating AI policies through regular review cycles, incorporating stakeholder feedback and evidence from ongoing monitoring, treating policies as "living documents."

Iterative Refinement Sprints: Short, focused periods of activity dedicated to addressing specific issues or implementing improvements for AI tools, processes, or policies, based on agile principles.

Key Performance Indicators (KPIs): Specific, measurable values that track progress towards achieving key objectives. In AI pilots, these can relate to efficiency, performance, adoption, or learning improvements. In Pilot Evaluation Context: Specific, measurable metrics defined prior to a pilot, used to track and evaluate the pilot's performance and success in achieving its objectives. *(Merged definition with context)*

Large Language Models (LLMs): Advanced AI models, typically based on deep learning architectures like transformers, trained on vast amounts of text data to understand, generate, and manipulate human language with a high degree of fluency and coherence. AI models, often a type of GenAI, trained on vast amounts of text data to understand, generate, and manipulate human language. *(Merged definition)*

Lessons Learned Document: A formal UOD compiling the key insights, challenges, successes, and actionable recommendations derived from the AI pilot experience, intended to inform future projects, refinements, and strategic decisions.

Lifelong Learning: The ongoing, voluntary, and self-motivated pursuit of knowledge for either personal or professional reasons, recognized as crucial for adapting to changing social and technological landscapes.

Lifelong Learning Strategies (AI Context): Plans for continuous AI-related professional development.

Living Document: A document designed to be regularly updated and revised as information changes or initiatives evolve (e.g., goals lists, plans, Team Charter, Stakeholder Map). Planning documents (e.g., Pilot Implementation Roadmap, Change Management Plan, AI Impact Map) that are designed to be dynamic and iteratively updated based on feedback and evolving circumstances throughout a project. Key pilot documentation (e.g., Risk Register, MVP specifications, Feedback Summaries, Benchmark Reports, Dashboards) that is continuously updated throughout the implementation phase to reflect new data, insights, and decisions. Key institutional plans, strategies, and guidelines (e.g., training roadmaps, policy manuals, improvement playbooks) that are regularly reviewed and updated to reflect ongoing learning, changes in the AI landscape, and evolving institutional needs. Strategic institutional documents (e.g., AI vision, ethical frameworks, policies, roadmaps, training plans) designed for regular review, evidence-based updates, and continuous adaptation to remain relevant over an extended horizon. A key concept for ongoing reports. *(Merged definition)*

Maturing/Scaled AI Initiatives: Key terms defining the scope of Phase 3. *(Definition requires further clarification from source or context)*

Meta-emergence: A higher-order emergence where new properties, structures, or systems arise from the interaction of previously emergent systems, often leading to compounding, interactive, and unpredictable transformations.

Minimum Viable Product (MVP): The simplest version of a new product or tool (like an AI application in a pilot) that includes just enough core features to be usable by early adopters and to gather initial feedback for future development. The most basic, functional version of an AI tool or system launched in a pilot to test core assumptions and gather initial user feedback efficiently before developing more extensive features. *(Merged definition)*

Multimodal AI: AI systems capable of processing, understanding, and generating information from multiple types of data, such as text, images, audio, and video, and integrating these different modalities.

Multi-Phase AI Expansion Plan: A key UOD from the UX Workbook.

Natural Language Processing (NLP): A subfield of artificial intelligence focused on enabling computers to process, understand, interpret, and generate human language in a way that is both meaningful and useful.

Needs Assessment: The process of understanding stakeholder challenges, goals, current workflows, and contexts related to potential AI applications, often employing human-centered research methods.

Needs Assessment (AI Literacy): The systematic process of identifying current AI literacy levels, skill gaps, attitudes, and institutional readiness among diverse stakeholders to inform the design of targeted AI literacy initiatives.

NIST AI RMF (National Institute of Standards and Technology Artificial Intelligence Risk Management Framework): A framework for managing risks associated with AI systems. A voluntary framework developed by the U.S. National Institute of Standards and Technology to help organizations manage risks associated with AI systems, structured around functions: Govern, Map, Measure, and Manage. (Referenced in prompt context). *(Merged definition)*

Nonviable Synergy (in Human-AI Collaboration): A human-AI interaction that fails to produce the intended beneficial result or function effectively even in the short term, often characterized by malfunction, immediate failure, or creating more problems than solutions.

Ontology: In philosophy, the branch of metaphysics dealing with the nature of being, existence, or reality. In information science, a formal naming and definition of the types, properties, and interrelationships of the entities that really or fundamentally exist for a particular domain of discourse.

Opportunity Statements (AI Context): Specific statements that articulate the anticipated benefits or positive impacts of AI integration in key institutional areas, such as teaching, learning, research, or administrative operations.

Organizational Agility (AI Context): An institution's capacity to sense, respond to, and capitalize on changes related to AI technology and its impact, adjusting strategies, structures, and processes effectively.

PALS (Personalized Adaptive Learning Session): The guided approach used in this workbook series to facilitate learning and partnership with GenAI using the HCASR framework. A structured approach using GenAI as a partner, guided by the HCASR framework, to enhance research, customize insights, and reflect on human-AI collaboration skills for specific tasks or learning objectives. *(Merged definition)*

PALS Prompts: Specific prompts provided in the PALS workbook (Step 5) designed to help users partner with AI to complete tasks related to the corresponding UX workbook section.

Persona Refinement: Adding detail to preliminary personas based on stakeholder analysis regarding needs, goals, AI interactions, and concerns, specific to a team's scope.

Personas (in UX): Semi-fictional, research-based profiles representing key user or stakeholder groups, used to guide human-centered design and decision-making.

PIER Phases: Planning, Implementation, Evaluation, and Refinement (mentioned in HCASR context).

Pilot Conclusion Recommendation: The evidence-based recommendation (e.g., Scale, Refine Further, Discontinue) formulated by the Smart Team following the initial pilot evaluation, provided as input for Phase 3 decision-making.

Pilot Data Collection & Feedback Report: A comprehensive UOD that compiles, summarizes, and presents all the quantitative and qualitative data gathered during an AI pilot, including analyses of stakeholder feedback.

Pilot Program/Project: A small-scale, controlled initial rollout of a new system, tool, or process (such as an AI application) to test its feasibility, effectiveness, and user acceptance, and to gather data before wider implementation.

Policy Review Trigger: A predefined event or condition (e.g., significant AI update, recurring ethical concern, new regulation) that initiates a formal review and potential revision of relevant institutional AI policies.

Priority Gaps vs. Quick Wins Matrix: A tool used to categorize and prioritize identified gaps and opportunities based on their importance/need versus the institution's readiness/feasibility to address them.

Prompt Engineering: The process of designing, refining, and optimizing input prompts given to generative AI models (especially LLMs) to elicit desired, accurate, and relevant outputs. The process of designing and refining inputs (prompts) given to AI models to elicit desired or optimal outputs. *(Merged definition)*

Proprietary AI Solution: An AI system or tool that is developed and owned by the institution itself (in-house) or a specific entity, as opposed to a generally available third-party AI model.

Qualitative Data Analysis (Pilot Context): The process of systematically organizing, coding, and interpreting non-numerical data from a pilot (e.g., text from Daily Feedback Logs, interview transcripts, open-ended survey responses) to identify themes, patterns, and insights into user experiences and perceptions.

Quantitative Data Analysis (Pilot Context): The process of statistically analyzing numerical data collected during a pilot (e.g., KPIs, usage metrics, survey ratings) to measure performance, identify trends, and compare outcomes against benchmarks.

Recognition & Excellence Framework (AI Context): A key deliverable; an institutional system for acknowledging and celebrating achievements in AI competency development, ethical leadership, and responsible innovation.

Regulation Synergy & Augmentation (RSA): An operational framework designed to guide the ethical and effective practical application of GenAI by integrating principles of Self-Regulated Learning (SRL), Co-Regulated Learning (CRL), and Intelligence Augmentation (IA), with an emphasis on Hybrid Human-AI Regulation (HHAIR). A core principle focused on mitigating AI challenges and augmenting human capabilities through evidence-based strategies. (See HAICS Guide). *(Merged definition)*

Risk Register (Initial Pilot): A document created during pilot design that identifies potential AI-related risks (e.g., data privacy, bias, security), assesses their likelihood and impact, and outlines planned mitigation strategies and responsibilities.

Role-Specific AI Literacy: AI competencies tailored to the distinct responsibilities, tasks, and ethical considerations faced by different roles within the HEI (e.g., faculty, researchers, administrators, IT staff).

Rollout Roadmap: A key UOD from the UX Workbook.

RRR (Rapid Research & Review): Strategically using AI for research synthesis with human validation. The process described in HCASR C3 for using AI to efficiently gather, synthesize, and review information, always coupled with human validation. *(Merged definition)*

Scaling (AI Context): Transitioning from pilot AI projects to broad, systematic, and sustainable integration.

Scaling Laws (for AI): Empirical observations in AI research suggesting that as model size (parameters), dataset size, and computational resources for training are increased, the performance of AI models (particularly LLMs) tends to improve predictably.

Scaffolding (in instructional/project design): Providing temporary structured support to help learners or teams build upon foundational work (like previous project phases, governance frameworks, or established policies) to achieve a goal. In this context, ensuring pilot designs logically extend from Phase 1 work.

Scenario-Based Testing: A testing method where users are asked to perform tasks or navigate workflows using the AI tool within realistic, context-specific scenarios to evaluate its effectiveness and usability in practical situations.

Self-Regulated Learning (SRL): A proactive process whereby learners set goals for their learning, and then attempt to monitor, regulate, and control their cognition, motivation, and behavior, guided and constrained by their goals and the contextual features in the environment.

Short-Term Wins: Tangible, early successes in a project or change initiative that help build momentum, validate efforts, and sustain energy, as emphasized in Kotter's change management model.

SMARTIE Goals: A goal-setting framework extending SMART (Specific, Measurable, Achievable, Relevant, Time-bound) by adding Inclusivity and Equity. Applied to team goals within charters. Framework: Specific, Measurable, Achievable/Ambitious, Relevant, Time-bound, Inclusive, Equitable. Reinforced heavily in Chap. 8 for assessment objectives. Its specific application to ongoing monitoring objectives could be reinforced if needed. *(Merged definition, noting context)*

Smart Teams: Diverse, cross-functional groups, often composed of faculty, staff, students, and technologists, formed within institutions to collaboratively manage AI deployment, innovate, solve problems, and monitor evolving needs related to AI integration. Diverse, cross-functional AI guiding teams established to lead specific AI integration efforts within HEIs. Diverse, empowered groups or coalitions within an institution responsible for guiding and implementing strategic initiatives, such as AI adoption and pilot programs. *(Merged definition)*

Sociocultural Context: The specific social, cultural, economic, and political environment influencing an institution and its stakeholders.

Stakeholder (AI as a conceptual): Viewing AI systems not merely as tools but as influential participants within an ecosystem whose actions and outputs can significantly impact institutional strategies, human behaviors, and outcomes, thus requiring consideration in governance and ethical frameworks similar to human stakeholders.

Stakeholder Engagement: The process of involving individuals or groups who have an interest in or may be affected by a project (like an AI pilot) in its planning, design, implementation, and evaluation. (See also Dynamic Stakeholder Coordination/Engagement and HCASR C4).

Stakeholder Journey Mapping: Creating visual representations of stakeholder experiences relevant to an AI project or the planning process, focusing on interactions and touchpoints.

Stakeholder Mapping/Analysis: Methodologies for systematic stakeholder identification, analysis of attributes (e.g., interest, influence, stance, concerns), and visualization (e.g., Power/Interest Grids, relationship maps) to inform engagement.

Strategic AI Plan: The comprehensive blueprint guiding an institution's approach to AI, typically encompassing the AI vision, strategic goals, governance framework, implementation roadmap, resource plans, risk management strategies, and monitoring/evaluation plans.

Strategic AI Planning (Long-Term): The process of developing a 5–10 year (or longer) institutional vision and roadmap for ethical AI integration, alignment with mission, and adaptation to evolving standards and technologies.

Strategic Alignment (AI Context): The deliberate process of ensuring that AI initiatives, strategies, and governance frameworks are consistent with and supportive of the institution's overall mission, values, strategic priorities, and other key operational plans (e.g., IT roadmap, budget cycles, DEI commitments).

Success Metrics / Thresholds: Clearly defined quantitative and qualitative indicators, along with specific targets or benchmarks, that will be used to evaluate whether an AI pilot has achieved its objectives.

Sustainability (AI Context): Ensuring AI initiatives have the long-term resources (financial, technical, human) and governance for continued effectiveness.

Sustainability & AI Competency Framework: A key UOD from the UX Workbook.

Sustainable Synergy (in Human-AI Collaboration): A human-AI collaboration that is not only viable (functional in the short term) but also ethically sound, equitable, and beneficial for the long-term health and development of individuals, institutions, and society.

Symbiotic AI Collaboration (Mode of HAIC): A mode of human-AI collaboration characterized by a balanced and interdependent partnership, where humans and AI actively contribute complementary skills and co-create outcomes, with shared control and mutual adaptation.

Synergy (in Human-AI systems): The phenomenon where the combined output or effect of humans and AI working together is greater than the sum of their individual, separate efforts, leading to enhanced capabilities or novel outcomes.

Synergy of Design Thinking and Change Management: Utilizing DT's human-centered methods to determine *what* needs changing and CM's frameworks to guide *how* change is implemented effectively.

Systematic Data Collection: A planned and organized approach to gathering pilot data, using defined methods and tools, to ensure comprehensiveness, reliability, and ethical handling, forming the basis for valid evaluation.

Team Charter: A formal document co-created by a Smart Team defining its mission, scope, SMARTIE goals, roles (including AI Ambassadors), responsibilities, operating norms, decision processes, and ethical commitments.

Thematic Analysis: A common method in qualitative research for identifying, analyzing, and reporting patterns (themes) within data, such as feedback from Daily Logs or interviews.

Think-Aloud Protocol: A usability testing method where participants verbalize their thoughts, reasoning, and perceptions while interacting with an AI tool, providing insights into their cognitive processes and user experience.

TLARA (Teaching, Learning, Assessment, Research, and Administration): An acronym used in the source materials to categorize the five main areas or levels within higher education institutions where AI integration and its impacts are examined. Used as a structure/scaffold for AI ethics.

Triangulation (Data Analysis): The process of using multiple data sources (e.g., quantitative metrics and qualitative feedback) or methods to examine the same phenomenon, aiming to increase the validity and comprehensiveness of findings.

Unsustainable Synergy (in Human-AI Collaboration): A human-AI collaboration that, while potentially offering short-term viability or benefits, leads to negative long-term consequences, such as skill erosion, ethical violations, inequity, or depletion of critical resources.

UODs (User Outcomes and Deliverables): The tangible outputs, results, or products created through the activities in the textbook and workbooks. Tangible products, documents, or results expected from completing a set of activities or a project phase. In the pilot design phase context, the key outputs. *(Merged definition)*

Upskilling (AI Context): Providing targeted training and development to enhance existing stakeholders' competencies related to understanding, using, and managing AI technologies effectively and ethically.

Usability Testing: A method for evaluating how easy and intuitive a product or system (like an AI tool) is to use by observing actual users as they perform tasks with it. The process of evaluating how easy, efficient, and satisfying an AI tool is to use by observing actual users interacting with it to perform specific tasks. Plans outline participants, methods, and metrics. *(Merged definition)*

User Experience (UX): The overall experience a person has when interacting with a product, system, or service, especially in terms of how easy or pleasing it is to use. UX design focuses on optimizing this interaction from a human-centric perspective. UX strategies in pilot design focus on ensuring the AI tool is user-centric. *(Merged definition)*

User Journey Mapping: A UX technique to visualize the end-to-end experience a user has when interacting with a product, service, or process, identifying pain points and opportunities.

User Onboarding: The process of introducing and training new users on how to effectively and ethically use an AI tool or system implemented during a pilot.

UX Workbook (User Experience Workbook): The companion workbook focusing on content-specific activities, strategies, and producing the primary UODs for each chapter's topic. Chapter 7 focuses on activities like drafting team charters, defining roles, stakeholder analysis, and initial deliverable scoping.

Viable Synergy (in Human-AI Collaboration): A human-AI collaboration that produces a positive, functional outcome in the short term, achieving an immediate goal or effectively solving a current problem.

Workflow Optimization: Evaluating and refining how an AI tool is integrated into existing tasks or processes to improve efficiency, reduce errors, or enhance human-AI collaboration.

References

Attridge, M. (2024, April 24). *Campus-GPT: How 2 University of California campuses are designing their own specialized AI tools. BestColleges.* https://www.bestcolleges.com/news/uc-san-diego-irvine-design-specialized-ai-tools/

Brighteye Ventures. (2024, January 30). *The European Edtech funding report 2024.* https://www.brighteyevc.com/post/the-european-edtech-funding-report-2024

Cameron, J. (2024, January 7). Integrating artificial intelligence into higher education governance. *Association of Governing Boards of Universities and Colleges (AGB).* https://agb.org/blog-post/integrating-artificial-intelligence-into-higher-education-governance/

Coffey, L. (2024a, January 19). Arizona State joins ChatGPT in first higher ed partnership. *Inside Higher Ed.* https://www.insidehighered.com/news/quick-takes/2024/01/19/arizona-state-joins-chatgpt-first-higher-ed-partnership

Coffey, L. (2024b, March 21). Universities build their own ChatGPT-like tools. *Inside Higher Ed.* https://www.insidehighered.com/news/tech-innovation/artificial-intelligence/2024/03/21/universities-build-their-own-chatgpt-ai

Culotta, A., & Mattei, N. (2023, November 20). Use open source for safer generative AI experiments. *MIT Sloan Management Review.* https://sloanreview.mit.edu/article/use-open-source-for-safer-generative-ai-experiments/

Fore, P. (2024, November 21). OpenAI partners with Wharton for a new course focused on leveraging ChatGPT for teachers. *Fortune.* https://fortune.com/education/articles/openai-chatgpt-university-of-pennsylvania-wharton-course/

Grajek, S., Pelletier, K., & Freeman, A. (2024, March 11). AI procurement in higher education: Benefits and risks of emerging tools. *EDUCAUSE Review.* https://er.educause.edu/articles/2024/3/ai-procurement-in-higher-education-benefits-and-risks-of-emerging-tools

Inclusion Cloud. (2024, February 28). What you need to know before choosing open-source or proprietary AI models. *Inclusion Cloud Blog.* https://inclusioncloud.com/insights/blog/open-source-llm-vs-proprietary-models/

Katsamakas, E., Pavlov, O. V., & Saklad, R. (2024). *Artificial intelligence and the transformation of higher education institutions [Preprint]. arXiv.* https://doi.org/10.48550/arXiv.2402.08143

Kotter, J. P., Akhtar, V., & Gupta, G. (2021). *Change: How organizations achieve hard-to-imagine results in uncertain and volatile times.* Wiley.

McIntosh, A. (2024, October 23). EDUCAUSE 2024: Institutions embark on cross-campus generative AI initiatives. *EdTech: Focus on Higher Education.* https://edtechmagazine.com/higher/article/2024/10/educause-2024-institutions-embark-cross-campus-generative-ai-initiatives

Nguyen, H. V. (2024, January 9). Open source vs. proprietary AI: A comprehensive guide to making the right choice. *SmartDev.* https://smartdev.com/open-source-vs-proprietary-ai/

O'Connell, A. J. (2024, February 7). How (and why) the University of Michigan built its own closed generative AI tools. *EDUCAUSE Review*. https://er.educause.edu/articles/2024/2/how-and-why-the-university-of-michigan-built-its-own-closed-generative-ai-tools

Paykamian, B. (2023, November 16). Should universities build their own custom AI tools in-house? *Government Technology*. https://www.govtech.com/education/higher-ed/should-universities-build-their-own-custom-ai-tools-in-house

Rispens, S. (2024, June 4). OpenAI builds ChatGPT system for colleges and universities. *EdScoop*. https://edscoop.com/openai-chatgpt-edu-colleges-universities-artificial-intelligence/

Rutgers Business School. (2024, November 21). Rutgers Business School partners with Google to enhance teaching and classroom learning with Generative AI. *Rutgers Business School News*. https://www.business.rutgers.edu/news/rutgers-business-school-partners-google-enhance-teaching-and-classroom-learning-generative-ai

Index

A
Academic integrity, 107, 108, 111
Adaptation, 283–294
Advanced learning prompts, 37, 52
Agile, 4, 21, 133, 211
AI adaptation, 283–285, 287, 290, 292, 293
AI Ambassadors, 124, 133
AI-Centric modes, 42, 43, 45, 47, 50
AI Champion, 332–334, 338
AI competency, 267, 269, 272, 273
  frameworks, 269–273, 277, 321, 324
AI Council, 123, 133–136
AI ethics, 7, 8, 20, 30–33, 35, 36, 50, 71, 103, 108, 127, 130, 138, 139, 146, 155, 156, 161, 164, 167, 272, 277, 283, 287, 290, 292, 293, 297–316, 323, 329, 330, 332, 333, 336, 338, 339
AI evaluation, 243–244, 255, 258
AI governance, 92–93, 103–116, 245, 253–254, 258, 267–271, 273, 277–279, 286–288, 294, 297–316
AI Governance Council, 306–308
AI impact assessment, 194–196, 258
AI implementation, 207–209, 213, 221
AI institutionalization, 277
AI-Integrated Curriculum, 277
AI integration, 103–104, 106, 108, 115–116, 123, 130, 132, 133, 139, 140, 145–167
AI leadership, 267–271
AI Learning Hub, 333
AI Literacy, 7, 8, 12, 16, 32, 36, 47, 50, 87, 108, 140, 156–158, 183, 194, 199–200, 209, 222, 240, 245, 247, 258, 260, 267, 269, 275, 277–279, 285, 286, 293, 297–316
  culture, 333–336, 339
AI maturity, 286
  models, 177, 286, 315
AI performance evaluation, 247–252
AI pilot, 209, 211–214, 216–217, 220, 235, 239
  design, 191, 193, 196
  projects, 189, 196, 229
  strategy document, 192, 194, 196, 199, 202, 203

D. Woodring, *Advancing Higher Education with Human-Centric AI Partnerships*,
https://doi.org/10.1007/978-3-032-05205-6

AI readiness, 115, 145–167
evaluation, 145, 146, 165, 167
AI risk management, 111
AI roles, 300, 302, 315
AI scaling, 266
AI sourcing strategy, 164–165, 179–180, 184, 185
AI sustainability, 315
AI vision, 30–37, 133, 145, 158, 166, 213, 278, 299, 326
statement, 172, 174, 176
Alt text, 8, 17
Artificial General Intelligence (AGI), 8–10, 12, 21, 23
Artificial Superintelligence (ASI), 8–10, 12
Assessment, 4, 12, 18, 29, 37, 41, 47, 63, 69, 70, 72, 79, 83, 85, 86, 88, 90, 91, 93, 103, 104, 108, 115, 116, 130, 138–140, 145, 148, 151, 152, 154–167, 172–175, 179, 180, 191, 195–198, 202, 207, 209, 210, 219, 233, 237, 238, 243, 247, 248, 252, 254, 256, 260, 269, 287, 290, 298, 302, 303, 310, 323–324, 329–331, 334, 335
Augmentation, 3, 33, 41, 83, 89, 91
Authentic learning, 88
Authentic High-Impact Practices (AHIPs), 79, 82, 87–93, 104, 122, 146, 172, 191, 208, 244, 266, 285, 298, 340

B

Barrier Analysis, 200
Baseline metrics, 195, 199
Benchmarking, 247–252, 255, 259, 260
Benchmark report, 234
Bias (AI), 17, 20, 32, 47, 48, 52, 60, 63, 107, 108, 111, 115, 131, 160, 164, 201, 214, 234, 236, 245, 267, 286, 305, 327
Build *vs.* Buy Analysis, 115

C

Capacity building, 266, 273, 278, 279
Change Management (CM), 4, 16, 18–21, 29, 103–106, 109, 115, 121–123, 146, 158, 159, 171–173, 181, 183–185, 189–191, 200, 207, 208, 213, 221, 229, 236, 244, 245, 251, 254, 259, 265–267, 273, 278, 284–287, 289, 292, 300, 310, 312, 324, 326, 328, 332, 333, 337, 339
Change Management Plan, 200
Change readiness, 158–159
Chief AI Officer (CAIO), 269–271
CLEAR framework (Prompting), 79
Co-creation, 13, 15–16, 127, 190, 210, 290
Cohort effects, 68–72, 74
Collaborative intelligence, 121, 124, 173, 184, 185
Collaborative partners, 31, 37
Collective Intelligence (CI), 30, 83, 246
Communities of Practice (CoPs), 271, 273, 277, 284, 331–335, 338, 339
Competency development, 283, 286
Competency framework, 172
Competency mapping, 156, 158
Compliance, 265, 269, 271, 277
Continuous assessment, 209, 210
Continuous evaluation, 243–245
Continuous improvement, 19, 20, 33, 190, 209, 233, 246, 255–259, 266, 269, 271, 293, 294, 300, 301, 305, 309, 310, 334
Continuous iteration, 33–34
Continuous monitoring, 246, 250, 259, 260
Co-regulated learning (CRL), 50, 83

Co-regulation, 31, 34
COVID-19, 69, 70
cohort effects, 68–72
Critical evaluation, 321, 322, 324, 328, 336, 339, 341
Cultural adaptation, 287–288, 293
Curriculum integration, 329
Custom frameworks, 302, 307

D

Daily Feedback Logs, 207, 209, 210, 215, 217, 218, 220, 221, 231, 232, 237–239
Data analysis, 229, 233, 237
Data collection, 229–233
plan, 231
Data-driven decision making, 257
Data governance, 155, 165
Data infrastructure, 154–155
Data integration, 232
Data privacy, 107, 111
Data synthesis, 233
Design Thinking (DT), 4, 18–21, 29, 41, 56, 103–106, 115, 122, 145, 146, 148, 171, 172, 181, 184, 185, 189–191, 207, 208, 211, 221, 229, 244–246, 251, 254, 259, 265–269, 275, 278, 284, 285, 288, 289, 292, 300, 309, 310, 324, 326, 333
Distributed moral responsibility, 33
Distributed responsibility, 37
Diverse stakeholder engagement, 103–116
Diversity, Inclusion, and Equity (DEI), 88, 181
Dual system (Kotter), 123, 133, 213–214

E

EDUCAUSE, 145, 160, 164
Emergence, 55–74
Empathy mapping, 23, 105, 129, 152, 154
Enhancing Academic Outcomes (EAO), 79, 82, 86–89, 92, 93, 104, 107, 108, 122, 146, 156, 172, 191, 208, 244, 252, 266, 285, 298, 340
Enhancing Human-Centric Skills (EHCS), 79, 82, 85–87, 89, 91–93, 104, 107, 108, 122, 146, 156, 172, 191, 208, 244, 245, 266, 285, 298, 340
Equity, 321, 323, 336, 337, 339, 341
Equity & Accessibility Review, 166
Ethical AI, 30–33, 35, 45, 243, 244, 246, 267–272, 276, 277, 297–316, 337, 339
Ethical AI Governance, 110–111
Ethical AI Pledge, 126
Ethical AI Policies, 139
Ethical AI Testing, 201
Ethical compliance, 234, 238
Ethical considerations, 234, 239
Ethical frameworks, 32, 297, 298, 302, 309, 315–316
Ethical governance, 190, 196–198
Ethical guardrails, 32, 214
Ethical oversight, 285
Ethical principles, 110, 112
Evaluation, 229–234, 237–239
framework, 244, 256, 260
Evidence-based decision-making, 230
Explainable AI (XAI), 32

F

Feedback analysis, 213
Feedback loops, 58–60, 63–67, 69, 131–133, 135, 138, 190, 191, 200, 202, 208, 209, 215, 219, 221, 246, 255, 285
Fifth Industrial Revolution (5IR), 3–23, 321
Foundational training, 326–329

G

Gap analysis, 157, 256
Generative AI (GenAI), 3, 6–9, 12, 15–18, 20–23, 29–37, 41–52, 55–74, 79, 92, 103, 107–108, 115, 116, 146, 148, 151, 154, 155, 158–160, 164, 165, 167, 172, 173, 176–178, 180, 183, 185, 186, 190, 194, 202, 208, 210, 222, 230, 231, 236, 240, 247, 254, 260, 274, 275, 279, 283, 285–287, 294, 297–299, 301–303, 305, 308, 310, 312, 314, 315, 321, 331, 340, 341
Goal setting, 21, 23, 30, 33
Governance, 7, 12, 15–18, 21, 34, 44, 49, 58, 92, 93, 103–116, 124, 131–136, 139–140, 145, 154–155, 159–160, 166, 167, 197–202, 212, 222, 240, 245, 247, 248, 252–254, 256–260, 265, 267, 269–271, 273, 274, 276–279, 283, 286–288, 291–294, 297–316
  frameworks, 115, 145, 154, 189
  refinement, 291, 293
  structures, 132, 134, 136
Guiding Coalition, 121–124

H

HAIC Factors (Goals, Interaction, Task Allocation), 47–48, 52
HAICS Guide, 104, 105, 107, 108
HCASR framework, 121, 122, 127, 130–132, 146, 158, 171, 172, 184, 185, 208
Higher education, 303
Higher Education Institutions (HEIs), 103, 113, 115, 116
High-Impact Practices (HIPs), 85, 87, 88, 90
Human-AI collaboration (HAIC), 3–23, 34, 36, 37, 41–52, 56, 61, 65–69, 71, 80, 92, 122, 130, 131, 146, 174, 190, 191, 201, 208, 217, 235, 297, 298, 305, 307, 313–315, 321, 322, 327, 328, 330, 332, 336, 338, 340
Human-AI Interaction (HAII), 80, 84, 89, 307
Human-AI partnerships, 29, 30, 33–34, 41, 42, 50
Human-AI synergy, 30–31, 36, 59, 81, 89
Human-AI Teaming (HAIT), 80
Human-centric AI (H-CAI), 21–22, 29, 41, 45–46, 50, 82, 91, 103, 207–222
Human-Centric AI Synergy (HCAS), 3, 7, 30, 50, 79–83, 88, 89, 91, 92, 103, 104, 122, 146, 172, 191, 208, 217, 218, 244, 246, 253, 266, 285, 298, 302
Human-Centric AI Synergy & Regulation (HCASR), 3, 7, 21, 22, 33, 37, 51–52, 104, 105, 107–109, 112, 115, 116, 121, 122, 124, 126, 127, 131, 132, 136, 139, 146, 160, 171, 190, 191, 194–196, 200, 201, 208–211, 213, 215–219, 233, 234, 237, 238, 244, 246, 254, 256, 257, 266, 268, 273–276, 284, 285, 287, 288, 298–301, 303–306, 308, 310, 312, 314
Human-centric approach, 121
Human-centric design, 210–211
Human-Centric mode, 42, 46, 121
Human Oversight, 190, 194, 196, 197, 200, 201
Hybrid human-AI intelligence, 29–37
Hybrid Human-AI Regulation (HHAIR), 83

I

Implementation, 33, 36, 44, 49–51, 83, 103, 104, 107, 152, 154, 158, 159, 164, 171, 180, 181, 183–186, 189–191, 199–203, 207–211, 213, 214, 216, 219–220, 222, 230, 235, 236, 243, 244, 253, 257, 258, 271, 276–277, 284, 290, 297, 305–307, 309, 310, 314, 324, 338
  roadmap, 171, 180
Inclusive (SMARTIE), 30, 32, 34–35, 49, 126, 127, 149, 151, 184, 210, 219, 269, 288, 289
Institutional AI Strategy & Roadmap, 184, 185
Institutional culture, 178
Institutionalization, 265, 277–279, 309, 315
Institutional mission, 106, 116
Institutional readiness, 255, 256
Intelligence Augmentation (IA), 80, 83
Interdisciplinarity, 31
Iterative development, 208, 219
Iterative improvement, 237
Iterative piloting, 211–212
Iterative policy development, 284, 285, 287–289, 293
Iterative Refinement, 209, 210, 215, 292, 294

K

Key Performance Indicators (KPIs), 194–196, 230, 232–234, 237, 245, 248, 252, 256–258

L

Large language model (LLM), 15, 17, 68, 84, 297
Leadership, 299, 310, 314, 315
Leadership Commitment, 178
Lessons learned, 230, 234, 237–240
Lifelong learning, 83, 85, 88, 89, 269–273, 278, 279, 284
Living documents, 130, 138, 164, 167, 186, 202, 218, 219, 221, 237, 240, 257–260, 277–279, 287, 291, 294, 300, 315, 316
Long-term strategy, 297–316

M

Metacognition, 328
Meta-emergence, 59
Metrics, 86, 88, 93
Minimum viable product (MVP), 208, 211, 213–215, 218, 219, 221, 222
Mixed methods, 232, 233
Monitoring, 243–248, 254, 256, 258–260
Monitoring & Evaluation (M&E), 180

N

National Institute of Standards and Technology AI Risk Management Framework (NIST AI RMF), 108, 111, 160, 196, 215, 221, 236, 248, 258, 301
Needs assessment, 145–149, 152, 154, 162–167

O

OECD AI Principles, 196
Onboarding, 326, 327, 338
Opportunity statements, 174–176
Organizational change, 283
Organizational culture, 158–159

P

PALS Workbook, 103–105, 107–109, 112, 115, 116
Participatory design (PD), 104, 332
Pedagogical innovation, 82
Pedagogy, 31, 45, 87, 92, 105, 238, 269, 327, 328
Persona, 36
  development, 129, 152
Personalized Adaptive Learning Session (PALS), 7, 23, 36, 37, 50–52, 73, 74, 93, 104, 108, 122, 124, 127, 130, 132, 135, 139, 140, 146, 148, 171, 172, 195, 200, 208, 210, 213, 216, 217, 219, 222, 244, 247, 252, 256, 257, 259, 260, 266, 268, 271, 273–276, 278, 279, 285, 287–289, 291, 293, 294, 297, 298, 301, 303, 308, 310, 312, 314, 315
Pilot data, 229–240
Pilot design, 189–191, 196–199, 202
Pilot evaluation, 229, 230, 234, 238, 244–245, 259
Pilot objectives, 192, 194, 195, 202
Pilot projects, 19, 175, 194–195, 265, 309
Policy, 327, 329–332, 335, 336, 338, 340
  development, 269, 302, 308
  mapping, 303, 312, 315
  refinement, 250–257, 259, 260
Power/Interest Grid, 128–130
Preparatory Exercises, 104
Priority gaps, 162–164, 166
Professional development, 326, 329, 339
Prompt engineering, 30, 32, 33, 37, 49, 85, 91, 156, 272
Proprietary AI Systems, 222
Proprietary *vs.* third-party AI, 134–136

Q

Qualitative data, 232, 233
Quantitative data, 232, 237
Quick wins, 163, 164, 166

R

Rapid Research & Review (RRR), 51–52, 299, 303
Refinement, 283–294
Reflective hybrid intelligence, 31, 33, 37
Regulation Synergy & Augmentation (RSA), 3, 31, 50, 79, 82–85, 88, 89, 91–93, 103, 104, 107, 109, 115, 122, 146, 190, 191, 208, 217, 244, 246, 253, 266, 285, 298, 300, 306, 340
Resource allocation, 180, 183
Risk management, 108, 111, 115, 196–198, 202, 214, 258, 259
Risk Management Plan, 180
Risk mitigation, 235–236, 239
Risk Register, 198, 199, 203, 218, 219, 236, 239
Roadmap, 314, 315
Role-specific training, 329, 338

S

Scaling, 265–268, 271, 273–279
Scenario-Based Planning, 180
Self-regulated learning (SRL), 50, 83
Self-regulation, 31
SMARTIE goals, 126–127, 150, 151, 171, 176, 178–180, 184, 185, 192, 194, 195, 215, 247
SMARTIE objectives, 146, 151
Smart teams, 16, 19, 103, 115, 116, 121–140, 145, 148, 158, 173–174, 181, 184–186, 200, 203, 207, 221, 238, 239, 260

Sociocultural context, 115, 116, 130, 139, 167, 186, 202, 207
Specific, Measurable, Achievable, Relevant, Time-bound, Inclusive, Equitable (SMARTIE), 171
Stakeholder, 7, 12, 13, 17–20, 22–23, 31, 34–37, 41, 42, 49, 51, 52
analysis, 127–132, 138
coordination, 284, 289, 293
empowerment, 208, 210
engagement, 16, 22–23, 29, 33, 42, 103–116, 128–132, 140, 148, 190, 191, 199, 200, 203, 219, 238, 246, 283, 289, 290, 292, 298, 299, 310
mapping, 41, 121, 122, 127, 132, 139, 171, 174
needs, 145–148
Stakeholder Engagement Plan, 131, 132, 138, 140
Strategic alignment, 173–174, 181, 183
Strategic planning, 115, 171–186
Strategic roadmap, 189–191
Success metrics, 192, 194–195
Sustainability, 4, 56–74, 80, 82, 88, 92, 93, 115, 116, 139–140, 167, 202, 222, 240, 243, 247, 248, 250, 252, 259, 260, 273, 276, 277, 279, 294, 297, 309, 314–316, 321–341
Symbiotic mode, 41–45, 49–52
Synergy, 4, 6, 12–15, 18–23, 30, 31, 34, 35, 37, 43, 49, 50, 52, 56–59, 61–66, 68–74, 81, 84, 89, 92, 105–106, 122, 135, 136, 146, 171, 172, 184, 185, 189–191, 208, 217, 300, 326, 333

T

Talent & Capacity Building, 180, 183
Task allocation, 47–49, 52
Teaching, Learning, Assessment, Research, Administration (TLARA Framework), 92, 93, 116, 139–140, 186, 202, 240, 278, 279, 294, 298, 300–316, 340, 341
Team charter, 125–127
Triangulation, 232, 233

U

UNESCO recommendations, 196
Usability testing, 195, 213, 216
Use case, 37
Use Cases Inventory, 138
User experience (UX), 16, 22, 23, 36, 37, 41, 50–52, 68–69, 73, 74, 93, 104, 105, 108–110, 112, 115, 116, 123, 130, 132, 139, 140, 148, 152–154, 166, 167, 171, 172, 186, 190, 191, 194, 195, 199–203, 208, 210, 213, 216, 217, 219, 222, 232–234, 238–240, 244, 245, 247, 252, 256, 257, 259, 260, 266, 268, 271, 273–279, 284, 286–289, 291, 293, 294, 297, 298, 301, 303, 308, 310, 312, 314, 315, 334, 341
research, 148
User Experience (UX) Workbook, 107
User journey, 148
User outcomes and deliverables (UODs), 115, 116, 122, 137–140, 146, 162, 165–166, 171, 172, 183–184, 191, 199, 202, 203, 209, 218, 220–222, 233, 237, 239, 245, 258–260, 276–278, 290–293, 297, 314–315, 322, 338–339
User story, 36
User support, 222
User training, 222

V

Viability, 56–74, 244, 248, 250, 265–267, 276, 278

W

Workflow mapping, 213
Workflow refinement, 254

GPSR Compliance
The European Union's (EU) General Product Safety Regulation (GPSR) is a set of rules that requires consumer products to be safe and our obligations to ensure this.

If you have any concerns about our products, you can contact us on

ProductSafety@springernature.com

In case Publisher is established outside the EU, the EU authorized representative is:

Springer Nature Customer Service Center GmbH
Europaplatz 3
69115 Heidelberg, Germany

www.ingramcontent.com/pod-product-compliance
Ingram Content Group UK Ltd.
Pitfield, Milton Keynes, MK11 3LW, UK
UKHW020955220926
475831UK00002B/91
9783032052049